Commentary on Dr Rudolf Steiner's Agriculture Course

Enzo Nastati

A revised transcript of 30 meetings lead by Enzo Nastati
between January 2001 and May 2005.

English translation by Mark Moodie with assistance from Sofia Mancini Griffoli, Claudio Zanette, Mike Atherton, Jane Brown and Jonathan Code. This publication has been generously assisted by the BDAA's Grange Kircaldy Trust, the Hermes Trust, and private donations.

Published by Mark Moodie Publications, Oaklands Park, Newnham, Glos, GL14 1EF, UK
www.moodie.biz

ISBN 978-0-9517890-6-3

CONTENTS

COMMENTARY ON RUDOLF STEINER'S AGRICULTURE COURSE

Led by Enzo Nastati

First introductory meeting (January 13, 2001)

In this series of meetings we will consider ecological agriculture from a new angle. Both biodynamic and modern organic farming spring from a series of 8 lectures that Rudolf Steiner gave in 1924 at Koberwitz, in Count Keyserlingk's mansion. This lecture series was published as: '*Spiritual-Scientific Foundations for the Renewal of Agriculture*', but is widely known as the 'Agriculture Course'. It contains directions that are essential for those who want to practice agriculture in harmony with the laws of the Earth and the cosmos.

It is clear to anybody who has read even one of Steiner's many publications (which cover virtually all fields of knowledge) that his books and lectures penetrate to the deepest truths and that, precisely because of the intricate and profound nature of the things that he says, his works do not yield their fruits to a person who is not motivated by a genuine and resolute desire for knowledge. Therefore one should not lose heart the first time one does not immediately understand something he says. We are convinced, however, that it can be productive to strive to understand the precise messages in these lectures.

Steiner based the course on very clear concepts, the first of which understands *nature* to be a threefold reality composed of body, soul and spirit: in addition to the sense-apparent aspect of nature there is a soul aspect that might be described as the *feeling sense of the world,* as well as a spiritual reality consisting of beings that are more evolved than humans. These three aspects, along with their activity and impulses, will determine how we seek to grasp the forces, laws and meaning of nature. Out of this interpretation an approach to agriculture can arise which is responsive to the reality of creation.

Dr Rudolf Steiner had a highly prepared audience. They were all students of his spiritual science or *anthroposophy*. Most modern people have not studied this discipline sufficiently to grasp Steiner's thoughts right away. So we will begin our work by clarifying some basic concepts that form the foundation for what we will build later.

First, one should know that Steiner was often asked to give a course on agriculture and that he had declined to do so because he considered that, with a bit of goodwill, his students could find enough guidance from his fundamental works, such as 'Esoteric Science', to develop a revitalised agriculture based on sound spiritual-scientific principles. Only in 1924, thanks to the insistence and hospitality of Count Keyserlingk, did Dr Steiner decide to offer the cycle of lectures to which we now refer.

Dr Steiner died in March 1925. This series of lectures was one of the last he presented during Pentecost when the forces of the Holy Spirit powerfully affect the world and people. All of Steiner's pentecost lecture cycles are of a special hue, but the 'agriculture course' is even more important since it represents the last time that Divine Wisdom, an attribute of the Holy Spirit, found voice and expression through this great man.

The stated primary aim of the course was to address three problems that had often been brought to Rudolf Steiner's attention:

- the degeneration of plants,
- the diminishing nutritional value of food,
- the lowering fertility of farm animals.

One can experience a certain wry amusement to think that these were pressing concerns in the 1920s when we would now be happy to have those plants and their nutritional value. Evidently, however, there were already early signs of a situation that was destined to worsen dramatically to the point that, in our opinion, it is now no longer remediable with the methods of organic farming alone.

Although we shouldn't expect to find explicit answers to these three questions the answers are *implicit* in the complex mosaic that Steiner composed.

Moreover, we have noted many times that Steiner often veiled the truths which he shared, not to guard the secrets out of jealousy nor for the pleasure of increasing the effort required by those who desire to learn, but mainly because the laws of nature which he exposed could be interpreted awry and used inappropriately. For example, during the course he was asked what one could do against pests, weeds and parasites. The response, that we can all read, essentially suggests that it is adequate to incinerate a sample of the creature in question when certain planets are in front of a particular constellation and later to sprinkle the ashes on the soil. A positive result is not always evoked in the first instance, and it is one of those practices that is easily mocked as witchcraft by those unfamiliar with the concepts.

In fact, whoever knows that each living being is connected to a spiritual archetype also knows that destroying the form of a being causes an equally violent separation from its substance, and a message is sent opposing the incarnation of the spiritual entity which had manifested in that being. This is especially powerful if at that moment the corresponding constellation is 'active'. In practice the umbilical cord through which the creature draws spiritual sustenance and strength is severed between the incarnated being and its 'mother', so the ashes scattered on the ground make that field uncongenial.

Not everyone knows that Hitler decided to incinerate the Jews after forcing his 'best' scientists to study the works of Steiner. Evidently he hoped to make Europe unfit for Jewish people. In this light we can also gauge the gravity of the continuous burning of the victims of foot and mouth disease in England. At some point or other the incineration would have to coincide with the constellation that corresponds to the cow and this would be experienced by the archetype of the cow as a gesture of rejection.

Be that as it may, a long time has passed since Rudolf Steiner delivered these agriculture lectures and it is time to lift the veils, at least as far as we are able, so that the truths can be available to those seriously committed to help nature find a new equilibrium. This will be the guiding theme of the course we are about to begin. It won't be a purely technical course but will give us a way to familiarise ourselves with the laws of nature. From there it is hoped that concrete initiatives may come forth: if knowledge remains at the level of thought and is not carried to the field of action it is worse than useless.

Before we begin reading the text of Steiner's lectures it is worth studying the architecture of the course itself because the sequence in which the concepts were placed already presents a profound riddle, the resolution of which adds meaning and perspective to each part.

If we had a musical score in front of us we could dwell on the individual notes and we would see that the first, for example, is a '*do*', the second one was a '*fa*'. If we could also contemplate the score as a whole then we can find the melody of the piece that transcends the individual notes. Anyone who has some familiarity with Steiner's books and transcripts knows that one can derive satisfaction from the details. Perhaps though we might also see that the most profound truths are uncovered when seen within their overall context. Then details that may otherwise seem trivial are revealed in their fundamental relevance. We are convinced that there is a very important message even in this course outline so we will begin with an examination of it.

First, the Course is divided into 8 lectures and we should try to understand why there are just so many by building a table that we will follow together. This will also bring some knowledge into focus that can bring us to a deeper understanding of Dr Steiner's message. We can sketch a table with 8 columns and 11 rows (see below).

The first header row outlines the evolution of our planetary system so it is labeled: *phase of evolution*. In Rosicrucian-Anthroposophical esoteric schooling the first manifestation of the Earth is called '*Old Saturn'*. This was a huge sphere of differentiated warmth filling the orbit of the current planet Saturn – its remnant. Time first appeared in this phase.

This sphere of warmth condensed to form the '*Old Sun'*, a reality consisting of gas whose existing remains are the planet Jupiter. During the old Sun, light manifested in the universe and space was born.

The second row of the table is dedicated to the etheric forces. We have put warmth and light corresponding to old Saturn and old Sun respectively. For a deeper grasp of these concepts please refer to "Esoteric Science" by Rudolf Steiner or the publications "The Quality of life", or "Nutritional Quality" available to members of l'Albero della Vita.

The third phase of the Earth's evolution is the '*Old Moon'* which consisted of liquid which has left the planet Mars as a kind of memory. Finally, we have the solid manifestation that is called Earth.

Up to this point, evolution had been an ongoing and phased condensation with a continual reduction of the volume occupied. The point of greatest condensation was reached approximately 2000 years ago when the entire process began to reverse its course, and this will eventually completely re-enliven or 'etherise' the Earth. The turning point was the 'event of Golgotha' when Christ introduced an impetus of resurrection into the evolution of the Earth. This opposed the process of death represented by continuous and progressive hardening. (We can add that radioactivity first emerged when the event of Golgotha occurred, which reveals the natural process through which matter 'becomes'.)

We consider the process of condensation to be a result of the 'fall' precipitated by Lucifer. But then came the moment when another spiritual entity, the Son or the second person of the Holy Trinity, entered Earth's events and activated the process of resurrection.

The current Earth, before starting on the ascending path, briefly retraced the steps above as indicated on the third row of our scheme. The repetition of the warmth being, ancient Saturn, has been called the *Polarian* epoch, the repetition of the Sun is the *Hyperborean* epoch. Then the repetition of the old Moon was *Lemuria* followed by the *Atlantean* epoch in which the Earth experienced a condensation close to the current one.

PHASE OF EVOLUTION	OLD SATURN	OLD SUN	OLD MOON	EARTH	FUTURE JUPITER	FUTURE VENUS	FUTURE VULCAN
Body	Physical body	Etheric body	Astral body	Preparation of the 'I'	Spiritual self	Life Spirit	Spirit Man
Ether	Warmth	Light	Chemical	Life	First new ether	Second new ether	Third new ether
'Repetition' on the Earth **Approx start**	Polarian epoch - 67,000 B.C.	Hyperborean epoch -52,000 B.C.	Lemurian epoch - 37,000 B.C.	Atlantean epoch - 22,000 B.C.	Post Atlantean Epoch - 7,000 A.D.	Seven seals + 8,000 A.D.	Seven Trumpets + 23,000 A.D.
'Repetition' in post-Atlantean cultures	Old Indian - 7,200	Old Persian - 5,067	Egypto - Chaldean - 2,907	Greco-Roman - 747	Germanic + 1,413	Slavic + 3,600	American + 5,787
Sign of the Zodiac	Cancer	Gemini	Taurus	Aries	Pisces	Aquarius	Capricorn
Spiritual leadership	7 Holy Rishis	Zarathustra	Hermes Trismegistos	Aristotle	Jesus Christ	Jesus Christ	Jesus Christ
Type of therapy	Spiritual	Polarity of good - evil	Alchemical	4 humours	Spiritual Scientific	Spiritual Scientific	Spiritual Scientific
Type of Agriculture	Nature	Light and Dark	3 processes	4 elements	New synthesis	New synthesis	New synthesis
Lecture		First	Second	Third	Fourth and fifth	Sixth	Seventh and eighth
Content		Plants, silica and calcium	13 nights, clay (+ calcium and silica)	Zodiac, Protein Cosmic nutrition Philosophers stone	Field sprays and compost preparations	Overcoming diseases	Agricultural organism (New Jerusalem)

During the repetition of these previous phases, the Sun, then the planets Mercury and Venus, and finally the Moon departed from the single mass that had initially encompassed them, leaving behind our Earth.

The first row has three remaining boxes to fill that allow us to cast our gaze towards the future of our planet. The next phase of the process will be the *Future Jupiter* in which there will be no more physical manifestations and in which everything will be raised up a plane. The *Future Venus* will follow and the seventh and final phase will be *Vulcan* where even the ethereal and astral planes will have been abandoned to be replaced by higher levels.

In this scheme, the Earth is central to the whole developmental process, but since everyone once was and everything must return again to being 'one', we can consider the later stages again as a kind of repetition of the three previous phases. In this way but at a higher level, the future Jupiter will repeat the stage of the Moon, the future Venus will be a repetition of the Sun and so on.

It can be enlightening to note that if each earlier stage of the Earth is joined with its repetition by an arc, we get the picture of the Hebrew calendar - an old image of the cosmos. This immediately suggests that what we are showing is not just dreamt up by Rudolf Steiner but belongs to an ancient body of knowledge that is all but lost.

The second line of the table shows the four etheric forces that have characterised the stages of evolution until now. Although we will not discuss much about this topic because it is largely handled elsewhere[1], we would like to point out that the ether linked to the Moon is preferably called the '*alchemical ether*' or the '*ether of chemism*', instead of the more familiar '*chemical*' ether. This is because it is an expression of a broader concept than orthodox chemistry that studies the interactions of chemical elements through the exchange of electrons. Material alchemical interactions are a profound interpenetration that allows the exchange of neutrons that are part of the atomic nucleus and not the 'suburban' electrons. It is clear that such a transformation requires much more powerful forces of combination and dissociation.

The four etheric forces emerge from successive transformations of the Earth. Each emerging new ether has retained the characteristics of the ethers that have preceded it whilst adding its own specific qualities. Therefore the Life ether, which is the youngest, contains the characteristics of all three other ethers. In other words the ethers have succeeded on the phylogenetic principle (the latter is the offspring of the former), but also the ontogenetic principle (the child retains the characteristics of the parent). We will not now discuss the new ethers and those acting upon future manifestations of the Earth because we will discuss them later.

Our third row tabulates the eras of our Earth which have also retraced the path of creation. It all looks like a set of Russian dolls. First there was the Polarian epoch, then the Hyperborean, then the Lemurian and then the Atlantean - the epoch in which humanity began to take on our current configuration.

The present era is the Post-Atlantean epoch. To suggest names for these future eras one can point to what John wrote in his Revelation and say that the next epoch will be the 'seven seals' and that will be followed by the 'seven trumpets'.

Even within the Post-Atlantean epoch we find a smaller repetition of what was

[1] 'The Etheric' by Ernst Marti, and many other l'Albero della Vita publications

already repeated! On the fourth row of our table we can write the names of the ages of civilisation.

The ancient Indian cultural era is the recapitulation of Old Saturn. The ancient Indian had a unified vision of reality and believed that only the spiritual world was real because it was absolute and unchangeable. The world that could be perceived by the senses was considered an illusion or *maya*. The consequence of this understanding is that the physical world was considered undeserving of our attention and that only the relationship with the spirit was cultivated.

The second period was that of the Assyrian-Babylonian or Persian civilisations, which was followed in turn by that of Egypt, and then the Greco-Roman. The present is the Germanico-Saxon. The Slavic and then the American stages will follow this period.

Each of these past periods of civilisation had a principal spiritual 'guide' as noted in the fifth row of our table. The Indian period was guided by the seven Holy Rishi. The teacher of the seven Rishi was the last of the Atlantean initiates who began and led a huge spiritual culture in Tibet. No one person was able to contain all the wisdom of the previous period so this was divided into seven parts, corresponding to the seven planetary mysteries, one for each Rishi.

The second or Persian era presented a vision of the cosmos as a struggle between Ahriman, lord of darkness, and Ahura Mazdao, the Lord of Light and the Sun. This shows a transition to duality from the Old Indian unity. The Persian era had Zarathustra as its spiritual guide.

The third period was led by Hermes Trismegistos whose philosophy was based on three basic principles: *Sal, Mercur* and *Sulfur*[2]. This survived in the work of the alchemists until the nineteenth century. The name of Egypt was 'El Kemi, or *the hidden one*. The three processes occur in condensation, exchange and in expansion.

The fourth or Greek period had Aristotle as a guide. He brought the theory of the four elements: Earth, Water, Air and Fire. The memory of this lasted up to 150 years ago when doctors still formed their diagnoses in terms of the famous four humors: blood, phlegm, clear and black bile.

Inevitably, after this fourth era we come to the present where memories of what has been are almost completely lost. But now we must decide whether we shall continue the trend from the Indian unity that has reached the four-fold conception. We could continue the fragmentation and move towards a vision of the world inspired by 5, 6, 7, 8, etc. Perhaps, however, we would prefer to work towards a new synthesis.

We cannot ignore the fact that the Greco-Roman era was not only the era of Aristotle but also the time of Golgotha[3]. Christ came to bring unity. We must start looking for a new synthesis and work so that the future can lead to unity.

The initial task will be to revive the three-foldness of the Old Moon but, this time, brought to a higher level that we could call the three fruits of a new alchemy. Only then might we go forward to a new duality and then a new one - a new unity with the cosmos.

We can also accelerate this timetable, thanks to the Christ in us all, and bring the evolutionary process described to full expression in our present lives. Remember the parable of the lost sheep but consider the interpretation of Thomas, who says that

---

[2] Italics are used for the elements of this 3-fold model. The spelling 'sulphur' is reserved for the chemical element, *sulfur* for the expansive pole of this trinitarian archetype.

[3] The four classical elements can be represented by the four arms of the cross. The inscription INRI, made by Pilate and put on the cross, shows the initial letters of the four elements in Aramaic.

when Christ went looking for the wandering sheep he actually went looking for His favorite sheep because it was the only one that had the courage to jump the fence. But who prevents us from jumping the fence during this life rather than waiting for thousands of years? One of the stimuli of the agriculture course is in just this direction.

Let us reconsider the structure of the Agriculture Course and assert, if it is not already clear, that the structure of the course is not arbitrary. If we consider that the human of the Indian era was not a farmer but a gatherer of the fruits that the divine made available, we understand that the first lecture of Rudolf Steiner's agriculture course is not aimed at India since agriculture did not then exist. The first lecture is, however, relevant to the work of Zarathustra, ie when people began to domesticate and breed plants. One can read this lecture as describing the duality between silica and limestone, between light and darkness.

The second lecture speaks of Egypt: we are introduced to the concept of clay as the mediator between silica and limestone, as *Mercur* mediates between *Sulfur* and *Sal*. The third lecture tackles the world of protein and substance; the four elements of the Greeks.

Steiner was not explicit in the various lectures that he was talking about the various agricultural initiatives in successive eras of civilisation, but everything can be argued this way with a careful reading.

This brings us to the point in which Dr Steiner describes the biodynamic preparations, which represent the new synthesis, and which were probably too important to squeeze into one lecture: he dedicates the fourth and fifth lectures to these preparations. The preparations are a synthesis between the various kingdoms - made by humans of materials from the mineral, vegetable and animal kingdoms. These are then buried so that they can be enriched further by contact with the Earth within which the impulse of Golgotha has been active for 2000 years. In this way a process takes place that might be defined as one of *Christianisation.* This allows one to have preparations that bear the first new ether of Christic redemption.

When one understands this new synthesis one can add, on the second line in the sixth column of our table, the first new ether which is arising from the transformation of the Alchemical ether.

Then comes the sixth lecture that corresponds to the evolutionary stage of the future Venus where the Slavic people will champion the highest spiritual development. Because Venus is linked to love, this lecture concerns the application of the forces of love. This is made evident when one considers victory over evil as a vital understanding of new ways of dealing with 'pests'.

In the seventh and eighth lecture the return of unity is envisaged, but at a different level than that of old Saturn. We know that mankind was expelled from an earthly paradise that the Bible describes as a garden, and we must be clear that the goal was not to return but to enter into the New Jerusalem - a city. This means that we must build a social life based on a new model.

Dr Steiner speaks of agriculture here too when he talked about the laws that govern how one plant will collaborate with another plant, one animal with another, how ants work with mushrooms and so on. Finally there is a detailed discussion on cattle that suggests that this animal will become the source of all the new forces of life.

In this way we can grasp Dr Steiner's complex design: presenting the evolution of agriculture in the light of the spiritual evolution of human beings, and understanding why the content of this presentation has been veiled with 'seven times seven seals'.

Under the guidance of Dr Steiner we can grasp the essence of Zarathustran agriculture and understand how this great initiate, whose name means *shining star*, acted to domesticate plants and animals. We can then understand the second phase of development of agriculture that corresponds to the Egyptian period through the meeting with the clay and the subsequent formation of humus (flooding of the Nile), and so on.

But we shouldn't attempt to grasp the laws of life in order to bring back the Zarathustran or Old Greek agriculture, because this wouldn't make any sense today: these laws are always the same and, once understood, they can be carried to a new, contemporary system which is adapted to the present earthly and cosmic conditions.

We must, however, begin from the assumption that our work is useless if its only fruit is the acquisition of a series of notions or points of view, striking as they may be, if it doesn't also provoke a deep transformation in our way of feeling and being. The book that we are going to study together is a book of initiation through work with and upon the Earth. Today farming is considered the most humble work but it was once the discipline that was taught only to the best of the best of Zarathustra's students - precisely because it gives access to the laws of life.

If we consider the whole lecture cycle we could say that the man who lived by gathering the natural fruits of plants and animals preceded it: Abel. The path from the first to the eighth lecture represents the journey of Cain who can redeem himself by following this path.

Cain means '*he who thinks for himself*' and is the free part of humanity that can raise all the kingdoms of nature, including Abel. Abel was not murdered; the murder is only an allegory to describe how Cain has freed himself from a thinking tied in to old relationships and how he begins to use his independent thinking. The same theme is touched upon in the Bhagavad-Gita when Arjuna is forced to fight his own relatives. If we consider that the last chapter of Steiner is mostly dedicated to the cow, we have a clear picture of the fact that at the end of his evolutionary trials Cain (who was a pastor) is able to pay the ransom for Abel. So if we were to look for a suitable title for this course, perhaps it might be: *The redemption of Cain.*

One can understand why we don't consider the term '*organic farming*' to mean non polluting. The true organic agriculture was that of Zarathustra or Aristotle where '*organic*' (or 'biological') meant the *logic of life* with its laws and its connections with the cosmos. Unfortunately those who practice organic farming today know practically nothing of this.

To continue our work we must now speak of the kingdoms of nature or, rather, better understand from where the animals, the plants and the minerals have emerged.

One who takes care of animals is called [in Italian] an '*allevatore*'. This word is composed of two parts: the pronoun *al* and the verb *levare*. *Al* is of Arab origin and signifies '*spirit*' while *levare* signifies 'to bring' or 'to raise'. Therefore *allevare* means to bring towards the spirit, or rather to help in its spiritual evolution. It seems, however, that the meaning of this word is not given much consideration in modern animal husbandry. This can also apply to other kingdoms of nature. But to work with these we must first realise the origin of what is around us.

If we now develop another table (below) we can add another set of information. We said that Old Saturn was a sphere of heat, so in the corresponding column we have written *warmth.*

We know that warmth is the only force that cannot be contained because it heats any walls around it which in turn continue to radiate the warmth. This can be confirmed by all of us because, if it were not so, it would be sufficient to heat a room once in winter and it would remain hot until spring if we are only careful not to admit the cold. In that reality consisting of substance in varying degrees of warmth, everything was *one*: us, the animals, the plants and the entire solar system.

We used the word 'substance' and not the word 'matter' because substance is the essence of matter - and matter did not really exist as such until the phase of manifestation called 'Earth'.

The second phase, corresponding to the old Sun and to the Persian civilisation, brings a separation of this unity. Therefore we have shown the duality between light and darkness.

In the third phase, corresponding to the Old Moon and the civilisations of old Egypt, Light is sublimated into *Sulfur*, Darkness falls in *Sal* and part of each form *Mercur*, the element of exchange or balance. We arrive at an image of the world as a series of processes.

We get to the fourth stage in which a part of *Sulfur* comes to rest as the *Fire* element, a part of *Sal* falls as an attribute of *Earth*, while combinations of *Sal, Sulfur* and *Mercur* form the elements *Water* and *Air*. We link *Earth* with the mineral world, *Water* with the plant world, *Air* with the animal world and *Fire* with man. This last Greco-Roman phase is the one for which the Tetraktys is an image of everything that we have proposed.

In the phase corresponding to the old Moon there was no distinction between humans and animals. These two kingdoms were merged together. It is not a coincidence that in the culture of Egypt, the repetition of Lemuria, there are many depictions of people with animal heads.

In Lemuria the conditions became impossible for human life on Earth, because the Earth still contained the Moon that was the bearer of the forces of death. In these conditions, humans withdrew from the Earth-Moon planet, with the exception of some who became set in a form and are today's existing animals. When the Moon was ejected[4] the conditions were mitigated allowing the re-descent and renewed earthly evolution of humans. This makes it clear that the human being is not an evolved animal; rather the animal is a derivative of the human being. One could better say that the animals are *human beings* that have sacrificed themselves by descending early to Earth.

During the Lemurian epoch there was also intermingling between the plant and animal kingdoms, and between the plant and mineral kingdoms. Present-day minerals were derived from this last intermingling as well as a few present-day plants. Therefore, there are existing plants that were derived from an evolution of *Sal* towards the element of Water, and others from the fall of *Mercur* towards the element of Water. The plants that are the result of rising up are the existing grasses, while those who descend from *Mercur* are all the others.

Among those who descended there are some who have stopped their fall at an intermediate level between Water and Air and these are legumes that are more rich in vegetable proteins. Even closer to the element of Air are the plants rich in alkaloids from which we derive drugs.

---

[4] This was a result of Mars colliding with our planet, whose evidence today is the Pacific Ocean.

TYPE OF AGRICULTURE	NONE	LIGHT AND DARKNESS	3 PROCESSES	4 ELEMENTS	NEW SYNTHESIS	NEW SYNTHESIS	NEW SYNTHESIS
DESTRUCTION					FRAGMENTATION (*CLONING*)		
DIVISION OF THE KINGDOMS OF NATURE AND THEIR REUNION AFTER GOLGOTHA				*FIRE ELEMENT* (HUMAN)			
			SULFUR (ANIMAL-HUMAN)		HORN SILICA (501)		
		LIGHT (DIURNAL BEINGS)		*AIR ELEMENT* (ANIMAL)		NEW PREP	
	WARMTH (IMMATERIAL PHYSICAL MAN)		*MERCUR* (VEGETABLE-ANIMAL)		COMPOST PREPS (502→ 507)		NEW (IRON)
		DARKNESS (NOCTURNAL BEINGS)		*WATER ELEMENT* (VEGETABLE)		NEW PREP	
			SAL (VEGETABLE-MINERAL)		HORN MANURE (500)		
				EARTH ELEMENT (MINERAL)			
DESTRUCTION					*FRAGMENTATION (ATOMISAT^{N}.)*		

Also between the mineral and plant kingdoms there exist intermediary conditions such as the world of *crystals*, which are the only minerals that 'grow' by an external accretion process.

Another intermediate sector between the mineral and the vegetable is that of the viruses, which do not have a life of their own, but use the life of the cells that host them to reproduce. In the world between the animals and man are all the monkeys and other animals with semi-erect bearing. It is vital to understand this information about animals and plants with which we must interact in our farming.

Obviously if we shift our attention from the trinitarian Hermes Trismegistos to the dual perspective of Zarathustra which corresponds to Hyperborea, we find more profound intermingling between the mineral, plant and animal kingdoms so the human being is very close to the animal and plant beings. Therefore, during the Zarathustrian times, human beings found it much easier to modify plants.

Having presented these considerations we must take note that modern people have arrived at a pivotal moment in which we have forgotten everything that the knowledge of the previous epochs represents. We can choose to take an ascending path guided by the impulse from the event at Golgotha thus creating a new synthesis, or we are condemned to further fragmentation that would lead to destruction.

This fragmentation (which is already under way today with the hell-bent exploitation of nuclear power as well as the fragmentation of the human being itself, now understood as a set of organs to be transplanted and maybe even cloned in a laboratory) will lead to the disintegration of the world, perhaps beyond a point of no return.

If instead we seek a new cultural synthesis - an operation that would start with a new 'three-foldness' – we could do a lot worse than see what is shown in the fourth and fifth lecture. Here is precisely where we are given the biodynamic preparations, where the synthesis of Fire and Air is the horn silica preparation, between the Earth and Water horn-manure, and the synthesis between water and air is shown in the preparations derived from yarrow, chamomile, nettle, oak, dandelion and valerian. In this way human beings can become the means through which a new evolutionary development of the Earth and its kingdoms can take place.

It will then be possible to reach the second and third syntheses depending on the strength we will have.

First Koberwitz Lecture

Second Meeting

Now that we have identified the structure of the 'agricultural course' held by Rudolf Steiner in June 1924 at Koberwitz, we can begin to look at the contents of the course itself. '*Spiritual-scientific Foundations for the Renewal of Agriculture*', to give it its longer name, begins with an introduction which is a transcript of the talk given by Dr Steiner in Dornach on June 20th, 1924 after his return from the Koberwitz course[5].

We will not analyse this introduction but recommend that everybody finds time to read it. Instead we will begin with the first lecture which was given by Dr Steiner on June 7th, 1924 and which has been entitled: '*Preliminary speech and introduction to the course. The Emancipation of human and animal life from the external world*'.

We will identify the different paragraphs of the Agriculture course transcript by referring to their first few words[6]. It wouldn't be practical to refer to page numbers because different editions of the agricultural course have different page numbers.

Keep in mind that the course was held at Koberwitz and therefore in Silesia. In 1924 this region was part of Germany whereas today it is part of Poland. This is a significant fact because there exist three cultures in Europe which may be defined as the cultures of the past, of the present and of the future. The culture of the past is the latin culture which refers to Rome. The culture of the present is German-Saxon, and the future culture is Slavic. Therefore, Rudolf Steiner chose a place which is on the border between the present and the future. We might, therefore, consider whether the course we are to study is a course for the future which will only manifest its full importance in times to come.

It is important to remember that the first lecture corresponds, though not explicitly, to the agriculture of Zarathustra practiced during the Persian civilisation. This period is an echo of the Hyperborean epoch and, further in the past, of the second phase of the Earth's manifestation which we refer to as the '*Old Sun*'.

We are, therefore, in a period of time in which a dualistic, polar or two-fold vision of the world and of life is predominant. Within this chapter we find references to the polarity between full and new Moon, between the upper and lower parts of the plant, between silica and calcium, and between the influence of the external planets and that of the internal planets.

The course begins with expressions of warm gratitude to the family, thanks to whose hospitality and committed organisational work the course was made possible. The introduction does not present particularly important points until the sixth paragraph. Only a phrase in the middle of the third paragraph ("*I am sure Frau Dr. Steiner ...* ") needs clarification, where Steiner acknowledges the impulse which Count Keyserlingk has given to agricultural initiatives that have their roots in another movement called 'Kommende Tag', literally 'the coming day' or 'the coming time', though we might also translate it as '*the future epoch*'. This movement was founded

[5] This is said for the Italian edition but is also true of most English editions

[6] This English translation will refer to the Creeger Gardner translation of 1993. ISBN 0938 250 353

mostly by young anthroposophists inspired by a model of society based on three-foldness. This way of conceiving human social life was particularly dear to Rudolf Steiner who felt that it was a necessity of the times. However, the movement collapsed, partly because of the disastrous post-war situation in Germany characterised, amongst other things, by the very high inflation which frustrated any economic initiative. We will not elaborate here on the three-fold social order, but would like to suggest "Basic Issues of the Social Question", by Rudolf Steiner as further reading for those interested.

Approximately in the middle of the fifth paragraph ("*Whether you will go away...*") we find the statement that agriculture is the central concern of human life. We would like to point out that Steiner did not intend to refer only to the question of nutrition because it is clear that the production of food, both vegetable and animal, holds a primary role amongst all other human activity. Agriculture and animal husbandry are also the activities through which we are most directly involved in the stewardship of life. Furthermore the farmer has the responsibility of furthering the evolution of the Group I's, encouraging positive experiences by means of their manifestations on the Earth, whether they be mineral, vegetable, or animal. No other activity is vested with such a heavy responsibility. Let's recall that Zarathustra, who domesticated plants - a fundamental act in the evolutionary path of nature - taught agriculture only to his most evolved disciples. Such was the importance he attributed to this activity and the level of morality demanded from those who practiced agriculture. What we can gather from this phrase is a call to our conscience for agriculture not to be considered merely from an economic point of view, so it may begin to be considered as an opportunity for spiritual evolution which must spring from an acknowledgement of nature's own evolutionary requirements.

We then come to the sixth paragraph (*"In some respects, agriculture ...")* in which Steiner speaks prophetically of the destructive character adopted by modern culture. Let us recall that the year was 1924 when many types of pollution would have barely registered on today's charts and environmental conditions were such that today's growers would yearn for them. However, the problems whose negative effects Steiner was anticipating were primarily of a cultural nature, and he could already foresee what would result if these trends continued.

Steiner's vision was so wide that it could not be confined to the historical moment in which particular concepts were expressed. His impulses will retain their validity for a much much longer time in part because he wasn't conveying techniques but rather investigating laws of the physical and etheric levels in relation to spiritual laws. The efficiency of a technique is limited to specific conditions, whereas a living thought may adapt to any situation, proving itself much more practical than a technique alone.

Unfortunately, we so often prefer rigidly categorised concepts and laws to the point that some indications given by Steiner to farmers could not, for example, be reconciled with the discipline of modern organic agriculture. Let's give an example: he was asked what could be done for a very poor soil where nothing would grow, keeping in mind that organic matter might not be available. Rudolf Steiner's answer would seem anathema to a modern organic farmer, since he advised the abundant application of a chemical-mineral fertiliser. However, he gave two conditions: that the fertiliser not be chemical nitrogen (he advised the use of potassium sulphate) and that it be used along with a strong poison (such as digitalis).

In large doses digitalis is a fatal poison, but in appropriate doses it becomes a heart tonic, in other words it helps circulation. Therefore, it is possible to use digitalis in homeopathic dilutions to stimulate the 'digestion' of the mineral fertiliser in the soil. This indication was applied in northern Germany by planting digitalis plants on a soil where potassium fertiliser had been used - with excellent results.

The seventh paragraph begins with the words: "*In order that we may speak in concrete terms and not in generalites ...*". We therefore leave the introduction and begin to speak of agriculture specifically. Steiner immediately begins by clarifying a concept which is valid in all human activities and not just for agriculture. This is the concept of *competence*: only he who is competent has the right to speak. This is surely a 'law' of common sense, but in our times this law is continuously broken when, in the name of freedom of expression, anybody may pronounce upon any subject. We have even gone as far as to invent columnists who express general thoughts and who are invited to speak on any subject even though they might not know anything about it.

Rudolf Steiner also defines what it means to be competent and states that competence must come from direct experience. He says that "*... people have no right to talk about agriculture, including its social and organisational aspects, unless they have a sound basis in agriculture and really know what it means to grow grain or potatoes or beets ... ... the only ones entitled to an opinion on agriculture are the people whose judgement derives directly from the field, the forest, and the stable*". Today, therefore, we have agricultural engineers who are inherently incompetent because their studies did not include a single hour of practical work in the fields.

This concept is restated in the eighth paragraph ("*The reson that all kinds ...*") where it says that it is not enough to analyze a plant from every possible point of view (Steiner uses the beetroot as an example) but we must also understand its vital links with the soil and with the season in which it matures. For us this approach is especially important. In fact throughout all our work we should always try to bring to mind that competence is born out of true knowledge which is itself founded on two pillars: 1) study, which allows us to acquire concepts, and 2) perception, which obviously must be adequately educated to grasp the essence of things. This knowledge can truly allow us to comprehend or understand the plant world, its relationship to the other kingdoms of nature and to the laws of life.

In the subsequent two paragraphs Rudolf Steiner gives an example of a study method which is also educational. He says that to understand anything it is necessary to start with a general vision and then to come down to the particulars. He then proceeds to give two examples, the first of which belongs to the inanimate physical world and the second to the plant world. The first example is the seemingly obvious statement that the cause of a magnetic needle's orientation lies not in the needle itself but in its relationship with the entire planet. The second example suggests by analogy that neither can the plant be considered as an end in itself and that its processes of growth are determined by factors existing in the entire cosmos.

This comparison is sufficiently good to highlight the fact that the plant, like the magnetic needle, is polarised. One end looks toward the Sun and the other end toward the centre of the Earth. In other words, just as the magnetic needle - which belongs to the inorganic world - is inserted in an invisible force field which directs it, so the plant in the organic world is inserted in another system of forces - invisible as well - which directs its growth. Therefore, if we want to understand the plant we must also understand this system of cosmic and earthly forces.

When considered in this light, a plant is simply an indicator or 'bio-assay' of the force field which envelops it, and so it will manifest different forms and colours depending on the qualities of this force.

The consequence of what we have just stated is that in order to improve our farming and gardening we would benefit from the ability to recognise the dominant forces within the environment in which we work. Otherwise we would not be able to understand which plants are appropriate for the environment - or if we did decide to cultivate plants which were not in harmony with the environment we would not be able to sustain these crops by supplementing the missing forces.

In the final part of the tenth paragraph ("*To many people, however,* ..."), Steiner makes an observation of fundamental importance. He states that modern culture has taken a path to knowledge which is based more and more on analysis, losing all connection with general and universal laws and thus creating a series of problems which are typical of our civilisation. Fortunately, he says, human beings still have a little instinct left which allows us to be aware of our real needs, beyond what modern science believes. In other words it was possible to do without any indications derived from the scientific world because in the end our instinct is capable of revealing our real needs to us. The problem is that we have lost many of the instincts which guided our ancestors. Exactly because of this we should be seeking, "*... to make use of a deeper spiritual insight in order to discover what these increasingly unreliable instincts are now less and and less able to supply*". We do this by becoming more and more intimate with the laws of nature through properly educating our perception.

In fact the force field of which the plant is a manifestation does not only give the plant its form, but also governs its internal processes such as those which - borrowing the term from the animal kingdom - we could call *instincts.* We could then recognise a plant's light and water requirements, its inclination to make a fruit pulpy or dry and all the other defining characteristics of that species. In all cases this is represented by appreciating the externally visible form as an expression of the force field that results in the manifest plant.

If we have grasped Steiner's thoughts expressed up to this point then we have grasped the basis for all agriculture! The rest of the course is the explanation of how to recognise these forces, how to understand what is lacking or what is excessive, and how to achieve a healthy balance.

The theme of instincts is also central to the twelfth ("*However, from an anthroposophical* ...") to fifteenth paragraphs ("*I cannot guarantee* ..."). We can consider instinct as the intelligence of the physical body in the same way as habit is the intelligence of the etheric body. Obviously there are other forms of intelligence such as those of the mind or even of the heart. Surely instinct is the most fundamental kind of earthly intelligence, that which conditions us the most, because it is determined by the strongest bodily needs of the human being such as hunger, thirst, reproduction and sleep.

There are, however, more evolved instincts which were incorporated into human beings thanks to the action of the great initiates, the first of which is Zarathustra. He had grasped the fundamental truths of existence but, considering the evolutionary level of his contemporaries, he had to use particular forms of communication such as sacred rituals in which mantra, chanting and liturgy played an important part. If repeated many times all of these acted powerfully on the etheric body of human beings until it imprinted itself in the physical body and created those new instincts which allowed one to grasp the messages of nature. However, very slowly, human beings detached themselves from the sense of truth carried by the great initiates and so even the sacred rituals gradually became repeated merely out of habit, until they became empty movements governed by superstition. In the end, having lost their initial meaning and become empty containers, even the superstitions were abandoned altogether.

Human beings are left with our basic instincts, and can no longer access those that were incorporated to help us in our evolution. In this way we are confronted by a *tabula rasa* and we realise that we have two choices: either build an artificial world according to the guidance of materialistic science which substitutes the white coat of the scientist for the black robe of the priest, or slowly re-establish a new relationship and alliance with the cosmos.

It's obvious that we intend to follow the second path, but to do this we must start from an intimate connection with the plant world which will allow us to formulate thoughts that we will experience as increasingly real, until they are imprinted first in the etheric body and then in the physical body, thus nurturing new faculties that will enable a renewed relationship with nature and the beings that manifest within her.

The eleventh paragraph ("*Before any science ...")* refers to what we have just stated. The old instincts were the result of an unconscious relationship with the beings that weave nature. Today we have a few remaining examples such as in those that have 'green thumbs' thanks to which any work that they do with plants is wonderfully successful. However, they do not know the reason for this great success. In contrast, the new faculties of which we spoke earlier will allow us to have a conscious relationship with nature. By this we mean that in the old relationship with nature we could grasp certain truths without penetrating them. For example we instinctively grasped the connection between binary leaves and the Moon. Even if this is enough for us, we should understand why the Moon governs binary leaves and this is the new discipline which we must develop.

Note that Steiner does not limit himself to say that we have to observe plants. He says "*... we have to commit ourselves to a much broader way of looking at the life of plants ...*". Evidently Rudolf Steiner invites us to make a different observation from that made by Linnaeus or that which would be made in a laboratory with a microscope and which would lead us to sterile cataloguing based on morphology. Rather he invites us to enter the etheric plane to grasp the action of the formative forces, recognising the gestures of the plants, understanding their transformations and metamorphoses. In this way we can understand what the plant needs and how to work with it. The best book could only give the best description of a plant and of its life, but to really grasp life we must come into direct contact with nature. Even the best description of the perfume of the rose will never give us the emotions egendered by smelling one.

At this point we will give an indication to keep in mind throughout the rest of the study of the agricultural course. When Dr Steiner uses the word '*life*' he always refers not to the physical plane but to the etheric plane and thus to the source of life which is the Sun. This supposes the development of superior qualities of consciousness within us.

In the thirteenth paragraph ("*It is quite right ...*") Steiner relates a diverting episode which should not, however, be interpreted merely as a digression to allow for a break. After having spoken of a conscious relationship to cosmic laws and of the etheric plane, he makes an example where he speaks of full Moon and new Moon in relation to rainfall, making a clear connection between these subjects. The Moon can be considered the fundamental *regulator* of the etheric plane, which is the support for all manifestations of life. Let's recall that during the period before Full Moon there is 7% more rainfall.There is another interesting observation: "... *science can be quite correct, but real life can't always afford to act according to scientific 'correctness...*" The explanation of this phrase lies in the fact that there are things that are correct but that are not true. An example can be found in Einstein's theory of relativity which is correct on a mathematical level, but not true. A while ago in the Cuneo region, we had worked on a therapy for chestnut tumours, obtaining very good results which had satisfied the local farmers. However, the statistical analysis showed insignificant results as if our work had had no effect - despite the evident success. Clearly, also in this case, the statistical calculations (which were correct) gave results which did not correspond to the truth of the situation.

In the sixteenth and seventeenth paragraphs ("*So by way of introduction* .." and, "*If we studied*..") Rudolf Steiner inserts the new concept that life is governed by rhythms which we must begin to grasp. He then says that the simplest forms of life, like plants, are complete expressions of these rhythms while the more evolved forms of life, like animals and human beings, although they follow the same periodicities, have internalised them. For example the menstrual cycle is connected to the Moon, but it does not coincide with the position of the Moon in the sky although it maintains the same frequency. Within the female cycle we can recognise a constructive phase that corresponds to the waxing Moon, and a destructive phase which corresponds to the waning Moon. The organs which have closed in on themselves carry within us the planetary forces to which they are connected, and they conserve their rhythms. The plant, on the other hand, does not have enclosed organs. It does not have a kidney, a heart, a liver and so on because the plant's organs are outside of the plant, in the planets. We could say that animals have internalised the cosmos while the plant lives by continuous connection to the cosmic periphery where it finds the forces that fulfil these functions. This is the reason that it is more challenging to recognise the animals' connection with the realm of the planets, while it is easier to grasp it within the plant.

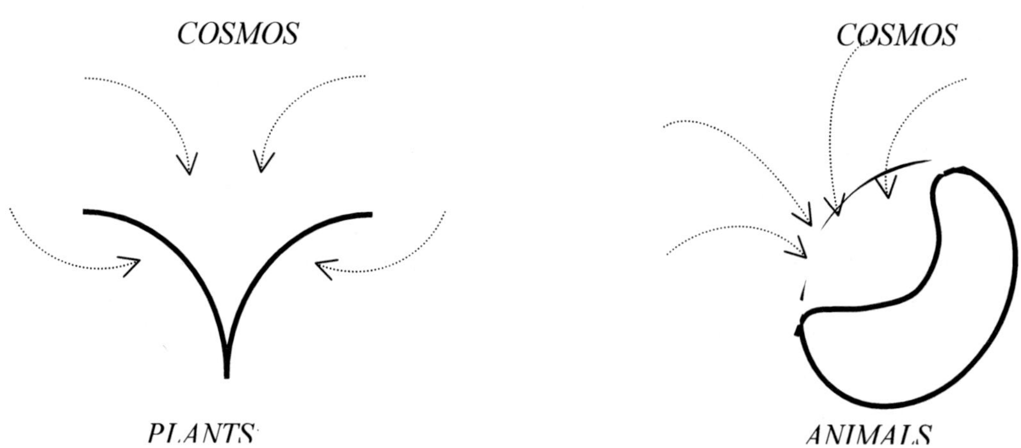

In paragraph seventeen ("*If we studied things more closely* ...") there is a reference to sunspots in relation to social illnesses. Steiner is referring to psychological illnesses which reach a peak every eleven years or so, a rhythm which corresponds to that of sunspots. He then refers to illnesses which manifest a cycle of fever over seven days, corresponding to the number of planets. In fact one of the illnesses which manifests a high fever is still today called influenza and the name itself indicates that we feel, or are influenced by, a particular planetary action which we don't manage to integrate. Those who have the ability to maintain their equilibrium when subject to such an influence do not fall ill even if they are exposed to the same risk factors. (The illness which we call influenza is related to the position of Saturn's rings with respect to the Earth. In fact it is easy to make the connection between the element of Fire, related to Saturn, with the manifestation of fever.)

We can observe that Rudolf Steiner, without making a point of it, used examples which first referred to the Moon, then to the Sun and finally to Saturn. That is he first showed the impact of an internal planet, then the central 'planet', and finally a representative of the external planets, thus embracing the entire solar system. With the first example he establishes a connection with the *etheric* world, the life world, especially related to water. With the second example he connects with the Sun and speaks about psychosis. For the American Indians, psychotic people were considered to be touched by the spirit and we can grasp a doorway to the *spiritual world* in the Sun which manifests within us as the higher Self. Finally Saturn and the world of fever is connected to a situation which we could define as *astral*.

The eighteenth paragraph ("*With regard to human life ...* ") is particularly important. Dr Steiner here clarifies the first principle of agriculture: everything that takes place on Earth is only a reflection of what happens in the cosmos, and since the plant is completely immersed in the life of nature and in its rhythms, it reflects what happens in the cosmos and is its indicator. The plant is a reflection of the cosmos.

The plant is surrounded by one life-force that we call Mother Earth and another that we call Father Sky. These embrace the plant and manifest as its functions, so that the plant is the direct expression of these forces. Therefore, we can consider the plant to be an image of these forces. Studying a plant can reveal many things about the planets that act upon it, but it is not enough to look at it distractedly. It must be studied deeply until the cosmic influences reflected in it are understood.

In the nineteenth paragraph ("*You see the Earth is surrounded ...*") we find the description of the solar system. It is important to note that in this paragraph Rudolf Steiner lists the planets in the order according to the Ptolemaic vision of the cosmos. With the advent of the Copernican revolution the names of Venus and Mercury were exchanged.

In the following paragraphs Steiner repeats the word '*life*' many times both in relation to the cosmos and the Earth. Within this repetition we must recognise a pressing invitation to look at the Earth, the Cosmos and their relationships from the etheric point of view. It will be impossible for us to farm in the way indicated by Steiner if we continue to hold a materialistic vision of the world. The plants and their life processes cannot be observed solely from the bio-chemical or the bio-electric point of view which modern science champions, because the world of life does not obey these laws of the modern materialistic point of view.

There are a few words which we mustn't miss because they are of central importance. Steiner speaks about "*a broad view of life on Earth...*" He asks us to widen our vision to the etheric plane, and also to the astral plane and even to the spiritual plane. Even in relation to the physical plane our consciousness should not stop at the Earth's surface but should embrace the globe right into its interior. Even in the bowels of the Earth there live spiritual beings who act forcefully upon the life of plants. Let's not forget that our whole approach to pests is based on an understanding of the various layers of the Earth and of the forces that emanate from there.

Also in this very important paragraph, Steiner states something which is in front of everybody's eyes but which, despite this, is almost unknown. He says that silica has a fundamental role in the life processes of the Earth. He points out that silicon constitutes 28% of all earthly substance. Later in paragraph 23 ("*Half, that is forty eight percent ...*") he specifies that if instead of silicon we consider silica - silicon oxide - we are considering 47-48% of what we find in nature. We should also note that 90% of the ashes of plants like equisetum is silica. Silica is opaque to light and this fact brings to mind the gospel of St John: "*...And the light shineth in darkness; and the darkness comprehended it not*". With these few words Steiner indicates the path to understand the laws of life on Earth - study silica.

We think it is worth examining how Rudolf Steiner approaches the subject of silica. First he refers to its form. He connects it to what he had already suggested about the existence of two poles in the plant - the cosmic and earthly poles. The key to understanding this duality is form.

Starting from these few words we can dive into an infinite sea. Form is the expression of a certain balance between cosmic forces and earthly forces and in the case of quartz this form is a hexagonal parallelogram with the same proportions as that of the bee hive cell. However, this form is found specifically in crystal quartz grown in Europe. The American quartz has a pyramidal shape wider at the bottom, and the Asian quartz is exactly its inverse with the wider end at the top. We all know that America is dominated by hardening processes of *Sal* and Asia by expansive processes of *Sulfur*. Evidently these forces are strong enough to modify the forms of minerals.

We must, therefore, understand the effect of silica and limestone on forms because they are fundamental variables to keep in mind when studying the forms that we will meet later on.

AMERICA *EUROPE* *ASIA*

At this point it is important to realise that, in this first lecture, Rudolf Steiner is describing the plant which manfiests within a macrocosmic polarity as with light and dark or the magnetic poles of a compass, and he describes silica and limestone as the poles of the plant. Silica is indicated as being the ambassador of the cosmic part of the plant, while limestone is that of the earthly part. The clarification that silica constitutes 48% of the Earth is of particular relevance if one is aware that silica is the bringer of cosmic forces to the plant via the roots, in the same way that limestone is the mediator of earthly forces.

In the twenty-second paragraph ("*From such things* ... ") we find the accusation that, despite the fact that silica is found in such great abundance on the Earth, it is practically unknown to the scientific world. It may seem incredible, but our scientists undertake detailed study of the effects on the plant world of bromine, of iron, etc which are only present in minute quantities while they completely overlook the most widespread element.

To understand silica's origin we must go way back in time to the Lemurian epoch. The Lemurian epoch is the pre-Atlantean repetition of *Old Moon.* Matter had not reached the solid state of aggregation but had at most reached the liquid state. We can imagine a liquid-etheric condition in which solid minerals could not have yet existed. The plant world had already manifested within this reality but we should not imagine it just as we see it today. In fact the plants of that epoch did not have stems and did not grow from the soil! They were only a flower which hovered in the sky like our modern butterflies. An effective representation of what we are describing was created by Botticelli in his '*Birth of Venus*' with the plant-flowers emanating from an androgynous, aereal being.

At a certain point the hardening process of limestone attracted these flowers to Earth which, in order to have a relationship with their changing environment, gradually had to develop leaves, stems and roots. The flower petals which had fallen to Earth condensed into silica so that, from this point of view, our planet, consisting of almost 50% silica, could be considered as being an immense flower. After all the earthly Paradise known as the garden of Eden dates back to the Lemurian epoch.

The part of the flower consisting of the ovaries, the stamen and the pistils has formed the feldspar and the better known clay, while the calyx allowed the formation of mica. Silica, clay and mica are the components of granite which can be considered to be the mother rock from which other rocks are born.

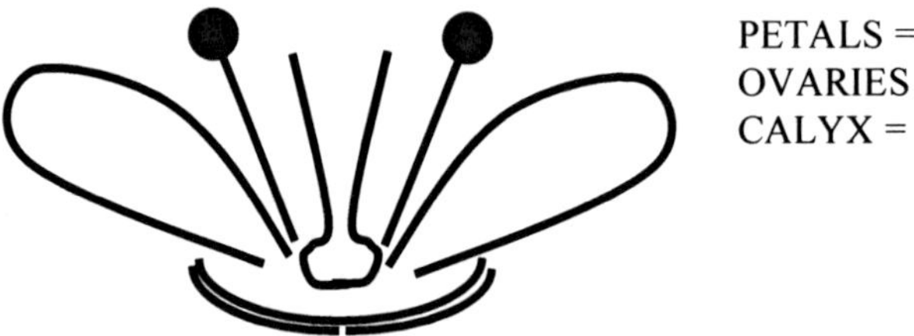

PETALS = SILICA
OVARIES = FELDSPAR
CALYX = MICA

The descent of flowers to Earth corresponds to the 'fall' when mankind was expelled from earthly Paradise. It was a fall to the physical level, the densification of Light into Darkness, the consequence of which is the modern opacity of carbon: diamond has become charcoal.

The idea that the plant was born from a flower, that it then gradually and successively produced the first leaf bud with the first leaves, then the stem and lastly the root, is undoubtedly unconventional but it is confirmed by the fact that if we make a sensitive crystallisation with the leaves of a plant, we discover that the base leaves, which emerge first on a modern plant, manifest a '*younger*' gesture than the apical leaves which are born later. This allows us to understand that the physical plant goes through the inverse process from that of the fall and, therefore, represents the resurrection from the fall itself. Because of this the plant represents an element of virginal purity which, when used as food, can sustain human beings in their evolutionary path. It is also the reason why we use plants as medicine.

In the twenty-third paragraph ("*Half, that is 48 per cent, ...*") Steiner tells us that if there were less silica on Earth we would have pyramidal-shaped plants (therefore of the 'American' type since the forces of limestone are dominant) and that flowers would atrophy. In this way Steiner is telling us that flowering is connected to silica.

Paragraph 24 ("*On the other hand ...*") speaks instead of the effect that a reduced quantity of limestone would have on the form of plants: by reducing the earthly forces, plants would have a very thin stem and a shaft similar to a vine, unable to support itself.

Steiner has mentioned two substances: silica and limestone. He will speak of clay only in the second lecture. Silica first emerged on the *Old Sun* and is carrier of *spiritual* qualities, while limestone has its origins on the *Old Moon* and carries *life* processes. Let's also say that clay originated on the *Old Saturn* and is the carrier of *astral* qualities. Obviously we are not referring to minerals which could not have existed in those epochs, but to real spiritual entities and to their relative forces which, later on Earth, will govern the three processes and the three material substances of the same name with which we are familiar. Dr Steiner speaks of processes and substances and explains that material substances are the residues of processes that have come to rest. It is always important to keep in mind that the plant lives in processes even though it uses material substances. This key will allow us to understand the reason why, if we want to act in a hygienic and therapeutic way, we do not believe it is appropriate to give substances directly to plants. Before inserting substances in the processes of plants we must release them from their stasis by dynamising at the opportune time.

Let's now attempt to go back in our imagination to the Lemurian epoch 35,000 years ago. In that epoch there existed a celestial body consisting of a united Earth, Moon and Sun. This entirety had a watery, albumin-like consistency and the plants, as we already said, lived in flower-form as very fine silica that floated around the Earth. This single planetary body which 'breathed' aerial plants became more and more dense so that the Sun, with its connected beings, detached itself to proceed unhindered with a higher development. Celestial bodies carrying less evolved entities - the Earth, the Moon, Venus and Mercury - remained together. At this point the atmosphere became albuminous and from the sky came a milky substance which condensed and became limestone. The Earth has hardened continuously and increasingly and the limestone intensified its contracting action, bringing plants lower and lower until they began to develop leaves and roots. Even animals breathed limestone and gradually they condensed it within until, at a later stage, the bones were formed.

After the Sun's exit, Mercury and Venus also detached themselves from the original celestial body because of the ongoing contraction, allowing the evolution of their connected beings. Therefore, there remained a reality formed of Earth and Moon. In that reality, plants, animals and human beings were distinct as they are now, but there were living forms in between plants and animals, as well as between animals and human beings. In the end the Moon also detached itself and the Earth we know was formed and continued to harden until its present state.

The function of limestone today is to hold life in living beings whether they be plants, animals, or human beings. Life itself, however, is connected to silica and not to limestone which represents death. Animal limestone is much more 'eager' than the plant limestone. The animal hides its reproductive parts while the plant does not. In the human being the reproductive function has emancipated itself even from the cosmic rhythm as a results of original sin, which lead to the loosening of the tie between the Moon (reproduction) and the Sun (representative of life in the cosmos).

After the fall man becomes aware of being naked because he has materialised and the limestone within him becomes the basis for death and hate. However, limestone can be redeemed and become a messenger of life. Particularly to symbolise this, all of Christ's life is tied to limestone (Nazareth, Gethsemane, Golgotha). After all, life on Earth develops in the spring because of the fact that the Sun warms the Earth, provoking the etherisation of limestone.

We can recognise three fundamental gestures in limestone which give rise to an initial classification:

- Lunar limestone which extinguishes life
- Earthly limestone which maintains life
- Solar limestone which transmutes life

From this initial classification we can come to recognise seven aspects of limestone:

- Inorganic limestone, lunar and dead: slaked lime (hungers for astral) - bone-mineral
- Inorganic limestone, earthly and dead: quicklime (hungers for etheric) - mineral
- Organic plant limestone, alive: corresponds to camomile
- Organic plant limestone, dead: corresponds to oak bark
- Organic animal limestone, alive: corresponds to the worm
- Organic animal limestone, dead: corresponds to the crab shell
- Organic limestone, resurrection: corresponds to the horn manure preparation '500'

We will come back to this subject but we can clarify a few aspects which were laid out above.

We all know that we place limestone in the compost heaps, more precisely slaked-lime in the plant material heaps and quicklime in the manure heaps. The reason is that the plant material heap is very rich in the etheric element, and so it is necessary to insert a substance in the heap whose hunger for life has been partly satisfied by the absorption of water, thereby favouring the absorption of the missing astral element. On the other hand the manure heap has a very strong astral element and is not very rich in the etheric element, therefore it is necessary to use quicklime whose intact ability to attract and maintain life ensures the maximum retention of the etheric element.

We will not dwell at the moment on chamomile, oak bark or the worm, but briefly clarify something about the crab-shell and its relevance with relation to limestone. Throughout its growth process, the crab must intermittently exceed the dimensions within which its exoskeleton has enclosed it by shedding its outer encasing and reforming it after having its growth spurt. Within this gesture we can grasp the capacity of limestone to transmute, or at least to dissolve and coagulate, a capacity which is used in anthroposophical medicine to dissolve kidney stones in particular.

The last and highest or Christianised type of limestone is the one carrying the forces of resurrection within - due to being buried underground. We are speaking of the horn manure preparation now known as 'preparation 500'.

Third Meeting

Together we have delved into the first 25 paragraphs of the first lecture of the Koberwitz agriculture course. We have appreciated how Rudolf Steiner characterised silica and limestone, completely ignoring the qualitative and quantitative chemical aspects and preferring to express himself in terms of form or shape. Understanding form is to understand the action of the *Spirits of Form* whom we now know correspond to the third category of the second spiritual Hierarchy. They are called the 'Powers' or the 'Dynamis' in esoteric Christianity and have their centre of action in the Sun.

In other words we could say that reading form gives us the possibility to pass through the 'gateway of the Sun' and to acquire the consequent initiation. We can also state that observation of form is a training that allows the modern human being to cross the threshold beyond which one may grasp the weaving of life. All of this will constitute our base from which to approach the higher hierarchies and the world of the 'group I' or the 'individual principal of each species'[7].

Our departure point for this journey is the Earth which, being a solid body, is impenetrable to light and is therefore internally permeated by darkness. In other words our feet rest on this world of darkness and clearly we must make a qualitative jump to begin to look higher. The problem lies in the fact that we have organised ourselves to make our life more pleasant in this material world to such an extent that we have not only fallen in love with the system that we created, but also find it difficult to come out of this dimension. Indeed, many consider it to be the only real one.

Starting then at the Earth and following the impulse of human beings to return to their cosmic origin, we first meet the lunar sphere. The Moon is the planet that administers life, so we can begin to familiarise ourselves with the laws of life and of life's rhythm via its celestial motions. The Moon also strongly influences the weather - which is so chaotic today - and here too it will be necessary to understand the imbalances in order to attempt to find a remedy.

Beyond the Moon we meet Venus and Mercury which strongly influence the entire realm of reproduction. We must also get a thoroughgoing understanding of these. Here also Steiner suggests which path to take since, as we have seen, after having spoken of limestone and silica he introduces the subject of form. He re-emphasises that study of the metamorphoses and variations in forms is the most important way in which we can be adequately prepared for reaching the solar threshold.

Let's go back to commenting on the course at Koberwitz from where we left off, from the twenty-sixth paragraph of the first lecture ("*With that, let us now look ...*"). Steiner is still presenting a two-fold view since we are still reading the lecture connected to Zarathustran agriculture and, while beginning to speak to us of the plant, he notes that the plant lives within a polarity between the lower part and the higher

[7] The I or self is that part of us which can learn from experiences and use these insights in future events. It is the I that endures and perseveres. The I plans and musters determination to see things through using its own will power. The I thus gets traction on its own destiny. Plants have an I but this is shared with all other plants of that species. Each dandelion, for example, has the one 'dandelion I' in common and we can call this the individual principle of the dandelion species.

part. The lower part looks toward the centre of the Earth and is connected to the "*inner force of reproduction and growth*". The upper part is dedicated to the formation of nourishment for the other kingdoms of nature. By 'force of reproduction' Steiner means the force of vigour and growth, or the vegetative forces that act from below - even though we connect the idea of reproduction with the flower. In order to avoid such a misunderstanding we will not use the term 'reproduction' when referring to these forces but rather 'vigour'.

The same polarity can be seen as an opposition between the world of light and the world of darkness. The former consists of an opening and giving, an expansion outwards so that the other kingdoms of nature may enjoy that which is produced by the plant such as smell, colour or food quality. The latter has a centripetal gesture, an expression of the plant's need to bring minerals, water and life for its own survival.

At this point Steiner clarifies that all that has to do with vegetative forces is related to limestone because limestone expresses a gesture that we could characterise as *greed.* Limestone dries, absorbs and takes. Now we can understand how a limestone soil favours the development of the plant's lower pole thereby enhancing the quantitative aspect of production while it wouldn't enhance the qualitative aspect connected to the nutritional value of the products. We will see shortly that the element that governs quality is silica. Let's remember that during the previous century grain crops have lost about 30% of their silica content, coinciding with a definitive decline in quality that manifested outwardly as the dwarfing of the plant. At the beginning of the last century wheat was as tall as the shoulder of a man while today wheat is about 50 centimetres tall and we are trying to make it even shorter. In this way we are convinced that we can counteract the phenomenon of lodging which is really symptomatic of a too great enhancement of the vegetative pole at the expense of quality. This comes about because of excessive and chemical nitrogen fertilisation.

The twenty-sixth paragraph concludes with another important piece of information: Rudolf Steiner indicates that the plant's lower element is connected to the internal planets, namely Moon, Mercury and Venus.

We may now summarise what has been presented up to this point. As we have seen, the limestone or silica environment determines the differing forms of plants. The forms connected to silica are the image of forces originating from the external planets that act upon plants through silica and cosmic warmth.

These planetary forces are further differentiated depending on whether the planet is in its 'regular' or 'retrograde' phase. The plant (like human beings and animals) observes the sky from its point of view which we can obviously characterise as 'earthly'. For all living beings, the dominant perception is to be at the centre of a system around which the Sun and the planets rotate. This system of reference is usually called the 'Ptolemaic system'. If we observe the movement of the planets from this point of view we will notice that at certain times the planets appear to be going backwards in relation to their overall orbital progress. This perception comes from the fact that each planet's centre of rotation is the Sun even though the Sun appears to rotate around us. In this way *epicycles* appear in the orbits of the planets (see figure with the example of Mars).

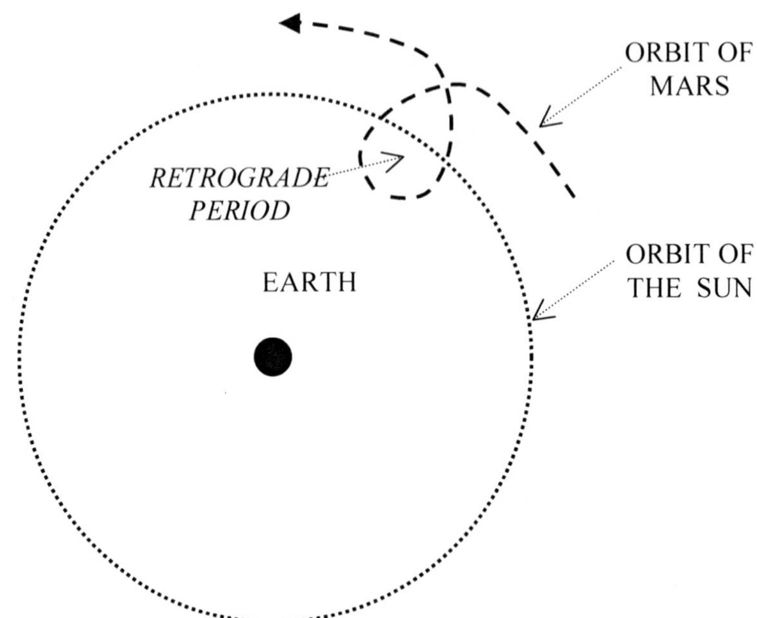

In spiritual-scientific terminology, the *primary influence* is defined as the action of a planet during its regular movement, and the *secondary influence* as its action during its apparent retrograde movement. The primary action can be characterised as '*incarnating*' or '*materialising*' and the secondary action as '*excarnating*' or '*dematerialising*'. The primary influence dominates during the first half of an organism's natural lifespan, and the secondary during the later half.

In this way each planet has at least two types of influence on the living world - the primary and the secondary. The following emerges from our studies:

- Saturn 1 brings the plant structure, verticality, genetic memory and memory of form;
- Jupiter 1 connects the plant with the cosmic form understood to be the planetary Group I, or consciousness;
- Mars 1 brings the plant growth within its form, asymmetry, and possibly encourages the formation of alkaloids.

All of this then establishes the formation of seed and the plant's quality that contributes to the sustenance of the human form - its capacity to nourish the human being.

As we have seen the earthly forms are connected to limestone and to the internal planets. In particular:

- Mercury 1: the formation of microvilli, the spiral insertion of leaves along the leaf stem (*phyllotaxis*), the ability to re-grow;
- Venus 1: brings the form into the seed and its development;
- Moon 1: brings the impulse to ramified roots, binary leaves, the basal rosette, and symmetry.

In this twenty-sixth paragraph we are told that if we consider the plant from the point of view of its capacity to nourish the animal kingdom and the human being, we must turn our attention to the action of the external planets, Saturn, Jupiter and Mars, which act through the silica element. This planetary influence is of an excarnating type and is therefore connected to the planets' retrograde movement.

Soon we will go back to these considerations which, we repeat, are of great importance. In the meantime let us turn our attention to the pivotal question asked by Steiner in the twenty-seventh paragraph: "*How can these forces be either intensified or somewhat restrained?*"

We will give some structure to our recent considerations with the help of a diagram. Let's remember the fact that the planets, even though they are always in the sky, are not always equally effective on plants. This means that the processes through which the planets act need a *mediator*. Depending on whether the influence in question originates from the internal or external planets, the mediator - whether a process or a substance - will be different.

At this point, having come to the twenty-eighth paragraph ('*When we observe how* .."), Steiner revives the example from the fourteenth paragraph when he speaks about water and the full Moon. We must get used to paying close attention to the examples brought up by Steiner during his presentations even if they might seem to be trivial diversions, because with each example he inserts observations or concepts which will be fundamental for the next subject. Here Steiner tells us: "*Water is much more than merely a chemical compound of oxygen and hydrogen. It is the ideal substance for bringing into the Earth those forces that come from the Moon for instance, for bringing about a distribution of the lunar forces within the Earth. There is a definite connection between the Moon and the water on the Earth.*"

In the twenty-ninth paragraph ("*Thus, we will need to consider* ...") Steiner asks whether it would be all the same to choose a rainy period before the full Moon or a random time for sowing seeds. The answer seems obvious because if the full Moon follows a rainy period it will fully activate the forces of vigour as well as the entire world which we have represented in our diagram as being underground. We should qualify this: if without further thought we were to work upon a soil that is still soaking wet the result would be the multiplication of lunar forces, encouraging the proliferation of weeds and fungi on compacted soil. On the other hand there could come a period of drought when water is scarce. In this case the full Moon could assist us since it could multiply the effect of any water that we irrigate on our fields.

We would like to digress briefly and note that all waters are not the same. In fact, looking only at the main distinguishing factors of water, we have water that rises up (spring water), water that runs (river water) and water that falls (rain). The water that rises sustains the ascending sap which, being rich in minerals, sustains the plant's vigour. Water that falls sustains the qualitative pole of the plant, while water that runs will have an intermediary value. (Stagnant water should be avoided because, being dead, it would only bring forces of death to our crops.)

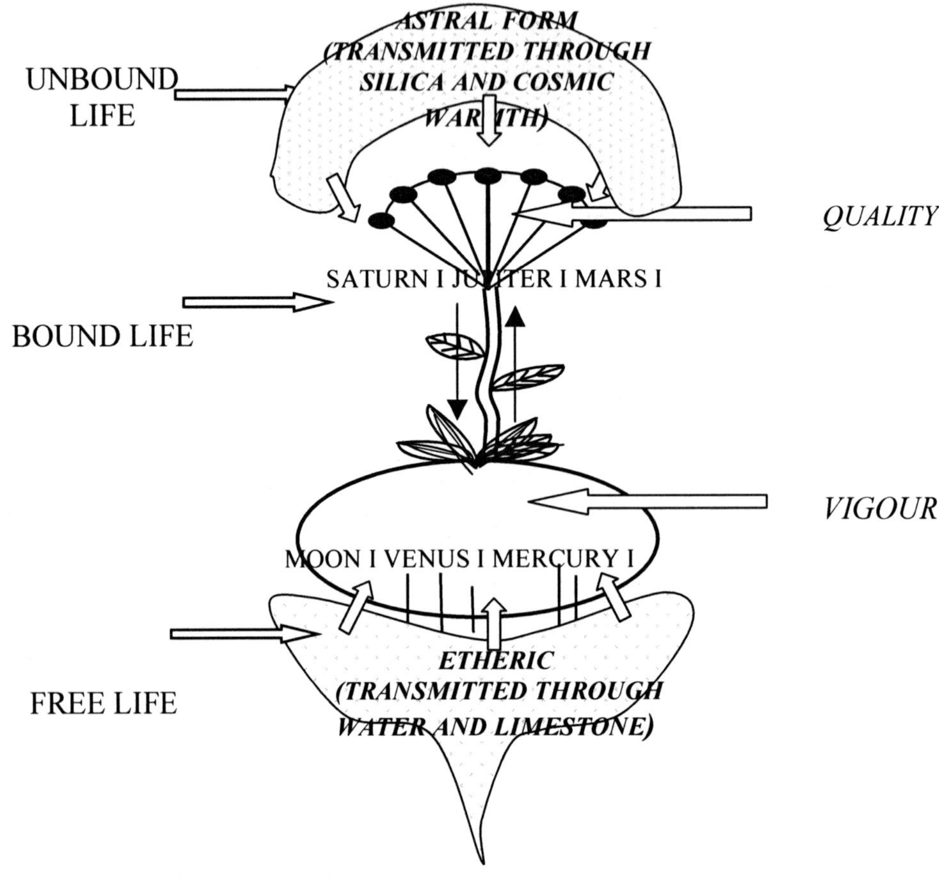

Having stated that water is the mediator between the forces of limestone and of the internal planets, we find the indication in the thirtieth paragraph ("*Let's continue* .. ") that the forces of the external planets are mediated by silica and the air's warmth.

From this we can understand that if the ripening processes come thanks to the forces of the external planets when the atmosphere is rich in warmth, these processes must take place when the fruits are still hanging on the plants. Evidently the techniques used today, which include harvesting fruits before they are ripe and storing them in a refrigerator until the market demands that it is the time to artificially ripen them, do not take the direction indicated by Steiner and can make us wonder about the quality of such products. The fruit must be 'cooked' by the Sun. It must be allowed to receive roundness from Jupiter, and its nutritional value from Saturn.

It's worth emphasising that Steiner also refers briefly to thunderstorms in this paragraph. We could present the most recent diagram in a different way. Let's construct a lemniscus and place our plant inside. The lower orb of the lemniscus represents the vegetative pole of the plant in which the contractive forces of vigour are strongly expressed, while the upper represents the expansive pole of nutritional value. At this point we can say that the lower orb also represents the etheric where life expresses itself the most, while the upper curve expresses the astral related to the external planets that, in theosophical terms, can be called *lower devachan.*

ASTRAL *POLE OF NUTRITION*

ZONE OF EXCHANGE

ETHERIC POLE OF VIGOUR

If we move from the plant world to the animal world, the upper curve is flipped over and comes 'into' the lower one. The animal is named such because it is the carrier of '*anima*' or soul and it is defined by the fact that the astral or soul element comes into the etheric element. In this way we obtain a figure that reminds us of the phenomenon of invagination:

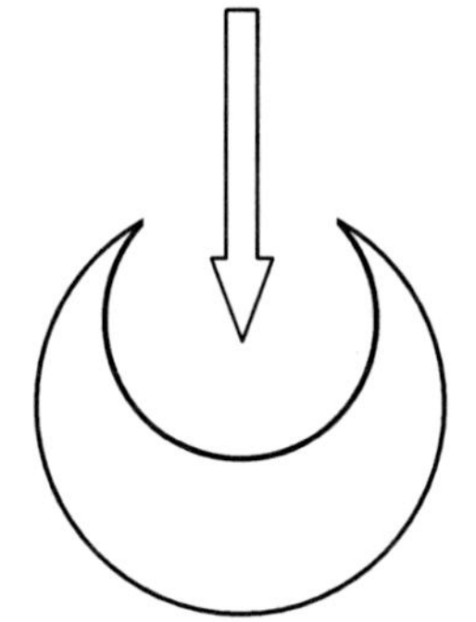

Now bring a typical thunderstorm to the mind's eye: an image of a dark sky out of which abundant rain falls. Suddenly lightning splits the darkness. We are spectators of this scene from the Earth in an atmosphere that is high in ethericity because we know that water supports the etheric. Above us, invisible, is the astral world that we place beyond the gateway of the Sun inasmuch as it is connected to the external planets.

Lightning is therefore a sudden and momentary split in an imaginary drape which allows us to glimpse the astral light, the one which we will see when we reach the Solar initiation or after we reach death.

Note how Steiner speaks of thunderstorms after having spoken of the warmth of air, clearly referring to Saturn and Jupiter, and since the life that manifests in the etheric comes from the higher worlds, lightning represents precisely Life which is brought from the astral plane to the etheric plane. In fact, after a thunderstorm everything seems more alive. We should learn to understand the experience of thunderstorms intensely. We can add yet another thought: with a thunderstorm comes purification of the air. So much is this true that during the plague people eagerly awaited rain and thunderstorms. With thunderstorms comes the formation of ozone that is trivalent oxygen (O_3). Oxygen is connected to the etheric plane, while ozone is connected to the astral plane. During the transformation from oxygen to ozone there is a contraction in volume that provokes an increase in warmth, creating the right conditions for Saturn to act as we have said above.

A final diversion on the problem of the 'ozone hole'; as we have been told such a 'hole' mainly occurs above the poles, in other words above the coldest areas of the Earth. Once again we see the relationship between the warmth of the air, Saturn and ozone. To be complete let's note that the ozone hole is also strongly linked to the lack of 'soul warmth' or the 'power of love'. No comment necessary.

In paragraph thirty-one ("*Where then do we see* ..") Rudolf Steiner begins by observing that Saturn takes about thirty years to make a full orbit around the Sun so it is difficult to notice its action on annual plants. However, it will have a strong influence upon perennials and especially conifers. He underlines the interconnections between annual plants and internal planets, and between perennial plants and external planets. He then gives a few examples which highlight the importance of choosing the right moment for sowing the seeds of perennial plants, depending on the intended purpose of those plants.

In the last paragraph, the thirty-second ("*Although these things* ..."), he says that ignorance of the correct times for sowing seeds due to a break-down of knowledge and by lack of interest in the laws of nature, actually brings about a degeneration of the products in terms of quality. In paragraph thirty-one he briefly refers to "*...the so-called ascending period of Saturn...*". Let's attempt to understand why Steiner has made this comment.

The planets, including the Sun, are not always visible in the sky because they are not always above the horizon. The Sun, when it is visible, makes arcs across the sky that get progressively higher from the winter solstice to the summer solstice. If we connect the highest points (*zeniths*) of the arcs which the Sun appears to make in the sky over the course of a year, we obtain a curve.

If we make similar arcs for the lunar cycles in a year we obtain thirteen tighter curves, because the sidereal Moon has a 27.3 day cycle which reaches the highest and the lowest points approximately thirteen times in one year. (We don't need absolute precision to make this particular point. The important thing is to have an understanding of the phenomenon to which Rudolf Steiner refers.)

We can clearly see in the lunar diagram the ascending and descending periods of the Moon. In a similar way we can map out lines for Venus and Saturn which we would use depending on which phenomenon we want to understand. For example, if we wanted to determine a cow's fertile period, we would have to consider the ascending curve of Venus.

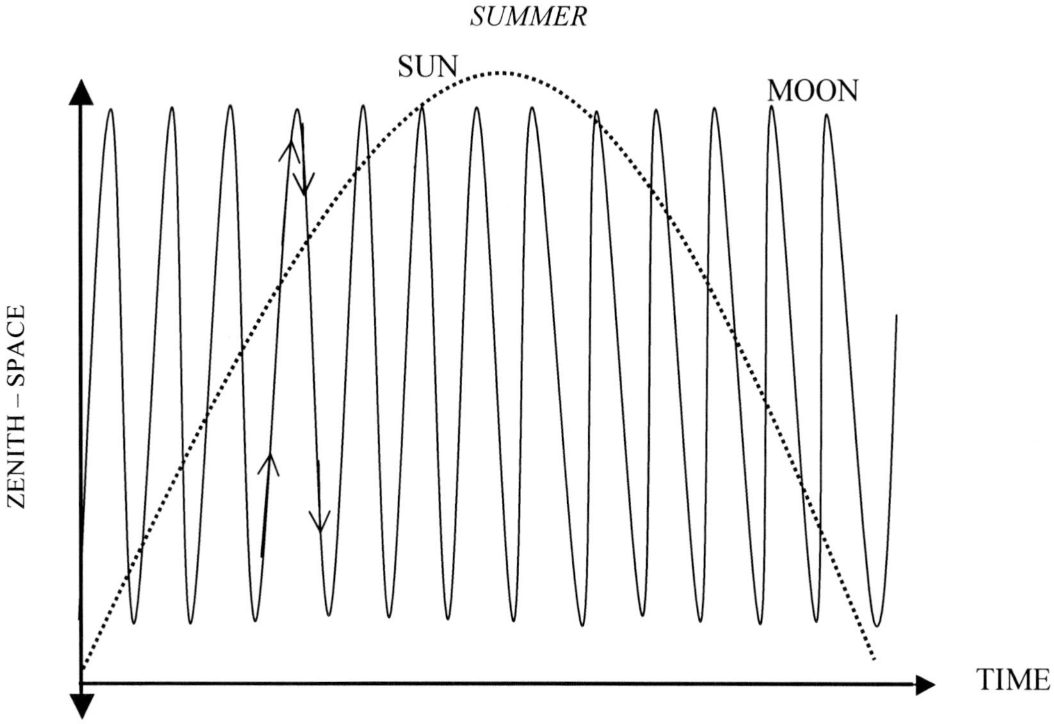

Annual cycle of Sun and Moon

To gain a better understanding of this concept we will leave Steiner's example because the Saturn-cycle is too long and would need a multi-year graph to see it as a whole. We will look more closely at Venus which, in 2001 (the year to which the graph below refers) made a significant path. Venus' diagram is useful to know for the sowing of all plants connected to Venus such as the *drupaceae*, in particular the peach tree.

We recognise four distinct situations in the diagram:

Venus ascending and above the Sun - A
Venus descending and above the Sun - B
Venus descending and below the Sun - C
Venus ascending and below the Sun - D

Obviously such a combination is also found if we refer to the Moon or to other planets and in the Moon's case, for example, these situations will be repeated many times a year.

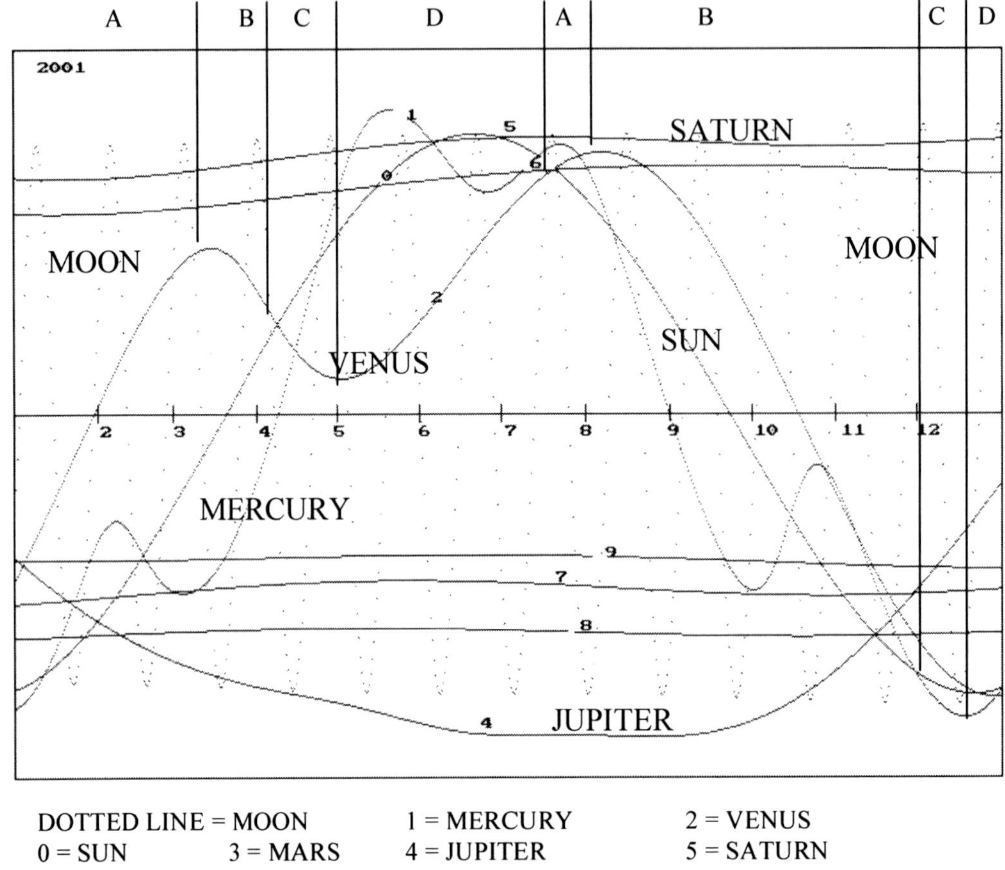

To understand what Steiner wants to communicate to us we must bring back to mind that the first lecture refers to Zarathustran agriculture based upon a two-fold vision of the world. We must also follow the path which Dr Steiner is taking us on; first he had spoken about water and the Moon (the example of the two professor's wives), then he talked about limestone and silica, then about quantity and quality, and finally about ascending and descending phases of the planets. Evidently he is still speaking to us of quantity and quality or of vigour and nutritional value, but this time in macrocosmic terms. From this viewpoint the period above the Sun's arc brings an impulse of vigour or quantity, and the period below the Sun brings an impulse of quality.

We note that during the autumnal and winter periods the Moon is more often above the Sun and brings the impulse of vigour or growth. We may also add that the action of the Moon is amplified when its ascending or descending phase corresponds to the similar solar phase. In other words the ascending Moon's above-the-Sun impulse of vigour is enhanced by the Sun in its ascending phase (from January to June) while the descending Moon's below-the-Sun impulse of quality is enhanced by the Sun in its descending phase (from June to December).

In this way one will have the most vigour during the first months of the year, coinciding with the ascending phase of the Moon, while the most nutritional value will come about during the last months of the year when the Moon is descending.

Obviously, when the Moon is below the Sun all its impulses will be weakened but the main principle of the Sun and the Moon's corresponding phases supporting each other will still hold. In this way we can understand that an ascending Moon in May, though it won't bring the same vigour of an ascending Moon in February, will be stronger than an ascending Moon in September, considering the fact that in May the Sun is in its ascending phase while in September it is descending. We can adjust the same reasoning to resonate with the macrocosmic impulses of quality.

What we have just presented is very important because it is the basis for the 'sowing calendar' with which Zarathustra successfully domesticated the wild plants. In fact Zarathustra only used the limestone-silica polarity of the soils as appropriate environments to enhance the specific planetary forces.

All of the above will help us to understand the fourth lecture and the use of the horn-manure and horn-silica preparations. In fact these two preparations correspond to the influence of the Sun. More precisely, the horn-manure represents the descending Sun and the horn-silica the ascending Sun. Therefore, in effect, the use of these preparations allows one to 'move' the solar forces, as if the curve of the Sun were moved lower or higher according to need.

According to this reasoning we may also modify the curves of any planet since Steiner gave us a preparation for each planet: yarrow corresponds to Venus, chamomile to Mercury, nettles to Mars and so on. At this point we may grasp the action of seed baths with the homeodynamic products that successfully create a microcosm around the plant able to benefit its growth.

Today, however, it is no longer sufficient to use Zarathustra's techniques of 7,000 years ago because the active forces have changed. In fact all of nature has lost the integrity it had at that time. Not even cosmic space has been spared from man's pollution. The sky has been violated an infinite number of times with satellites, by computers, robots and nuclear reactors, and it continues to be penetrated by myriad electro-magnetic waves caused by man's desire to communicate. Before our very eyes the weather has been completely altered.

But we haven't only seen negative events. 2000 years ago humanity witnessed an event of cosmic significance known as the mystery of Golgotha. When Jesus' blood was poured onto the Earth it brought with it the forces of the whole cosmos and these are now available to us, here on Earth.

Second Koberwitz Lecture

Fourth Meeting

The second lecture was held on 10 June 1924, at Koberwitz, and has been given the title: *The forces of the Earth and the Cosmos*. In this lecture and the two surrounding ones Rudolf Steiner expands on "*the conditions that are necessary in order for agriculture to thrive*".

It is now time to strive for a deeper understanding of the influence of the planets that will prove to be very useful when we study the sixth lecture. If one grasps the basic concepts now one has a foundation for deeper understanding of the subject later.

Let us continue to bear in mind that we are now commenting on the second Koberwitz lecture which corresponds to Egyptian agriculture and the threefold vision of the world dominated by the processes known as *Sal*, *Mercur* and *Sulfur*. While Steiner speaks of the key polarity between the internal and external planets in the first lecture, he now introduces the Sun which represents the *Mercur* of our system. Rudolf Steiner speaks of the Sun as the source of life and how its influence is modified by the interior planets as carriers of processes associated with vigour, and the external planets as carriers of the processes of quality that point to the stars or to the macrocosmic image of the plant.

In the sixth paragraph ("*Now, why do I say ...*") Steiner mentions - almost in passing - the parallel between the influence on the growth of the plant of what takes place beneath the soil and the influence of the human head upon the growth of the human being, particularly as a foetus. Indeed, the great cosmic images that shape the child descend directly into the head through the soft spot, which today tends to close ever earlier. In adults this process continues through breathing. By mentioning this Steiner wisely presents a first relationship between the root of the plant and the head of the human being: the plant, exactly like human beings, is 'nourished' by cosmic images as well as earthly substances, and this relationship will be explored more deeply in later meetings of this course. We may, however, consider the fact that nowadays the subtle processes that take place under the soil are being completely destroyed by modern techniques of agriculture and their use of synthetic chemicals. One major consequence is that these have expelled the cosmic images from the soil thus making it impossible for the plant to receive them. This is particularly severe when using herbicides. In light of this it becomes very clear that the most serious consequence of the use of chemicals in agriculture is not so much that food crops are contaminated and become carriers of substances harmful to human health, but rather this spiritual aspect which we have just mentioned which, if meditated upon, will appear extremely serious. This is the main reason why an urgent priority must be to restore this forcibly interrupted stream.

It is worth recalling that in the human being the same type of damage can result from the use of alcohol, or more generally from all substances that contain alkaloids: marijuana, coffee, cocoa, nicotine, etc. However, many common foods also contain alkaloids, such as potatoes, aubergine, peppers or pulses. Clearly the effect of beans is not comparable with that of marijuana, but let's not forget that in his school of initiation Pythagoras banned the consumption of legumes. Another task for humanity is therefore to reduce the level of alkaloids in plants, an undertaking which we have already initiated.

To understand the seventh paragraph ("*In addition we can say ...*") we will draw a diagram with our familiar plant in the centre. On the left we put the internal planets and on the right the external planets. Following Steiner's directions we show that the internal planets (Moon, Venus and Mercury) act on the upper part so we place them above, while Mars, Jupiter and Saturn are placed below.

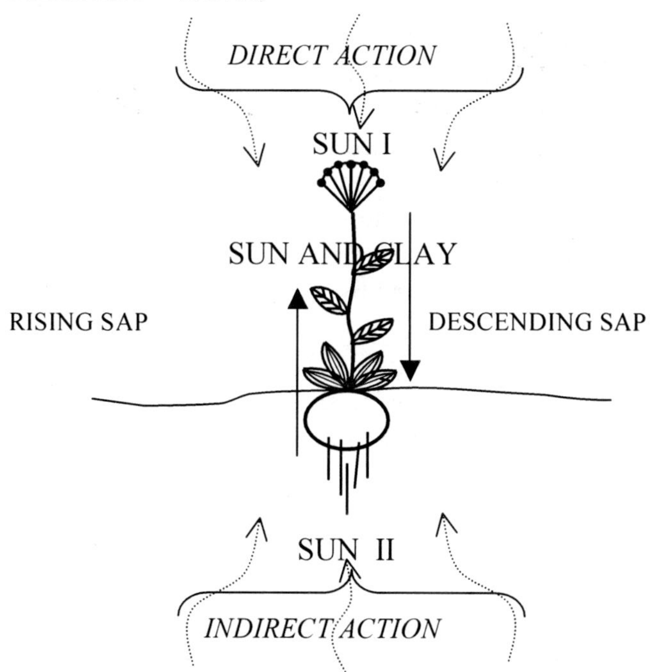

The sixth paragraph highlighted the reciprocal action of what is above and what is below the ground. This allows us to understand clearly how two opposing currents penetrate the plant, one rising and the other descending - anchored in the flow of sap. The ascending sap brings the influence of the external planets to the upper part of the plant whilst the descending sap brings the forces of the internal planets to the lower part.

Before proceeding we should clarify that the planets act continuously upon the Earth even when we do not see them in the sky since they are on the other side of the world. The difference in the forces that we receive, however, is crucial because when the planets are visible they radiate directly onto the Earth and the plants, while when we don't see them their forces reach us through the mass of our planet and are therefore modified by the body of the Earth itself. We can therefore say that both internal and external planets radiate directly and indirectly. In the passage that we are reading Steiner refers to the direct action of the internal planets and to the indirect action of the external planets. It is also important that we understand the way Steiner uses, in this context, the terms earthly and cosmic; instinctively we would think that cosmic is what comes from the sky while earthly is what comes from the earth. Instead Rudolf Steiner refers to direct irradiation as earthly because it flows *towards* the Earth, while that which flows upwards after having penetrated the Earth is referred to as cosmic. In other words the meaning attributed to the words refers to the destination rather than to the point from which they radiate.

To clarify, we use the word 'direct' to mean that which falls directly and 'indirect' to mean that which comes out of the Earth after having penetrated it. Before discussing the contents of paragraphs seven and eight it is worth noting that the two apparently forgotten currents - the direct one from the external planets and the indirect one from the interior planets - have been dealt with in Steiner's first lecture. We recall that we were invited to consider the indirect forces of the internal planets (which therefore act from below the ground) in connection with the reproduction of plants, whereas for food quality the direct forces of the outer planets were indicated.

At the end of the seventh paragraph ("*In addition we can say ...*") Steiner introduces the variable of soil and says that its make-up (siliceous, limey or clayey) determines the effects that the above-mentioned forces may have. In fact the life forces are brought to Earth by the outer planets which act as close mediators of the source of life, and such forces can be abundantly welcomed by siliceous land, which, in turn, has a strong relationship with the periphery of the universe. Note carefully that Steiner does not say that life is preserved by silica, but that silica allows the abundant entry of life forces.

About halfway through the eighth paragraph ("*To begin with ...*") Steiner uses the term "*it's own chemistry*" that could give rise to some confusion, but if we read it as a process of transformation we believe the concept becomes clear.

We have now read and commented on seventeen pages of the Agricultural Course and we can observe that Rudolf Steiner has not yet said a word about the biology of the plant. Instead he spoke about planetary influences and of forces linked to the life of the soil and we shall see that the rest of the Koberwitz course will proceed in this same direction. Steiner thought that the plant was only the result or manifestation of the forces in which it is immersed and could not express anything independent of these forces. From another point of view the observation of plant forms allows us to grasp the quality of those forces which the plant expresses. Mainstream science's approach is obviously completely different - it maintains that the plant expresses the forces that are in its seed and so to modify the plant it is necessary to modify the seed. Genetic modification techniques are a logical consequence of this point of view.

Towards the end of the eighth paragraph Rudolf Steiner use an expression that seems difficult to pin down; Steiner says that "*... and with plants whose roots are of particular interest or importance, we should never forget that silica is indispensable ...*". We must pay attention to the fact that he does not say 'specifically for their roots' but "*... plants whose roots...*". In other words the being, the essence of the root, is to allow contact with the world of spiritual cosmic images and these images find their highest expression in a particular form, which is that of the single stout root. We don't use the expression '*carrot-like*' because the linear action of Light has entered this particular form. Instead we use *single* and *stout* to include in this expression spherical roots such as turnips, radishes, beets and others.

Once again we note how Steiner ties the manifestation of the root in its essence to the existence of certain conditions in the soil composition, which in this case is the presence of silica in the soil.

There is another important indication in the eighth paragraph: in order for silica to exert the influence that we have mentioned it must be imbued with life or - even better - immersed in the etheric. Otherwise it is nothing more than an inert medium.

The ninth paragraph ("*The next thing ...*") introduces clay as a *mercur* substance that allows for the continuous exchange between the lower part and the upper part of the plant, or between the 'head' and what takes place above the soil that can be compared to the 'belly'. The balanced development of the plant requires that what comes from the cosmos is bound to the root through living silica so it can then be brought upwards through the ascending sap. *"Substances like clay support the upward flow of the effects of the cosmic entities in the ground."* The concept just presented is discussed further in the next paragraph to underline its importance.

We have always spoken about clay as an amphoteric substance that acts as an acid with basic elements and as a base with acid elements and which can, therefore, connect with both the upper and the lower worlds. Furthermore we should be clear that dried clay can become as hard as stone (Earth element) and, on the contrary, when it is wet (Water element) it can become malleable (Air element). We also know how it easily withstands cooking (Fire) to then reach a rigid form (Earth).

The mediating action of clay may also be included in the connection between the inner and outer planets and in this way clay can be considered to be the bearer of the forces of the Sun.

Going back to what Steiner said earlier we know that the internal planets support the influence of the Sun, and we may add this to our drawing above by placing a Sun above the plant indicating its primary force. However, the external planets also modify the Sun's influence, bringing to light a secondary force that we can indicate in the lower part of the drawing. Steiner has also told us that there is a third action of the Sun related to clay that we include in the diagram.

We have represented three Suns with which we should become increasingly familiar because they correspond to three aspects of life. If we wish to offer an alternative method of practicing agriculture to today's 'conventional' one, we must know the laws of life intimately and appreciate that which allows us to connect with life such as living silica, the warmth of air, humidity, water, and the three Suns which we have just seen.

We believe it is interesting to note that in paragraph ten ("*Once we get down to practice ...*") we find the following words: "... *we must treat it* [clay] *so that it* [soil] *becomes fertile ...*" This phrase seems to recognise the fact that the way clay is today does not allow it to fulfill its task completely, because the two poles between which it should mediate have been under continuous and increasing assault from human activity. In fact the soil today is incapable of carrying life because of the continuous use of substances that, by their nature, block the flow of life. Furthermore, even space is troubled by strong turbulence from the launching of thousands of space vehicles into Earth's orbit, some atomically powered, thereby obstructing the forces which we are studying.

It is up to the human being - to a human being with a different consciousness than that which has ruined the natural equilibrium of millennia - to become active and treat clay in a certain way so that life may begin to flow anew in the soil. First, however, it is necessary that this human being becomes familiar with and fully understands the laws of life.

At this point we must note an inaccuracy in the [Italian] translation of the last phrase of the tenth paragraph which says: ".. *the primary thing we need to know is that clay promotes the upward stream of the cosmic factor.* ". We believe it is more accurate to use the following expression: ".. *the primary thing we need to know is that clay brings life to the upward ascent of cosmic forces*". Within the expression *'promotes the upward stream'* there is only an impulse of movement, while in the expression we prefer, the element of life is added to the ascent and this is fundamental for understanding this passage. In fact life comes from the cosmos, particularly from the Sun, and in this way the function of clay as being the 'third Sun' is highlighted, in which it gives life to all forces that are currently languishing and dying.

Thus we come to the eleventh paragraph ("*But this cosmic stream ...*"). Here there is a clarification of the concept of earthly forces as forces that are generated above the soil by water and air and which must descend underground.

At this point Steiner uses the word 'digestion' frequently. Those familiar with alchemy will know that once upon a time, in order to transform substances, the alchemists used a technique called '*digestio*' which was achieved by placing substances in a vial and heating them with various sources of heat such as, for example, different types of dung or the heat of the Sun. Central to this technique was knowing the qualities of the different types of heat in order to cause predetermined transformations of the substances. Today everything has changed and the remedies made by the alchemists with their wise techniques do not work anymore. In addition to the lack of equilibrium in nature, the consciousness of human beings has profoundly changed so that even the temperatures used by the alchemists (about 32° to 37°C) are no longer sufficient. However, the technique of *digestio* itself is still valid. Steiner tells us that these transformations happen naturally in the upper part of the plant and that the result of these changes, which are brought about by the elements as carriers of planetary forces, must be brought down to the lower part in order to complete the exchange mentioned earlier.

We cannot ignore the significance of what happens in the plant during this digestion. In many plants something marvelous happens when the sap, which is mainly *Water*, is transformed into oil that is *Fire*. It is clear that what takes place no longer belongs to the laws of chemistry but is something infinitely larger. Let us stop to consider that *Water*, which has the capacity to extinguish *Fire*, is transformed into *Fire* itself. Only the shortsightedness of the present way of thinking and observing nature makes it possible that such a phenomenon does not awaken feelings of awe and wonder. In the past these things were fully comprehended by people that are today considered primitive by our scientists.

Also in this paragraph it is revealed to us that earthly forces are aided in their descent by the calcium contained in the soil or by the distribution of calcareous substances in homeopathic quantities directly above the soil. Note the clear reference to '*homeopathic doses*' that supports our beliefs and the direction we have taken with the work that we have been trying to develop for years.

In the twelfth paragraph ("*You see, everything ...*") Steiner speaks about warmth above ground connected to the internal planets, and warmth under the ground connected instead to the external planets. More specifically he says that warmth above the ground may be considered as dead warmth, while warmth below the ground should be considered to be live warmth. It becomes clear that the former is that which we define as *extinguishing fire* that finds its expression in measurable temperature, while the latter is the Warmth ether. When dead warmth is brought inside the Earth which is alive, it itself becomes alive. This is exactly what happens in the human being during breathing.

It is clear that the Warmth ether acts on the root of the plant because it is from the root that the plant develops. In us the Warmth ether is found in the head and it allows one to feel enthusiasm for life and to keep a young spirit even at an advanced age.

In the middle of the paragraph there is an imprecise translation of a phrase that we believe should say the following: "*The warmth under the ground, especially in winter, has united with something which is an internal factor of life*". This is precisely what allows the plant to have enthusiasm for life and to overcome the problems which otherwise would threaten its very existence.

The current problem lies in the fact that we have severed the plant's ability to connect with the world of life so that, faced with the tiniest obstacle, it no longer knows how to activate its processes of defense. Therefore the human being must continuously intervene with various products to sustain it. We believe it is clear that the solution is to work towards developing plants that are particularly rich in Warmth ether. At this stage of the course we do not have the tools to discuss how this may be done so let us be satisfied with pinpointing the problem. Later we may find the right solutions.

Also in the twelfth paragraph it is clearly stated that limestone has the ability to attract the imponderables. Note that when external warmth is attracted into the Earth this is attributed to the limestone content of the soil. Then if we think of the human being it may be easier to grasp this concept: in the human being the largest amount of the limestone substance is found in the head where it is even placed on the outside, as opposed to what takes place in the rest of the body where the bones are covered by fat and muscle. Well it is precisely in the head that we have the highest capacity to grasp the imponderables. Thoughts are first and foremost grasped in the head and within the head are also the teeth, which are simply bones in contact with the outside world, through which we may capture Light and the cosmic images.

Just before this, Steiner affirms something that may be difficult to understand in the first instance: "*If we human beings had to experience the warmth that is active down below, we would all become exceedingly stupid, because in order to be smart, we need to have the dead warmth around our bodies*". In this context, the terms intelligent and stupid are given the same meaning that we usually give them, yet considered from a different point of view. Intelligence is usually turned toward the comprehension of the physical world and the analysis of its phenomena, while stupidity is the loss of such an ability to comprehend. If we consider the same not from the point of view of the physical world but under a more subtle light, intelligence is the means through which we confront the physical world, which is the undeniable precondition for tackling the logic of the higher planes. If living warmth would enter us our ability to confront the physical world would be compromised and thus the possibility to access higher levels of consciousness in the right condition: after having sufficiently experienced and understood the logic of the lower planes. Such a state, seen from the physical plane, can only be perceived as stupidity or lack of intelligence. However, dead warmth may not enter us in any which way; it must enter "*…in a condition of slight vitality*".

We could say that if living warmth would bring us into direct contact with realities that are too elevated for our current state of consciousness, dead warmth would keep us in a state of death in which any experience would be impossible. What happens is that when dead warmth enters a living organism it is made alive, but only in sufficient quantity that it may light up intelligence.

However, for this dead warmth to enter us it must go through the necessary process of 'digestion' that allows it to enter the world of life. When this process of digestion is not sufficiently completed there come about situations of partial consumption that we have all experienced during our lives and which are called a cold or influenza. Therefore if our cold is the result of our inability to 'digest' the different external temperature, it could be due to excessive cooling or heating. Furthermore, the condition of imbalance which insufficient digestion provokes in us creates a condition where we are not able to relate sufficiently with the world of astral and planetary forces, the result of which is the source of illness - hence the name 'influenza'.

In the thirteenth paragraph ("*We are aware ...*") Steiner speaks, after the digestion of warmth, about the digestion of air. These are subjects that we are not used to discussing but they are essential if we want to cure the plants, the animals and even ourselves. We must understand that warmth is connected to the I. We have repeatedly stated that the I lives within the warmth of the blood. From these simple considerations sprouts the importance of acting from within the fullness of the I. When we spoke about how to understand and cure pest problems we stressed the fundamental importance of adequate meditation so that the forces of warmth that we activate in this way may be unlimited and even if we are enclosed in the home we may act on our plants. With meditation, which we may define as a 'maceration' of thoughts, we can achieve the therapeutic *digestio* that is beneficial for our fields.

Warmth and Air are attracted in the soil and vivified by the limestone that must be living. But for the limestone to be living we must activate specific processes within it which Steiner taught us to do by using a preparation based on chamomile. In this way the limestone in the soil is vivified and makes the live warmth and air - which in other instances we have called Warmth ether and Light ether - available to the plants.

In the fourteenth paragraph ("*It is different ...*") it is stated that the exact opposite happens for Water and Earth, meaning that when they are carried inside the soil they become even more dead. Steiner tells us that if they become even more dead, thereby losing part of the vitality which they had above the ground, they acquire the ability to take in cosmic forces of the more distant planets.

That which takes place underground also takes place under our skin where the elements of Water and Earth are represented by the lymphatic system and the bones. Our bones are so dead that death itself is represented by a skeleton. We can say that if bones are so dead it means that they have the ability to extinguish life, and we know that forces that kill life are ahrimanic forces in opposition to the forces of the Logos - the bearer of life. The reasoning that we have just taken allows us to conclude that human bones are the seat of ahrimanic forces and of hate, because out of hate life could never be born and life is essentially an act of love.

Water, which in our body is represented by the lymphatic, endocrine and glandular systems, is where the luciferic forces are expressed. We cannot avoid confrontation with these forces. We should even learn to recognise the symptoms that call for the necessity to confront them such as pain in the bones. Generally those who have this problem are people who have reached a certain age and who should have acquired, through the work done over the course of their lifetimes, the ability to confront Evil. This pain is none other than a reminder on the physical plane of the work that we should be taking up. Even osteoporosis has the function of making us conscious of the need to take on a spiritual development. It is not by chance that osteoporosis is more prevalent amongst women. It is an illness caused by luciferic forces and we know that women, being more developed as bearers of astral forces, are antagonised by Lucifer while Ahriman mainly opposes the spiritual forces of the I predominant in men.

At this point it may be interesting to note how the bones of the human body do not have a definite number, in fact there are between 203 and 206 bones because the sacral and parietal vertebrae may be welded together in different ways. It is no coincidence that in the Gospel of St John, in the passage where Jesus appears on Lake Tiberias, the boat on which the apostles are is said to be about 200 cubits away from the shore. This distance represents the bones of the human being. In fact the apostles who were present on the boat had not yet overcome the confrontation with Ahriman and had not yet reached the shore of the lake which represents the Zodiac, or spiritual world. Let us also remember that the number of stones forming the statue of the Zed inside the Great Pyramid is 205.

Coming back to bones we believe it is interesting to note how today children are highly prone to cavities. This is due to the fact that our children are forced to learn to read and write at too young an age (before 9 years of age). The head forces at this stage should be directed toward the formation of the organs of the body, but they are rerouted toward precocious intellectual development. Cavities are a sign that should make us understand that we are hindering the development of the organs. These problems, however, will come to light only after 60 years of age, but at this point nobody will be aware of their cause.

We are digressing from commenting on the Koberwitz course but it is good to bring these thoughts forward in the attempt to break apart the mental cages in which we have enclosed ourselves over the course of our lives. The task that Steiner suggests is to enter the laws of life, but we will never be able to do this if we don't throw out the baggage that stops us from lifting our thoughts toward truths that apparently seem devoid of meaning because of our obtuse minds and the conditioning to which we have been subjected.

In the same fourteenth paragraph Rudolf Steiner speaks about crystals and presents them as deserving of a prise for their extreme descent into Saline processes, as if being trapped into a form would make it possible to become transparent for cosmic images. This is why we say that the Earth is the bearer of cosmic images or that within the Earth lies the memory of the cosmos.

Just before this however, Steiner begins to be practical; he mentions two dates that have a precise reference to physical work. Between the fifteenth of January and the fifteenth of February the minerals emancipate themselves from the Earth and acquire the highest power of crystallisation. This is a way of saying that the minerals lose the 'opaque' nature of salinity and become transparent for the forces of the cosmos. In a better way we could say that they become antennae or a very powerful eye capable of perceiving the distant cosmos. Let us remember that in the middle of that period we celebrate Candlemas[8].

In this way we are presented with the fact that the Earth has the possibility to receive the forces originating from the far reaches of the cosmos and to conserve them for the entire year so that they may be used to direct the growth of plants. He then specifies that a period preceding the one mentioned above is particularly suited to the communication between minerals and plants, and the latter can therefore readily receive the forces radiated by the minerals. There will come a time in which we will understand that this is the only way to cure a plant which has been contaminated by genetic modification or which has undergone any type of degeneration.

By following these indications we will be able to direct the growth of plants. This is not to parade our power but to protect and allow the plants to develop qualitative characteristics that are adapted to modern human beings and will promote their

---

[8] See annual 'Agricultural Astronomical Calendar' from l'Albero della Vita

spiritual growth.

Let us temporarily leave reading and commenting on the Koberwitz agricultural course and let us turn our attention to the planets' influences on the plant world - without pretending to be exhaustive as this is a vast subject. Just consider that each planet acts not only on the physical plane but also on the etheric, the astral and the spiritual - and keep in mind that the plant also has a physical, an etheric, an astral and a spiritual component. For the time being let us concentrate on the physical and etheric planes.

Let us go back to the diagram that helped us at the beginning of the meeting and attempt to understand something more. We have represented the direct action of the internal planets and the indirect action of the external planets, but also the internal planets logically have an indirect action and the external planets have a direct action. For the Sun it is simple to understand that its direct action happens when it is visible and thereby directly sends us its rays, while the indirect action happens when the Sun is on the other side of the world and therefore sends its rays through the entire Earth.

It is a little more complicated for the other planets because in addition to what we have just said about the Sun, the direct or indirect action also depends on the apparent movement of the planets: in fact direct action is defined as the action that is exerted by the planet during its regular motion while indirect action is exerted when the planet is retrograde.

It is known that the planets revolve around the Sun which in turn seems to revolve around the Earth. If we turn just a little of our attention to this image we can grasp how, for a certain period of time, the planet moves - with respect to the Earth - in the same direction as the Sun. This is its regular motion. But the planet, at a certain point in its orbit, inverts its apparent motion and turns in the opposite direction as the Sun. This is what is called retrograde motion.

This consideration gives us the opportunity to understand how we may act therapeutically on plants.

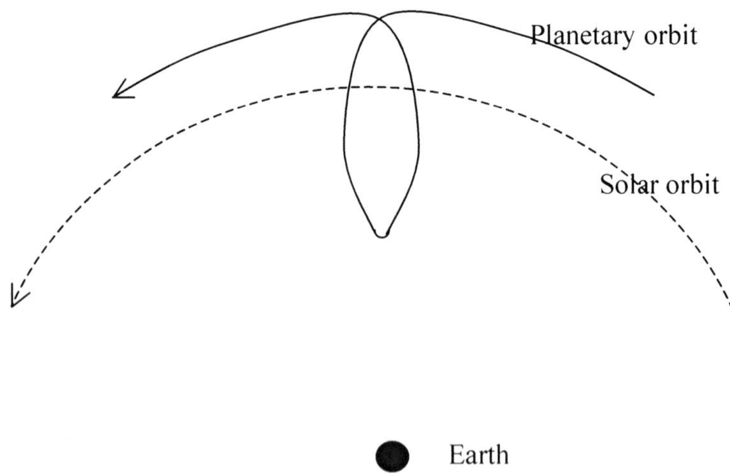

We know that the action of the planets may be incarnating or excarnating and that the incarnating action corresponds to direct action, and the excarnating action corresponds to the indirect action.

Thanks to the apparent reversal which we have just spoken about we have the possibility to cure a disease brought about by an incarnating action, for example unbalanced Moon 1 can bring fungal diseases which can be opposed by bringing forces of Moon 2 or Saturn 2. The choice between the two possibilities requires knowledge that we have not yet touched upon, so let us be satisfied with understanding that there are two possibilities to intervene.

We can add - assuming that each planet has a corresponding metal – that the use of the metal itself allows us to bring the incarnating action connected to that planet, while the use of a salt of the same metal allows us to bring the excarnating action.

We may also use homeopathic potencies in the sense that up to the potency of 15x (we are in the realm of the Earth and Water elements) we are calling on the incarnating action, and by increasing the potency (therefore entering the realm of the Air and Fire elements) we call more on the excarnating influence.

We will continue with this subject in the following meeting.

Fifth Meeting

Let us begin this meeting by resuming our study of the planets. Since we are working with the plant world we will attempt to deepen our understanding of the influence of the planets on the etheric level. (As for the other levels for now we will just say that the planets act differently within them.)

The process of life becoming manifest occurs in a journey through the successive planetary spheres starting at Saturn followed by Jupiter, Mars, the Sun, Venus, Mercury and finally the Moon. On the reverse path, which we may call the return to the archetypal sphere, we first meet the Moon and lastly Saturn. Our discussion about the world of planetary influences will follow the order that we have just listed.

Let us emphasise what we discussed in the previous meeting - that for each planet there is a primary influence linked to its regular progress and a secondary influence linked to its retrograde motion. The primary influence corresponds to the path toward manifestation which we will call *incarnating*, while the secondary influence concerns itself with the opposite *excarnating* path.

As we have seen the process of incarnation unfolds in seven stages that correspond to the planets whose spheres must be traversed to arrive on Earth. Referring to the plant world, the incarnating process begins with the seed connecting with the idea of the plant that it must incarnate within itself. This is, therefore, an action that happens in *space*. The plant makes the first step toward manifesting in space and therefore begins to lose its strictly spiritual character. Access to existence in the solar system inevitably involves a withdrawal from the spiritual world. This occurs in the sphere of Saturn and we can identify this primary influence with the word '*space*'.

The second step in the manifestation from the seed is to produce a *form* for the idea with which it has connected. Form is thus the characteristic of Jupiter in the incarnating process.

Then comes Mars, the active outgoing warrior, and thanks to Mars the form that has emerged receives the impulse to manifest or to enter the physical plane. We could say that Mars' primary influence is to *externalise*. This is the force that allows the seed to rise above the soil into the light.

At this stage the plant needs matter or substance, in other words it must *materialise* and put on plant substance. The Sun's primary influence is to 'materialise'.

Now the plant must begin to nourish itself. It must begin to bring nourishment inside itself in order to live and grow. The metabolic processes start at this point, regulated by Venus. We can summarise Venus' primary influence as *interiorising*.

At this point Mercury brings its impulse, which is *movement*: sap begins to flow. Mercury is considered the protector of merchants and thieves because of its characteristic of bringing movement: if the goods don't move it is a tragedy for these two occupations. Mercury is also the patron of medicine because it sustains all the healing processes through circulation or the movement of substances via the bloodstream, and because the breath moves in and out of our bodies it is also linked to Mercury's movement. We could also speak here of the movement of thoughts. Mercury's movement is evident in the strong force of re-growth that many plants exhibit when cut: think of alfalfa, of hazel trees or of quince trees.

Now the plant is ready to live in *time* – to manifest within existence - and this is a role of the Moon.

The plant has now passed over to its earthly existence – one that can be considered as an inevitable journey toward death, or dis-incarnation if we refer to the idea that lives within the plant. So we can now consider the excarnating path within which *space* is the secondary characteristic of the Moon. For the plant this is the space within which it establishes itself. We therefore find once more the characteristic space that we had already met as Saturn's primary, incarnating influence. That earlier space, however, was the expression of the capacity to 'enter' into space, while this secondary space is that into which the plants grows or moves through activity. When a plant occupies a quantity of space which is larger than the capacity which was determined during its passing through Saturn, this excess allows the manifestation of viruses. The transition from the Earth to the Moon is the transition from the physical plane to the etheric plane, from the mineral to the plant world; well, this imbalance of secondary Moon leaves room for those beings that are between the mineral and the plant world: that is, the viruses. Let us remember therefore, that when we fight against viral diseases we will always have an excess of secondary Moon.

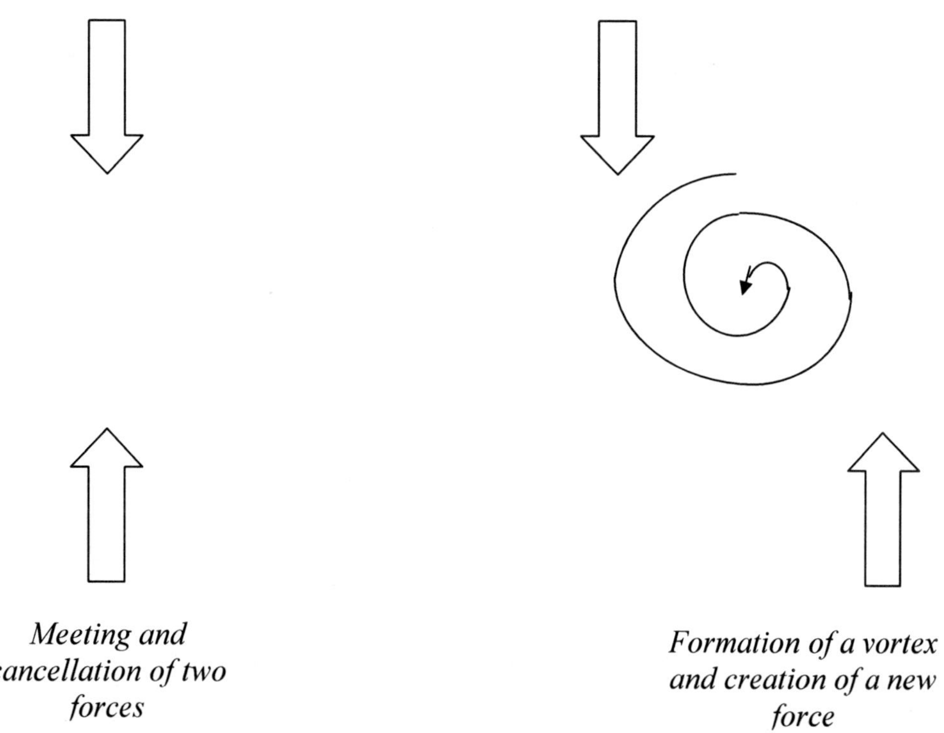

Meeting and cancellation of two forces

Formation of a vortex and creation of a new force

Let us now consider Mercury whose primary characteristic is movement. The opposite of movement is stagnation that is an over-consolidation of form. We have already found that form is Jupiter's primary influence. However, that is a form governed by an external influence that leads to the constitution of a structure: for example a joint. The *form* of Mercury's secondary influence is derived from movement. Think of what arises when two opposing currents of water meet. In order for both to be able to continue flowing they find an equilibrium that generates vortices. A vortex is a sustained empty form that enables the two currents to continue flowing in a dynamic balance. This type of balance can also be found in the social sphere such as within a cultural movement when two different streams form. Instead of annihilating each other in a head-on clash, they find a way to co-exist. Then, by mutually stimulating each other, internal arrangements are found which focus on building something new within the original single body.

EXCESS INFLUENCE	NORMAL ACTIVITY	CHARACTER-ISTICS	PLANET	CHARACTER-ISTICS	NORMAL ACTIVITY	EXCESS INFLUENCE
DOES NOT GO TO FRUIT	INCARNATION OF STRUCTURE	**SPACE** FERTILISATION OF THE COSMIC ARCHETYPE	**SATURN 507**	**TIME** MATURATION RESURRECTION	SEED INDIVIDUAL DNA	LACKING VERTICALITY VINES
RIGIDITY DEPOSITS DEFORMATION	FROM LIQUID TO SOLID, ARTICULATION	**FORM** SYMMETRY AND ROUNDNESS	**JUPITER 506**	**MOVEMENT** MOTILITY	SWELLING FRUIT	SAGGING SOFTENING
DESTRUCTION	GROWTH	**EXTERIORISATION** WILL TO INCARNATE	**MARS 504**	**INTERIORISA-TION,** CHLOROPHYLL, PROTEIN	FERTILISA-TION OF FRUIT	STERILITY
APOPLEXY HARDENING	COSMIC NUTRITION	**MATERIALISATION** SYSTOLE	**SUN 501 – 500**	**MATERIALISA-TION** GROWTH DIASTOLE	ETHERIC OILS SPROUTING	STRUCTURAL MALFORMA-TION
HYPERTROPHY	CELLULAR NUTRITION	**INTERIORISATION** NUTRITION	**VENUS 502**	**EXTERIORISA-TION** EXCRETION, LOSS OF SUBSTANCE	CELLULOSE MINERAL SKIN eg BARK	HYPOTROPHY
LOSS OF FLUID	CELL SAP	**MOVEMENT** ETHERIC COMES INTO MOVEMENT	**MERCURY 503**	**FORM** CONCAVE FORMS ASYMMETRY	REGROWTH LEAF, ADJUSTS TURGOR	OBSTRUCTED STAGNANT
TUMOURS	REPRODUCTION	**TIME** GERMINATION, EXPANSION	**MOON 505**	**SPACE** LIMITATION INTENSIFICA-TION	FROM OUTSIDE TO INSIDE	VIRUS
DIRECT STREAM INCARNATING OR PRIMARY INFLUENCE			**EARTH**	**INDIRECT STREAM EXCARNATING OR SECONDARY INFLUENCE**		

Venus' primary influence is internalisation and we have said that it is connected to metabolism. But that which is internalised, whether in plant or human being, is then processed in order to extract the nutritive elements destined for the construction of the body and that which is not used is eliminated. Therefore the secondary characteristic of Venus could be expressed by the word *exteriorising*. To make an example of what we have just said, think of a cork tree: the eliminated (or exteriorised) part of the tree is the bark. Therefore if we want to obtain thick bark we must bring secondary Venus forces to the cork tree.

If the primary influence of the Sun is materialisation, then the secondary influence is *dematerialisation*. We all know by now that the spring-summer Sun (and the day Sun) bring impulses of materialisation while the autumn-summer Sun (and the night Sun) bring impulses of dematerialisation. The ripening fruit of an annual plant is already the first step in the death of that plant.

If we take the earthly life of the human being into consideration we can divide it into seven year periods. The first seven are governed by the influence of the Moon, the second seven by Mercury and the third by Venus. The Sun rules the period between 22 and 42 years. During this period the human being starts to have the maturity to begin developing spiritually. Three times seven years later one reaches 63 years, which is the age of wisdom when, excarnated in Saturn (not dead), one's consciousness can expand to an understanding of the Zodiac, our spiritual homeland.

Mars' secondary influence is *internalisation*. Life requires alternating internalising and externalising influences such as inspiration and expiration.

Mars expresses itself with aggression but when humans reach a degree of maturity – usually above 42 years - they develop the ability to control their aggression and to focus their activity. What was at first aimed outward (Mars I) now works within oneself: a transformation may take place which could even modify one's character, limiting one's cantankerousness and emphasising, for example, one's tolerance. The result of this transformation is the ability to act in the outer world through the use of the word combined with the strength of transformed thought and the strength of the heart. This latter influence is the complement to the primary influence of Venus that is the ability to welcome and internalise. The word can be seen as the fruit of the human being, a gift to others.

The primary Jupiter influence, as we have seen, held the form. Static form, immobile in itself, becomes paralysis and can lead to disease. The form connected to life must have movement in its interior. Think of an apple, beautiful and red in its completed form whilst in its interior it nurtures the movement of its juices. The secondary Jupiter influence regulates such movements. Therefore Jupiter's secondary influence could be expressed with the word *movement*.

This movement is not the same as the one we identified as Mercury's primary influence. That was a movement directed toward the outside while this is an internal movement or circulation. When we speak about the movement of juices we refer to movements that we could define as being quantitative; however, there are movements (by this we mean a meeting of currents) which rather develop qualitative aspects, such as the process of ripening, the formation of gluten in wheat or of sugars in fruit, and that of oils. We are therefore speaking about the transformation of rough matter into finer and finer, nobler matter. The secondary Jupiter influence is the impulse that succeeds in giving that extra touch of quality to the ripening fruit.

The secondary influence of Saturn is *time*. The return to time is the exact consequence of the end of movement. This is the time of excarnation, or, rather, of resurrection; the time in which time doesn't end: we exit from the door of life and enter through the door of eternity.

We have just seen the fourteen main influences or activities of the planets. Now we will attempt to bring the manifestations of the influence of each planet into focus, as well as the consequences of an excess of these same influences.

Naturally we begin with primary Saturn whose influence, which we summarised with the term *space*, can also be understood as *cosmic fertilisation* or *spiritual death into space*. Such an influence manifests vertically, in incarnation, in the physical structure. The influence of Saturn is, as we have said, the first in the process of incarnation: Saturn is indeed the nearest classical planet to the Zodiacal sphere. When we are on Saturn - obviously meant figuratively - we experience a process of incarnation: its transition between intention and fulfillment. This is the moment of cosmic fertilisation when the impulse to manifest leaves from the spirit; and this, from a spiritual point of view, is the equivalent to a loss or to a death because the being which is incarnating is about to leave the spiritual world. That which is a birth from the material point of view is a death from the spiritual point of view, and vice versa.

The passage between the two realities manifests as the will to maintain a connection and induces a gesture of verticality. This gesture can be seen as the impulse to maintain a connection between its anchor in the physical world and its home in the subtle worlds. If we observe what takes place at the birth of any plant we see that, in accordance with the principle set out, first the radicle forms and the shoot immediately afterwards. As growth continues the plant diversifies but the plants that are strongly influenced by Saturn will maintain and amplify this vertical gesture - such as the conifers. Recently we spoke about physical structure: plants linked to Saturn express themselves mainly through the 'organ' - if we may call it this - that enables this connection, even by sacrificing other parts of the plant. If we think of the leaves of conifer trees this becomes clear. In this way the plant itself fulfills its task with the formation of large quantities of dry substance. Every time we want to accentuate a plant's inclination to produce wood we must amplify Saturn's influence on it. (If we shift our attention to the human being, the aspect of physical structure is expressed in the skeleton.)

We have said that the influence of a planet can also be excessive: excess Saturn causes a predominance of the structural part of the plant so that the higher forces do not manifest. The consequence is that the plant will not succeed, for example, in forming fruit.

The inability to form fruit can also depend on other factors such as poor fruit set but in these cases the problem will not lie with Saturn and the possibility to implement a correct and effective therapy will depend on our ability to deliver a correct diagnosis. If we are on a farm with excessive woodland we should be able to understand that this is a situation with an excess of Saturn and therefore, if we are to choose a crop for that area, we wouldn't plant an orchard as this would probably struggle. We could instead devote that area to horticulture, or even better, plant potatoes. (There always remains the alternative of chopping some of the trees.)

Let's move on to Jupiter who brings the *form*. (The typical form related to this planet is spherical, symmetrical.) The manifestation of this influence is found in activity that leads from a liquid to a solid state. To imagine this concept one could think of the hardening process that takes place in the structure of a growing child. At the moment of birth, the child can be considered to be related to the liquid element: it appears as a totally soft and round little thing and even the bones are relatively elastic.

Then, with the passing of time, the entire structure is consolidated, the cartilage hardens and the bones solidify. This process is the effect of Jupiter's influence. The typical parts of the organism governed by Jupiter are the joints. In fact within the joints one finds synovial fluid, cartilage and a bony part with the typical rounded form.

We also find Jupiter's primary influence in the movement of the muscles. Muscles are permeated by blood (liquid) but if they are flexed through movement in the joints, they take on a roundish form and harden; only when the muscle is relaxed once again does it soften and its liquid circulation is able to return.

An excess of this influence brings stiffness, deposits, malformations and cramps. This follows logically if we think of an excess of a process that brings liquid to solid. In an elderly person the joints become blocked, each movement becomes a challenge.

Mars was represented as an activity directed outwards which we defined with the single word *exteriorising*. Mars - the warrior - expresses the will to manifest, and by extension also to conquer the physical plane. We could say that Mars, with its will, allows for the intention existing on Saturn to move on to actualisation. This gives the physical plant the strength to grow upward and downward and gives the shoot the strength to push away the little rock which covers it. In another sense, the force with which the influence of fertilisation is concretised derives from the influence of this planet.

The excess of Mars' forces brings destruction, premature aging and a short life. Planting an apple tree with Mars in opposition coinciding with its regular movement will consequently result in a short lifespan and early fruiting.

The primary Sun *materialises* and therefore encourages the shift from cosmos to matter or from the imponderable to the ponderable. It is the idea that - in order to enter the ponderable world - must *wear* matter. Since the heart represents the Sun in the human being, we can say that this is the systolic phase, the contractive phase. The manifestation of this influence is found in the formation of substance.

Excess of this force brings hardening. Let's be careful not to confuse this influence with the stiffness that we identified with an excess of Jupiter, because the stiffness occurs between two parts (a typical example is the stiffness between two joints) while hardening means the sclerosis of a single part such as the bone, the muscle, wood. Excess of this influence also causes premature aging, degradation taking place too fast.

The Sun - unlike the other planets and indeed precisely because it is not a planet - does not have a regular and a retrograde motion. The primary and secondary influences are therefore determined by the alternating of the seasons, so that the summer Sun (or day Sun) performs an incarnating influence and the winter Sun (or the night Sun) an excarnating influence.

After the Sun in the incarnating process, we find Venus and its *internalising* influence. The main function of internalisation is without doubt to nourish, which we could define as acceptance and conservation. We are now talking about a coarse or ponderable nutrition stream that works from the roots, not the cosmic one which works downwards. The nutrition of which we speak occurs on the physical level and is the daughter of humus, water, and the soil-minerals.

In excess this influence causes hypotrophy that is the consequence of holding back excess nutrition and of the resultant block. It is the typical condition in which plants have stopped at a point in their development and cannot make progress nor regress despite our attempts to stimulate them.

Primary Mercury causes the *movement* of sap. To be a bit more precise we could say that we perceive a movement in the water element but in reality it is the etheric that is put into motion. We know that the etheric is connected to water and that water is the basis of life. For this reason we connect the concept of etheric movement with revitalisation. For a concrete example of what we have just said, think of alfalfa. When we cut alfalfa we take away the physical substance but the form remains attached to the etheric body of the plant for a while longer. The etheric immediately tries to 'fill' that form to bring new life (revitalisation) to that plant, which immediately begins to grow again and reforms its integrity: re-growth.

When there is an excess of this internal movement it cannot be contained so there is some loss of liquids and consequent dehydration. In the human being this excess can mean excessive sweating or an excessive loss of liquid secretions. Another type of excess loss may be provoked by the blood's inability to clot. (Keep in mind that within the influences that we are describing there can hardly be the isolated influence of a single planet. We are making examples and tend to create marked separations where in nature they are not so clearly defined.)

Let's come back to our excess of Mercury and, in this context, let's attempt to understand what may be a disease caused by an excess of movement. The fact that Mercury gives the possibility of revitalisation may lead to the repetition of re-growth even over a long period of time, almost to the point of boredom. This life that repeats itself can be found in the human being within the realm of work life where for 30 years one may be doing the same things again and again so that the monotony of repetitive work stifles even the enthusiasm to undertake other activities. The consequence will be the unconscious propensity to insert another life within one's own, in this way accepting bacterial infections. Bacteria in fact have a life of their own and do not follow the life processes of the human being. Naturally the same disease can arise in the plant world, for example with alfalfa when we rely on its re-growth too many times in a season.

We spoke about bacteria and not viruses because viruses are between the mineral and plant world while bacteria belong wholly to the animal world and completely fulfill the concept of autonomous life within our own. Secondly bacteria have an enormous capacity for reproduction that is another link with Mercury.

Mushrooms are linked mainly with the Moon and Venus. Mushrooms are divided into external and internal fungi. Examples of the former category are *oidium* and *botrite* and are mainly linked with an excess of primary Moon, while downy mildew is an example of fungi connected with an excess of primary Venus. External fungi appear via spores and settle in the plant while internal ones live right inside the plant, even during winter.

We defined the Moon's primary influence with the term *time*, a definition which we will fill out as '*manifestation in time*'. In this influence we find the impulse to germinate, to expand and to multiply.

Let's immediately clarify the concept of multiplication connected with the Moon by comparing it to the multiplication that we had connected to the concept of movement of the etheric. The multiplication of Mercury involves the repetition of a complex organism with many organs and many functions. When we cut alfalfa, the plant which regrows is a different plant with bigger and smaller leaves: it is complex. The Moon cannot differentiate and always reproduces the same identical model. An excess of such reproduction is a tumor.

We know that the liver is an organ that regenerates itself. Its regeneration is connected with Jupiter (form) and with Mercury (movement). In the case of liver cirrhosis one should bring in secondary Jupiter forces to remove rigidity (excess of primary Jupiter) as well as primary Mercury forces to reproduce the liver. The liver is a very complex organ with more than 3000 functions; not all of its cells are the same therefore we cannot help it with primary Moon forces.

The harmonious influence of primary Moon brings vigour, vegetative forces and growth. The Moon, like the Sun, does not display a retrograde motion and its two influences are marked by the position of our satellite with respect to the Earth and the Sun: the full Moon phase brings incarnating forces, while the new Moon phase brings excarnating forces.

This concludes the incarnating path so now we will go deeper into the planetary influences following the excarnating path along with their excesses. Secondary Moon, as we have said, determines *space*. Our space is defined by our skin; therefore the Moon *produces* skin. Generally when we speak of *skin* we also mean the external tissues of the organs.

The influence that we are examining is in polarity with primary Moon, so that if primary Moon was multiplication (in the sense of the reproduction of the same cell) then secondary Moon brings differentiation or diversification. The skin marks the beginning of something separate and different from what was there before (the organ or organism). The normal influence can be found in the fertility of the flower and also in our capacity to bring pregnancy to its natural conclusion.

An excess of secondary Moon is connected to diseases of the skin and to viruses. If we think of the animal world and we try to pick out those within which the principle of skin is at its maximum, we easily recognise the rhinoceros and the elephant. In the insect world all those with an exoskeleton are linked to secondary Moon such as the potato beetle and the bedbug. If we find these insects on our plants we are able to understand that a secondary Moon environment has developed.

Secondary Mercury is linked to a *form* that springs from two currents: the vortex. We speak here of concave and asymmetrical forms. (Forms related to Jupiter are symmetrical.)

The normal influence of secondary Mercury is reproduction meant as participation. The leafy network is also related to this force, as well as the turgidity of tissues. An excess brings stagnation of liquids and blockages. The blockage of sap is the primary cause of aphid infestation.

Let's move on to secondary Venus that *externalises*. All activities that involve yielding to the outside, a letting go, the entire world of excretions and secretions can be connected to this influence. The loss of substances and liquids such as hemorrhaging also belongs in this category.

The normal influence can be seen in the formation of cellulose and of bark. An excess of secondary Venus is hypertrophy in which the plant yields more than it succeeds in taking, leaving it suffering and sickly.

We may now shift our attention to the Sun. We had augmented the primary influence related to *materialisation* with the expression 'from the cosmos to matter'; therefore we may describe the secondary influence with the expression 'from matter to the cosmos'.

When moving toward the cosmos, matter refines its qualitative aspects more and more and frees its lightest parts. The formation of oils belongs to this dynamic.

An excess of this solar influence, which is an excess of dematerialisation, appears as weakness, fatigue when standing, or an inability to maintain an erect position (scoliosis, lordosis, kyphosis). In fruit trees, secondary Sun regulates fruit drop. This opens an interesting chapter for fruit growers because by using the forces of primary or secondary Sun one can moderate fruit drop.

Secondary or *interiorising* Mars produces chlorophyll, albumin and protein. Mars' secondary influence often goes hand in hand with primary Venus. Indeed, both interiorise. In this way Venus brings in the nutritive elements that we have called coarse nutrition, and secondary Mars transforms them. If for example we wanted to produce soy with a higher quantity of protein than that which we normally obtain, we would want to give secondary Mars forces together with primary Venus.

Let's attempt to make another step. We recently mentioned the transformation that can take place in the human being who carries Mars forces. We had said that upon reaching maturity one's character changes and that force which earlier had been focused on the outside is now transformed and becomes the force of the word that is useful for other human beings. In the plant we can recognise the same force in the formation of pollen that will fertilise other flowers and in the capacity that this pollen has to fertilise (male aspect).

An excess of internal activity can bring the paralysis of internal functions and in a spiritual sense, internal paralysis. It could be that an excess of secondary Mars could make one sterile.

Let's move on to secondary Jupiter whose influence is internal *movement*. It's typical expression is muscular movement. Perhaps the word that best describes it is *motility*. The normal effect of the influence we are examining is the swelling of the fruit. An excess of swelling means softening or sagging. Let's imagine a slight, puny person who gains strength and power by becoming bigger. If he becomes too large he is unable to be active and in the end he becomes flaccid.

Secondary Saturn is *time*. We could say "maturing" as the last step before disincarnating. A fruit, having concluded its cycle, can hold nutritional value and thus nourish others. Inside the fruit is also the seed that, in its interior, has the ability to reproduce life. The function of secondary Saturn can really be called "life within the seed". We could define the seed as the memory of itself.

An excess of secondary Saturn is the opposite of primary Saturn. Therefore it is a loss of verticality. This loss of verticality may be confused with the influence of secondary Sun. To clarify we may say that the influence of the Sun manifests as bad posture, in fact we had spoken of scoliosis, while excess of Saturn does not allow one to remain standing.

We have now deepened our understanding of planetary influences on the plant identifying the incarnating and excarnating influences. In future we will come back to this subject to try to understand how various plants are expressions of planetary forces. There are plants whose group I is connected to the Sun (*graminaceae*) or to Venus (*drupacae*) or to another planet - and they primarily manifest the forces of that planet.

It is also true that all the group I's of plants have a connection with Jupiter! This is linked to the fact that this planet is the memory of Old Sun, the second stage of Earth's manifestation, when the evolutionary level allowed the descent of 'plantness'. Let's recall that in the preceding stage, Old Saturn, the pre-material, physical element had appeared.

However, we know that plants are strongly connected to the physical Sun as the source of life for our solar system. We also know that the etheric plane, and therefore life itself, has found an indispensable mediator in the Moon which we can consider to be a regulator of life forces. We cannot say that the Moon brings life, but it determines some conditions of manifestation. For example, a human's or animal's gender is determined by the amount of time spent by the I on the Moon during the phase of descent towards incarnation.

In this way we have outlined the three key elements of the plant world, in which Jupiter represents the past, the Sun represents the present and the Moon is a point of regulation such as the dial on a radio. From this we can understand why it is so important to take into account the relationship between the Sun and the Moon to decide when to sow seeds.

However, what we have said, though true, is no longer sufficient because human beings, with our selfishness, have done many things which have resulted in the withdrawal of the group I's, first to the sphere of the Zodiac and then toward the Milky Way. Such activities include hybridisation, genetic manipulation and, worst of all, genetic engineering.

Let's resume reading Steiner's text by going to the fourteenth paragraph of the second lecture ("*It is different however ...*") to focus on the concept of the term '*death*'. Our way of looking at death is conditioned by the fact that normally we give an absolute value to our existence here on Earth. Therefore death, the severing of our relationship with this world, can only represent the worst evil.

However, within a broader view, precisely because death brings the end of vital and astral processes, it allows for a particular relationship with cosmic distances which, in this new condition, permeate the physical element as if it had become transparent. This is what happens with the crystal, the diamond and with all the precious stones that perhaps fascinate us precisely because they are bearers of such forces. Steiner even indicates a specific period of time during which the Earth is particularly dead and therefore is "*least reliant on itself, on its mineral masses*" and more exposed to the crystallising forces that exist in the cosmic distances.

If we were to use the term 'excarnation' instead of death it would be easier for us to understand how our essence on this planet is only following a path of experience and does not itself change at the moment of death - it only changes consciousness. In fact the process of excarnation begins at the moment of birth and accompanies the growth of consciousness which reaches its zenith at the moment of death because at that moment another dimension of consciousness is added to that already acquired during earthly life: consciousness of the other life.

In any case it is precisely consciousness of the world that distances us from consciousness of the cosmos and its laws. The plant world - which has a higher consciousness than the mineral world - easily connects with the planetary sphere, whilst the mineral world has the ability to connect with the cosmic distances, the world of the stars.

The etheric element hovers above the ground, permeating it, therefore the further we are from the Earth's surface the more we move away from life which is why the Earth is more subject to crystallising forces which come from cosmic distances. This is the reason why biodynamic preparations are placed underground.

Also in the fourteenth paragraph Steiner indicates the specific period during which the connection that we have just mentioned is at its maximum; on our part we must only try to avoid misinterpretation and set in our minds that he is speaking of minerals and not plants for which the dates are different.

We consider it important to note an expression used by Steiner in the third line from the bottom of the paragraph because it is full of meaning: he says that the interior of the Earth has the strange *feeling* of being less dependent[9]. He speaks of the Earth not as an inanimate physical body but as if he were referring to a being with a soul, one that can feel. The soul is connected to the astral plane that we know to be represented by the planets, and indeed it is from these planets that the forces that create form are derived. Therefore with that simple word Steiner wants to tell us that from January 15th to February 15th the Earth is capable of connecting with these formative forces to crystallise and therefore create forms.

In the fifteenth paragraph ("*The situation is this: ...*") we read that during this same period plants have maximum neutrality in relation to minerals. However, we can add here that if the plant does not communicate with minerals during the period in which the minerals enjoy a heightened communication with the cosmos, then the plants themselves do not speak with the cosmos.

To grasp the situation we must imagine all minerals and plants in a winter scene in which the minerals feel nostalgic for cosmic distances, thereby creating a bridge of light to the cosmos. At the same time the plants are as if cast under a spell until, waking up, they can receive from the Earth what passed on the planetary bridge which we mentioned. In fact plants are not able to establish contact with the cosmos by themselves because they have come too far from it, due to the human being.

We know that ancient wisdom transmitted spiritual truths through fairy tales and we may, if this helps us to understand, imagine what we have just said in the form of a fairy tale. In this way we could tell a story that during the entire year the beings of the earth called gnomes stay hidden in darkness, waiting for the time of mid-January during which a marvelous and enticing light reaches the Earth. During the same period, the undines - beings of the plant world - have been put under a spell and are unable to participate in this event. However, when the gnomes have filled themselves sufficiently with light they see the undines re-awakening and they give them part of the light that they had accumulated so that they too may live joyfully for another year, until the cycle repeats itself.

Also in the fifteenth paragraph we read how just before and just after the period in question, the situation is exactly the opposite. Perhaps it is worth remembering that in the middle of this period there falls the festival of Candlemas. If we were in the east we would call it Kundalini, but we could also call it Black Madonna since these names represent the same Spiritual Being. Before January 15th and after February 15th the plant is not spellbound and can connect to the cosmic forces through the mineral world and the period immediately preceding and following the mentioned dates is particularly important to direct the growth of plants. It is worth noting that the words "*particular importance for plant growth* " are repeated three times in seven lines and two lines later he adds that "*People will some day realise how important it is to take advantage of such things in order to be able to regulate the growth of plants"*. Clearly it was Steiner's intention to give particular emphasis to these words. In fact in these words is the secret to a new, ethical genetics.

In this way we have identified three periods, that is a period in which the plant is connected to cosmic planetary forces, followed by a period during which the plant is spellbound and finally a third period in which the plant is newly ready to connect. It's clear that the first and second connections are not the same because they are separated by a period in which minerals receive new forces from the cosmos. The second

9 Note that this is not so much the case in the English translation being more implicit within the phrase '*the inner nature of the Earth*'.

connection is that which brings to the plant the new life forces for the entire agricultural year. Before, however, there is that magical period of particular importance, the 13 Holy Nights, during which forces from the constellations and the planets descend to Earth from the cosmos to fertilise all the kingdoms of nature, starting with the mineral kingdom and through this continuing with the plant kingdom.

For those who want to attempt to direct the growth of plants with ethical genetics there are three important moments during which it is necessary to intervene: the first is the period of the Dead when the forces of death that embrace all of nature are dominant, the second is during the Holy Nights with the descent of the new forces, and the third is Candlemas when the new forces begin to be available to the plant world.

Thus we arrive at the sixteenth paragraph - which we consider to be of crucial importance - where Rudolf Steiner, speaking in 1924, sends a message to future human beings by saying: "*Let me remark here that if we are dealing with a soil that does not carry these influences upward during the winter period as it should, it is good to furnish the soil with clay, the dosage of which I will indicate later.*"

When he refers to *soil that does not carry these influences upwards* he means the condition of devitalisation of our soils, killed by repeated chemical interventions and especially by herbicides. Farmlands at the beginning of last century did not have more than the slightest hint of the problem of which Steiner is talking.

The sentence we have just read holds the promise of a clarification that is not given. Nevertheless implicit in this sentence is one of the fundamental secrets of agriculture. Clay is an amphoteric substance, in other words a substance that has the ability to act as a base with acids and as an acid in a basic environment, so that this ability involves it in a double circulation: it can carry upwards and can bring downwards. We can also say that clay allows the connection between the cosmos and the Earth to happen, a diminishing ability for our modern soils.

Steiner's indication may seem strange to us. However, research has shown that during the Middle Ages a mix of clay and saltpeter was used as fertiliser by spreading it on the soil in minute, almost homeopathic doses. According to our cultural bias the peasants of the Middle Ages were ignorant yet, because of tradition but also thanks to the ability to perceive from instinct and feeling, they were bearers of a knowledge which seems almost like science fiction, or at worst like witchcraft, to us.

However, by reading closely what Steiner explained at Koberwitz and attempting to connect various fragments of information we may understand something more and grasp the practical aspect of a teaching which at first glance seems to say very little.

In the preceding paragraph Rudolf Steiner spent a lot of time talking about three periods of the year and then he cites clay; we might infer that he is indicating the three moments during which clay should be used. Furthermore we must keep in mind that clay has a double action: there is an action that takes the crystallising forces that the plants received from the mineral world and brings them towards the heights so that they don't remain in the roots: and then there is another action that brings down what the plant developed in its upper part. In these few lines we can find almost all that we need to know to farm well. Indeed they hold the key to fertilisation, to cosmic fertilisation as well as an indication of the way in which this manifests in the growth of plants.

Now, however, we are left with the task of understanding how to use clay properly, bearing in mind that it must be a vehicle for descending and ascending forces. In antiquity these forces had a dual representation. The first was that of the horns of Isis with the solar disc in the centre, while the second was the horns of the stag with the cross in the centre. The first belongs to the Mediterranean world where Isis represents the forces of Mother Earth's fertility, while the second, typical of Celtic culture, grasps more the cosmic forces, in harmony with the Scandinavian feeling more in tune with Father Sky. (We will speak more about these two symbols when we explore the subject of humus formation.)

Coming back to the practical production of the clay preparation, we could consider the plant as a result of the union between the forces which we call Father Sky and Mother Earth and so, using the ancient wisdom expressed in the two mentioned symbols, we may draw the practical indication of placing clay inside a horn "of Isis" (cow horn) and bury it during the period in which the force "of the stag" is active, that is during the 13 Holy Nights.

To complete the image we can think of the Holy Nights as the force of the stag with the cross that descends to fertilise the Earth, while the passing year represents the fertility of the Earth, well represented by the cow; for the Egyptians this second aspect was grounded in their experience of the Nile.

The meeting of the two mentioned currents is also the origin of humus, but can also be recognised within other processes, one of which is without doubt bread-making in which the two currents are translated as ferments (Isis) and yeast (stag). It isn't a coincidence that once upon a time the baker made a cross on the bread before putting it in the oven, clearly it was a way to connect it with the cosmic current. Indeed we can remember that crosses were placed even inside the ovens as well as in the chimneys of the kitchens.

Returning to clay, however, we must remember that we need two kinds of clay, one capable of carrying upwards and the other of bringing downwards. Clay that carries upwards must have a higher degree of purification than the other because it must allow something earthly to rise to the cosmos and something impure would be unable to do this. This problem does not exist for the descending forces because that which is in the heights is already pure and only needs to find adequate reception.

Now we may consider the fact that by definition the purifying element is water. Therefore if the clay horn is buried in a wet hole it may undergo a process of purification and become a suitable upward carrier; if it were buried in a dry hole it would bring the opposite influence. Besides, in paragraph fourteen Steiner tells us that underground water becomes more dead and therefore more able to communicate with cosmic distances.

Regarding the use of clay we can refer to the three periods mentioned by Steiner and therefore we will use the purified clay preparation during the period of the Dead to encourage the withdrawal of the life forces from the Earth; we will use the same preparation at Christmas to encourage the descent of cosmic forces during Holy nights and finally we will use the clay preparation that carries upward on Candlemas to encourage the ascent of forces received from the mineral kingdom to the higher realms. What we have said - which can be further investigated elsewhere - can be represented by the three drawings below.

Look at the first drawing that refers to the period of November. At this time the plant is still alive although it is now ready to withdraw. Therefore we represent its etheric body with a lemniscate that is the expression of etheric circulation. Since the etheric body is related to water and therefore to sap circulation, it should be drawn in

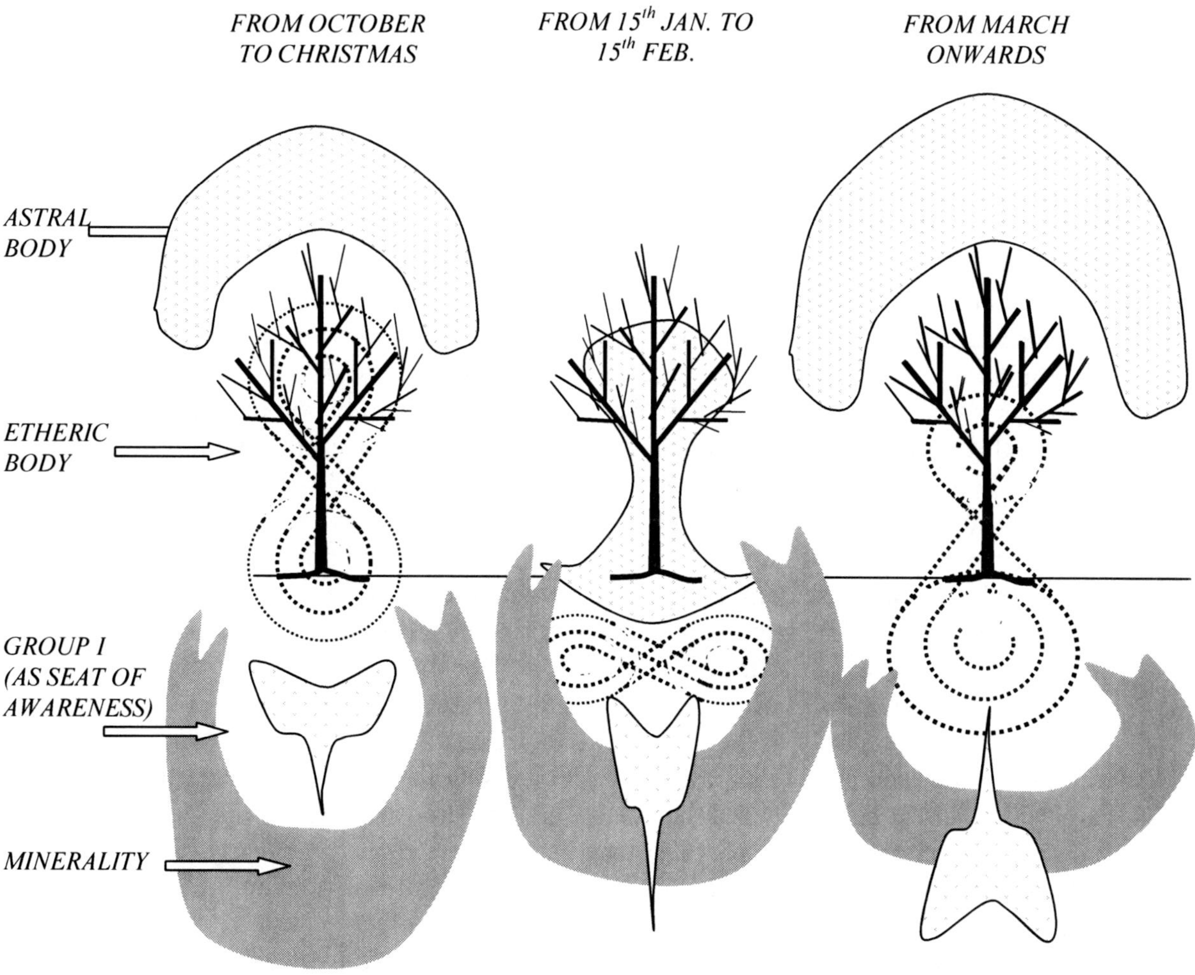

the colour blue. Above the plant we depict the astral body, similar to the head of a mushroom, imagining that it is coloured in yellow. We recall that the astral body hovers outside the plant and cannot enter inside otherwise we would be looking at an animal. We also recall that the astral body allows plants to go into flower.

Under the plant, also on its exterior, we depict the inferior I of the plant through which the group I of the plant is confronted with the I of the Earth, bringing about the birth of a process of consciousness. This part of the drawing should be imagined in red. Under the plant, in black, we also depict the mineral world that is ready to receive. We have said that during this period we may participate in this natural process with an influence that encourages the descent.

We now attempt to understand imaginatively the period that starts on January 15th and ends on February 15th. We are in winter and the plant is closed in on itself, therefore we represent the etheric body with a small, horisontal lemniscate placed underneath the roots since the tree does not have sap circulation during this period. The astral element descends into the trunk in a typical gesture expressing a situation of blockage. We could imagine the naked branch as if it were blocked by a cramp. The rigidity of the astral element is also demonstrated by the fact that in winter, branches lose their elasticity and if we try to bend them they inevitably break.

The I of the plant is welcomed by the mineral, while the latter, which has received the forces of cosmic distances, crystallises.

The third drawing is meant to represent life which manifests once again, therefore the etheric ascends back through the plant; this is represented by making the lower part of the lemniscate larger than the upper part. At the same time the mineral withdraws and the I shows interest once again in the plant, while the astral element is pushed by the etheric out of the plant with the result that the plant flowers: it is springtime.

We have not mentioned the 13 Holy Nights that we could link to the second drawing when the mineral consciousness of the Earth speaks with the plant consciousness.

We can become aware of this if we think about the fact that mineral consciousness is connected to the Sidereal Moon while the plant consciousness is connected to the Synodic Moon, but since the two lunar cycles have different durations, 27.3 days and 29.5 days respectively, normally they don't coincide. They only coincide during the Holy Nights, in fact a simple mathematical calculation demonstrates that throughout the year, 13.36 cycles of Sidereal Moon are completed and 12.36 cycles of Synodic Moon are completed: 365/27.3=13.36 and 365/29.5=12.36

It becomes clear that if during Christmas the two cycles are perfectly overlapping, at St. John's time they are completely out of synch, thus impeding any dialogue between the Earth and the plants. Indeed the period of St. John is particularly important for the care of the animal world and not for the plant world.

Returning to the subject of the use of the clay preparation it is important that the burial take place precisely in the period of communication between the two consciousnesses so that, encompassing the revitalising forces of Mother Earth and Father Sky, it may be considered the synthesis of all preparations.

From these considerations we must take the initiative to act practically on our land. Therefore during the period of the Dead we will spray the clay preparation which we have called 508; the one which brings downwards (dry hole).

During St. Lucia we will use the 508 once again to bring downwards, together with the well-known ‘Opening to the group I's’, based on yarrow prepared in the stag bladder (G01).

During Christmas we will spray 508 carrying upwards together with the generic plant I (O02).

During Candlemas we will repeat the treatment of 508 carrying upwards with the generic plant I.

The above-mentioned treatments replace all those which we were used to doing in other years during the Holy Nights, except for the specific ones such as Pro-humus, Anti-weeds and the production helpers (series I).

We add a few more details by saying that during St. Lucia there takes place the purification of the Earth's soul. During Christmas there comes about the meeting between all the kingdoms of nature with the "Lesser Guardian of the Threshold" who is the being of Jesus of Nazareth. At Epiphany there takes place the meeting with the "Greater Guardian of the Threshold" who is the Christ.

What we have said constitutes the practical aspect of these paragraphs of the Koberwitz course, namely how to direct the growth of plants. The clay preparation can also be used as a seed bath together with other preparations to improve certain functions of the plant such as enlarging the fruit, increasing the oil yield, etc.

Sixth meeting

In our previous meeting we read some paragraphs that we believe constitute the core of the whole Koberwitz course. We will remind ourselves that Rudolf Steiner emphasises the importance of what he is saying by repeating it three times in a few lines: that it has a "*particularly strong effect on plant growth*" and, after only three lines, adds "*how important it is ... in order to be able to regulate the growth of plants.*"

Because we attach a particular weight to these lines we will take them up again at the end of this series of meetings, when, after having gained more familiarity with the wonderful weave of the puzzle that Steiner is composing, we will develop our understanding of the influence of planets further. For the moment we can continue reading to dwell on the later lines of the seventeenth paragraph ("*So you see ...*") where we find a curious expression: "*the life of the soil is especially strong in the winter, while in a certain sense it tends to die down in summer.*

Normally we think that nature is asleep during the winter and revives during the spring, and this belief is based on the fact that life can be seen outside in the warmer season, while during the cold season we have the impression that life retreats. In reality, if we employ our *minds eye* which is not restricted to viewing the manifestation, we can understand that the Earth is more active when the plant closes in upon itself and this takes place in the winter. It can help us understand what we have just said by thinking of what all of us do when we concentrate: Whilst we can certainly jump and run around we need to pause and turn our attention within ourselves to really develop ideas. For the consciousness of the plant it is exactly the same during the winter, when there are no visible signs of life, the plant itself is awake and can perceive the descent of the cosmic forces.

We can imagine the alternation of the seasons as an expression of one cycle of a *breath of consciousness* of the Earth (and this gives us the opportunity to suggest that the four fundamental texts of Steiner are read in a different season. For example, "A Philosophy of Freedom" - which is the text to bring us into the world of thought - should be read in the period from St. Michael to Christmas, while during the winter it would be ideal to read "Knowledge of the Higher Worlds" and then "Theosophy" and "Esoteric Science" in the successive seasons.)

Be that as it may, let's continue our journey by reading the eighteenth ("*Now, especially ...*"), nineteenth ("*The truth of the matter ...*"), twentieth ("*So, if we ever ...*") and twenty-first paragraphs ("*Let us say ...*"). We will address them as a whole. Steiner speaks of the seed and sowing and begins by saying that to cultivate the soil it is first necessary to understand the circumstances in which the cosmic space and its forces can influence the Earth. Once more the cultivation of the land - and thus agriculture - is seen as enabling a dialogue between the universe and the Earth, or rather between the forces of Father Sky and Mother Earth. Only after this introductory word does Rudolf Steiner speak of the seed and of the plant. He opens this subject with the demolition of a widely held preconception that has adversely affected the understanding of the whole reproductive process. It is commonly believed that between the mother and daughter plant there is a continuity of life for which the seed is the vehicle. Thinking this way we consider that the development of the new plant is already present in the seed and so the seed must have a complicated structure. Well here we read that it is not so, because the new plant is formed by cosmic forces for which a seed is only an anchor point.

The clarification of the necessary concepts requires the help of some sketches to bring them together. We can show in the sketch that the plant reaches its completion with the formation of the seed into which it concentrates its life as the mother plant dies. In the seeds we can picture the terrestrial element caught and conserved in time as a latent extract of life.

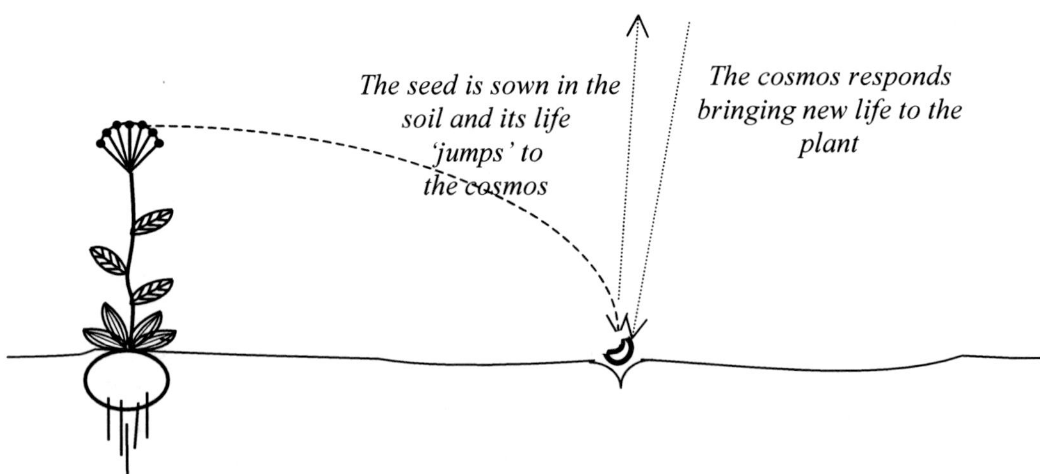

At some point we decide that it is time to sow and we make a tiny furrow in the soil. Now we might think that this space is filled only with air, but in fact there is a sea of undifferentiated ethers and the forces that come from the planets flow into this sea. From even greater distances the forces of the zodiac come through the planetary world to infiltrate the etheric and vivify the physical world.

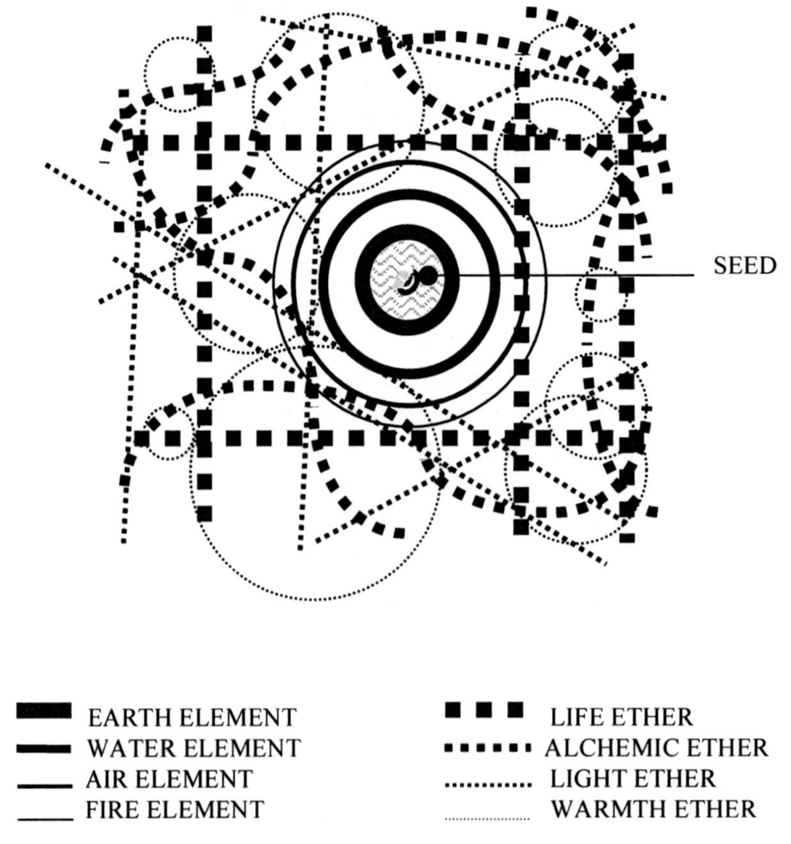

We put the seed into this generalised etheric situation as it is placed in its furrow. The seed is thus in contact with moist soil and awakens its life, inflates, and undergoes a process that makes its protein structure chaotic. But we have to consider the fact that the life of the seed, which awakens when in contact with water, is an individualised life which provokes a sudden retreat in the open sea of ethers (which is polar compared to the individualised ethers present in the plant) freeing space wherein the new life can arise. This 'retreating' action, however, causes echoes throughout the universe and is recognised and interpreted as an opportunity to manifest by the forces that resonate with the life of the dying seed. If we had put a carrot seed into the soil its call would have been noted by the forces acting through the planet Jupiter, which would have responded by pouring themselves into the free etheric space in which the new plant will be born. This does not represent a continuation of the preceding plant. It is a new life from cosmic distances.

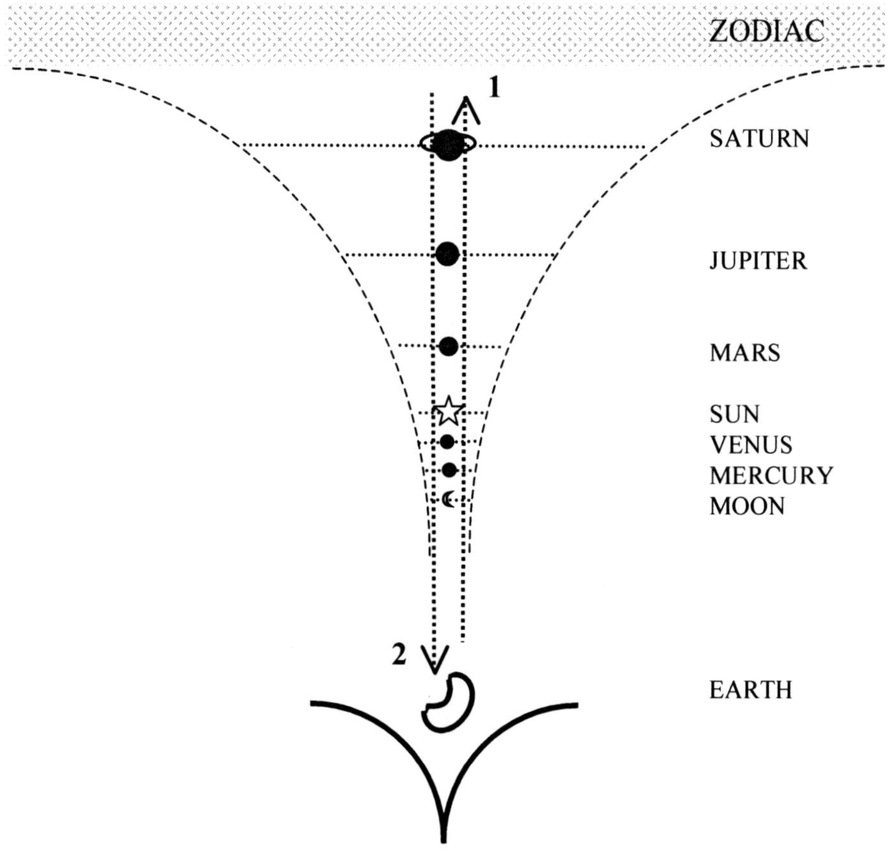

The seed 'sends' its life to the cosmos (1), which responds with new life (2).

Of course, if the reproduction does not occur through a seed all that we have just described is not the case.

But now we can sense the importance of choosing the correct time to sow the seeds if we want our plants to receive the maximum expression of cosmic planetary forces. Since different plants are different combinations of the various cosmic forces, we can imagine them as a musical harmony and the more the forces that make that harmony are in the sky at the time of sowing the more the plant will be their harmonious expression.

This fact is crucial for two reasons, primarily because the physical plant is the means for the group I to have experiences on Earth and to evolve through these experiences. Then, because plants are a source of food for us and for our animals and only if they are connected with their own spiritual matrix will they be valid for our subtle 'components', and thus suitable to nourish our higher faculties.

If we wanted to evaluate the importance of good seed compared to all our other inputs, we should reckon on 80-90% being attributed to the seed.

We make our best contact with the cosmic forces through the seed - through this chaos in the protein of the seed that appears to us to be the extreme of disorder. To a higher vision, however, it must be considered as the opposite - as the beginning of life. Moreover in the potentisation of homeopathic and biodynamic preparations it is crucial that chaos is created, because only in this way will we create conditions so that new cosmic forces can penetrate the preparation to be passed on to the plants.

It may seem strange but many scholars are committed to studying the so-called 'mathematics of chaos' and are trying to identify the higher order underlying the physical disorder that we perceive. These scholars would probably be surprised to know that they are dealing with the laws of the Father, which are the laws of life.

Returning to a concept just expressed in passing, we can reaffirm that the seed in its furrow creates a free space in the sea of the ethers and this will receive the characterisation given by the etheric body of the seed. This causes a vibration in this 'ocean' where the new cosmic forces that resonate with those of the seed can incarnate and, upon entering that space, undergo a reversal of action. This reversal is not only in the case of the seed but in any form of incarnate life - plant, animal or man. In this way we can understand how the head is the bearer of the warmth ether, but around it is the ether of Life; similarly the reproductive organs are bearers of the life ether and are surrounded by warmth ether.

INCARNATED		***FREE***	
ETHER	*ELEMENT*	*ETHER*	*ELEMENT*
LIFE	FIRE	WARMTH	EARTH
CHEMISM	AIR	LIGHT	WATER
LIGHT	WATER	CHEMISM	AIR
WARMTH	EARTH	LIFE	FIRE

This has important practical applications because it is clear that if we need to increase the Life ether in a plant we should bring about an increase of free Warmth ether. By acting outside we must remember that the extra etheric force will enter the plant.

From everything we have seen sowing is the most important moment for the life of the plant and anyone intending to commit to direct the growth of the plant cannot ignore this fact. Dr Steiner makes this explicit because, after having spoken of the important times of the year for the growth of plants, he moved right on to speak of the seed and the circumstances in which the cosmic space and its forces can act on earth.

It seems clear, at this point, that our main task is to create the conditions so that widespread life - *Zoe* - can transform itself into abundant incarnated life - *Bios* - in the plant, because only in this way do plants become strong, healthy, resilient and the bearers of high-quality food.

Of course everyone tries to find the best sowing conditions as defined by their understanding. So someone is concerned that the soil has just the right level of moisture, whilst another seeks to obtain organic seed because they believe one can easily establish a relationship with the world of life. Maybe this also means seeking to sow in accordance with the phases of the Moon. Some people can raise their gaze further and take into account the position of the planets in the sky, or even of the constellations. We use the seed baths because we are convinced that the seed makes its etheric appeal when it is in a position to receive a response of the cosmos, making the best of what lies in its immediate vicinity. In recent decades what the seed finds around it is electromagnetic pollution, radioactivity, smog, herbicide residues and chemical fertilisers. This dramatic condition did not exist before and makes the dialogue with the forces of the universe extremely difficult. For this reason we encase the seed, by means of the seed baths, with an etheric capsule that can bring the seed all the forces it needs, and protects it from environmental disturbances.

In the twenty-first paragraph ("*Let us say we plant ...*"), Steiner suggests the need for the seed, after establishing the dialogue with the cosmos, to be connected with the forces of the Earth so that it can manifest the formative principle that it received. But one must bring the plant to the Earth ... "*... by infusing the plant with the life as it is already present on Earth, that is, with life that has not yet reached the stage of complete chaos – the stage of seed formation – but has stopped at an earlier stage.*"

Steiner is talking about the importance of humus. Today we find it difficult to find soils with sufficient humus and so the pragmatic solution lies in replacing or augmenting the ponderable humus with the *forces* of humus.

We are fully aware of the fact that Steiner, in this context, uses the words "earthly" and "cosmic" to characterise the direction or goal towards which the forces are striving and not to their origin. Thus earthly forces are going towards the Earth and the cosmic forces are those that go towards the cosmos.

Having said that we can now understand that humus is the result of a process of vegetalisation of the Earth that has not reached completion. (If the process had been completed the Earth would turn into plants gushing up from underground.) The plant connects to this incomplete process from outside like a parasite, and joins with the predisposition to vegetalisation that it finds in the soil and brings it forward to its completion, or until the formation of the flower.

We could say that humus is an attempt to bring the mineral realm towards life but it is one that fails to individualise it and so remains general or widespread. This unindividualised or general life can, however, enter the plant which *does* have the principle of individuality and can thus embody this life.

From what we have said, however, it is clear that for the plant to complete the process of vitalisation, it must have a start. In other words the plant needs to find humus in the soil otherwise it will not be born and develop. The problem of the world today consists precisely in the fact that humus is increasingly scarce and so we must either find ways to increase it, replace it with something else, or stop farming!

There are a few lines of the twenty-second paragraph ("*Assume you have a plant* ...") that contain a decidedly unorthodox image of the plant. In fact we are accustomed to regard the plant as a result of a series of development processes that build the plant starting from the bottom and going up, because this is what we see of the growth of the physical plant. Steiner is proposing a completely different point of view. He says that what the plant develops horizontally[10] is the result of the terrestrial forces. It is as if these forces, coming from the sky, crash-land on the earth and splatter themselves around. By contrast, the cosmic forces move from the root vertically up along the stem to reach the seed, from where they can radiate throughout the plant.

Note that the terrestrial forces are the same as those found within humus. In this sense these do not 'push' the plant up, but act as a pole of attraction for the forces radiated directly from the planets. In their journey down these forces pass through the plant constructing it in a horizontal plane. We have, therefore, an image of humus as an organ that acts like a sponge that attracts and welcomes the forces of the cosmos.

For us an alternative remains of strengthening the presence of humus by strengthening its forces. In other words, working between the biological element and the dynamic element. Steiner is giving us directions to understand how the forces of the cosmos act on the plant and is indirectly providing a method for reading the plants' form. Through this we can begin to establish a diagnosis and move on towards remediation.

In the next paragraph ("*Just take a look* ...") the subject of colour is introduced with the indication that the colour of the leaves express forces radiated directly from the planets, while green is the direct result of the Sun. On the other hand the colour of flowers depends on the cosmic influence of the Sun reinforced by the cosmic influence of the external planets. Red is brought by the planet Mars, the white and yellow from Jupiter and blue from Saturn. It may seem strange but the colour of the flower arises - in harmony with the Goethean theory of colours – from the meeting of the light from planets with the darkness of the Earth and cannot therefore be the direct result of the radiation upon the plant.

It is known that there are plants, such as hydrangeas, which change their colour depending on the soil in which they grow so that one can work by enriching the soil with metals to get certain colours, such as adding iron on the ground to get blue hydrangeas. This way of working should make us realise that the colour comes from the forces of the earth, because if it depended on the direct radiation it would be more logical to spray the plants externally rather than to change the soil mineral balance.

In paragraph twenty-four ("*What is visible in* ....") Steiner says that what appears in the colour of flowers is what a superficial observer might think comes from the impact of external light. In fact it comes from the root, which has the clear function of making a connection with life from the distant planets. In an analogous way the substance of the flower is governed by the terrestrial elements, and this substance then reveals the cosmic element through its colour. The terrestrial element is expressed in the flowers in the differentiation of its parts. (Bear in mind that the term '*terrestrial*'

---

[10] This means away from the central and vertical line of growth.

means 'from the cosmos to earth' and '*cosmic*' means that which from the Earth 'rises to the universe').

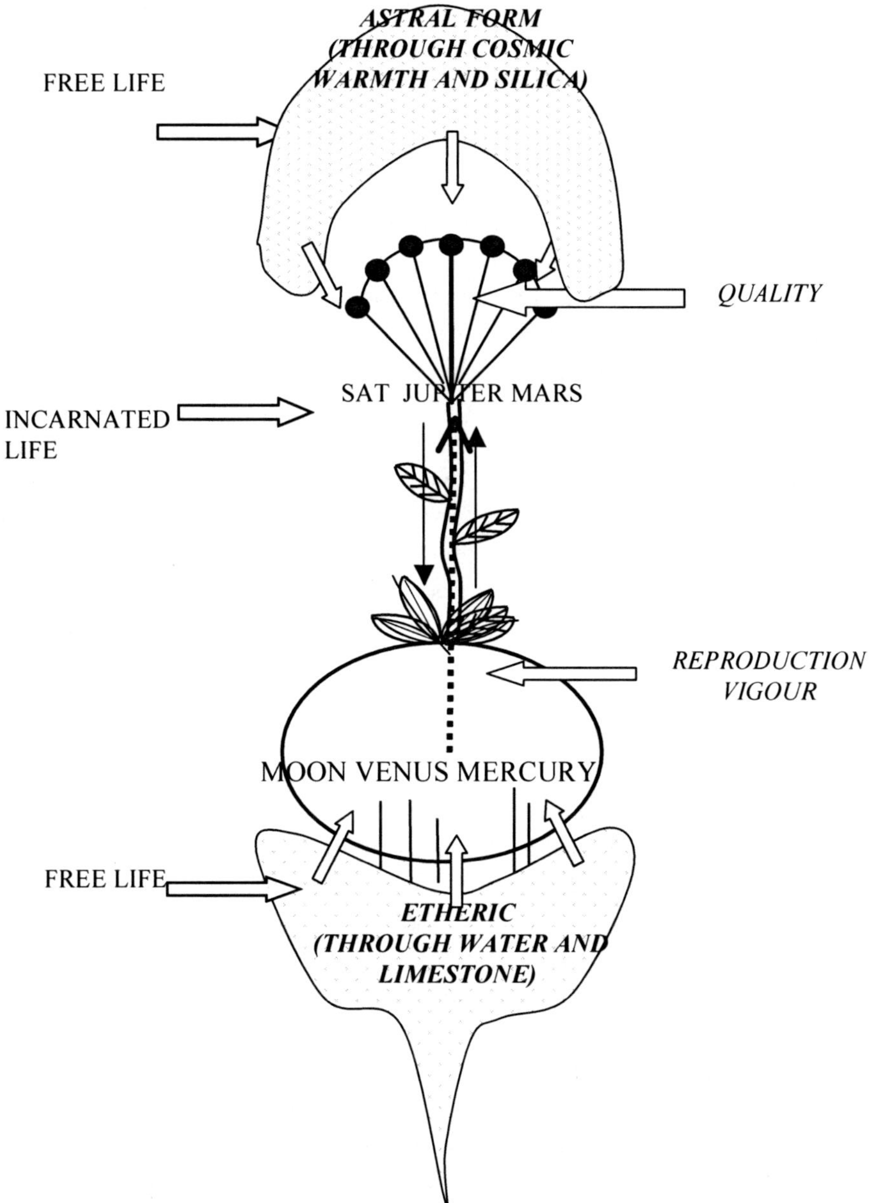

If the terrestrial element must live in the root exactly the opposite happens, and in this case it is the form of the roots that reveals that the impact of the terrestrial element by becoming fibrous.

The cosmic element, in as much as it is the bearer of life, tends towards unity (tap root, single apical flower), contrary to the Earth element which induces differentiation and therefore causes division or fragmentation (fibrous roots). We note, however, that the greatest differentiation can be seen in the flower, because whilst in the roots there is an almost infinite ramification these many roots are all essentially the same. On the contrary in the flowers we find petals, sepals, stamens, pistils, calices, ovaries. Orthodox botanical taxonomy knows this so all plants are identified by what can be observed of their flowers.

In summary, we can say that the plant is continually crossed by two currents: the terrestrial one that descends and which develops the plant horizontally and brings differentiation, and the rising cosmic current which strives towards verticality, unity and quality.

What enables this crucial and continuous interpenetration within the plant is clay, and so we have made two clay-preparations that allow us to promote either of these two currents and so direct the growth of plants.

If we consider a carrot we will see the typical result of the cosmic forces being retained in the root. The carrot is normally orange and carrot flowers are usually white because the carrot mainly expresses the forces of Jupiter. Well if we encourage the carrot with the clay-preparation that 'brings up', the cosmic forces should climb to the flower and we should get less colour in the root and a yellowing of the flowers. If we wanted to improve the root we should use the other clay-preparation that 'brings down' to maintain the cosmic shape at the root level.

Later in the paragraph Steiner talks about how we can bring down the terrestrial forces and here he mentions lime. We already know that lime tends to 'bring down' because it is the bearer of a contractive gesture but perhaps we had not grasped the practical implications of this dynamic. On an operational level, however, there arises a dilemma: should one 'bring down' the terrestrial forces with the lime or with the preparation of clay? Well, you use both whilst taking into consideration the fact that the action of the clay-preparation is more powerful. We can moderate the impact of our intervention by combining the two different preparations.

The plants and the environment were so much healthier in 1924. Now we have herbicides, the ozone hole, GMOs, artificial radioactivity and so on. Then it was enough to intervene with lime, but today everything is a little more difficult and so it is necessary to have a more powerful tool consisting of clay to which lime can be a valuable complement to calibrate the quality of any intervention.

Moreover we believe that it is no accident that Steiner was speaking explicitly about lime, whilst in talking about clay in paragraph sixteen ("*Let me remark here*") he postponed giving further details until '*later*'. In retrospect we might take this to mean not later in the course but at a future time when, due to the degradation of nature, lime would not be strong enough. The times in which we live are the ones covered by Rudolf Steiner. Today everything is losing life but fortunately with the clay we can now draw the necessary forces of the cosmos, especially when all lines of communication are open during the '13 holy nights'. It is clear that in this context lime takes a back seat.

Steiner has just finished speaking of lime and in paragraph twenty-five ("*Now, let us assume ...*") silica is brought into the reasoning when talking about the possibility of retaining the *cosmic* forces in the plant. In this way Steiner has wisely indicated the second variable at our disposal for being active in the development of plant growth.

In summary limestone can, to a greater or lesser extent, attract the terrestrial forces downwards, while silica allows control of the ascent of the cosmic forces.

All the work of Zarathustra is encapsulated in these few indications, and it is also the basis on which Martin Schmidt (who was present at the Koberwitz course) set up a method for the regeneration of seed. However, that method would take a hundred of years of thoughtful sowing. But now we have the clay-preparations available and these can accelerate this process and make similar progress within one plant generation.

Moreover today we are not interested in transforming one plant into another. We 'just' purify plants from transgenic contamination and other forms of pollution that affect the dialogue with the most distant cosmic forces from as far as the Milky way. In this way we can grow food that is more suited to modern times and our spiritual needs.

Steiner continues with concrete examples of how to direct the growth of plants.

Paragraph twenty-seven ("*From all this ...*") begins with a sentence that is repeated hundreds of times around here. It basically says that the ABC of the whole process of plant growth is to recognise what is cosmic or terrestrial. If one does not gain this ability it is useless to ask how to intervene to transform a soil that is condensing cosmic forces and retaining them in the roots or leaves, so that these forces will climb up to the fruit to improve the quality and taste.

After the observation that people have lost the "*instinctive primeval wisdom*" so it would be impossible to repeat the great work of domesticating plants produced by Zarathustra, Steiner says that we should always be grateful because without Zarathustra's work we could not now have the enormous variety of fruits we do have. But Steiner concludes the paragraph by saying that if we want to continue to be productive we cannot shirk the difficult task of penetrating the process of growing plants "*rationally*". The term "*rationally*" in contrast to "*instinctive primeval wisdom*" highlights the need to develop new knowledge, since the old one has been irretrievably lost.

The twenty-ninth paragraph ("*What our friend Stegemann ...*") can be surely called *prophetic*. Steiner actually describes exactly what is happening today. In this context we believe it is important to emphasise the fact that Steiner cites the end of Kali-Yuga. This is the so-called 'dark age' that lasted 5,000 years and ended in 1899 during which Man lost the perception of spiritual light. The end of this period marks the beginning of an era in which one can gain a new awareness, because the human soul is changed and changing and now has the latent ability to perceive nature, the cosmos, the divine for which one may have a new longing. But around us all of nature is changing too and this change requires a new connection.

To understand the extent of the change in nature we must remember that the end of Kali-Yuga was preceded - by about twenty years - by the start of the regency of the Archangel Michael. The evolution of our system is influenced by seven Archangels who take turns, about every 360 years, in bringing the new evolutionary impulses. Until 1879 there was the regency of the Archangel Gabriel. This Archangel is familiar to us as the one that has 'announced' to Mary that she would become the mother of Jesus. Of course the fact that the announcement of that happy event - for Mary and for humanity - has been made by Gabriel was not arbitrary. This Archangel is responsible for supporting the 'hereditary stream', or that which is transmitted from parents to children through reproduction. Michael, on the other hand, supports the 'individual stream' and therefore the free will and activity arising from the conscience of the individual.

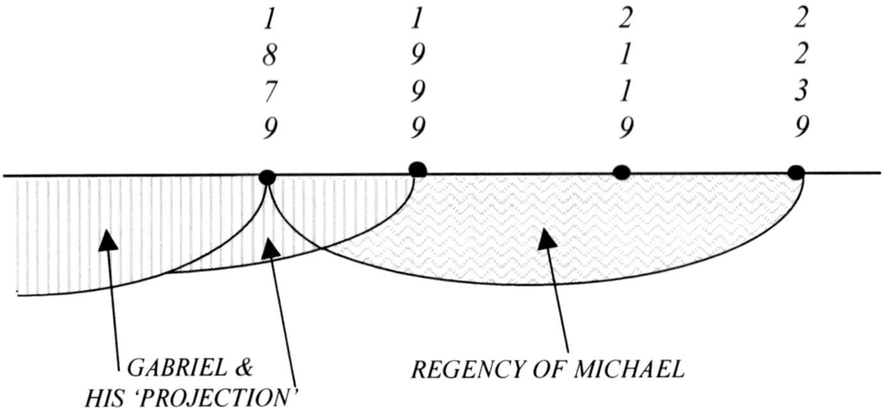

Given that since 1999 we are experiencing the full period of Michael's regency we can understand how individual initiative is to the forefront in this period and how this is a fundamental problem for the plant world. We know that the plant has incarnated physical and etheric bodies, and these are the ones that perpetuate the hereditary stream. Furthermore the plant alone is not able to find new ways to sustain itself with the cosmic forces. In this situation only mankind can help the plant to establish a new dialogue with the forces that bring life. We have this great opportunity that is also an equally great responsibility … and we also have the ability to ignore all this and to bring the plant kingdom to total destruction. Unfortunately, looking around, it seems that humanity has chosen the second path, because agriculture is increasingly carried away with techniques that divorce the plant from its spiritual matrix, starting with hybridisation right up to transgenic manipulation. Moreover it seems that the actions of man are no longer guided by ethical and moral principles, but only by the desire to maximise financial gain - at all costs.

Humanity must find the awareness that allows us to take a different road where it is essential to nurture the renovation of our souls in harmony with the laws of the spirit. It is not a coincidence that Steiner cites the transformation of the psyche.

In the thirtieth paragraph ("*People today have a ...*") Steiner completes his discussion about humus by saying that silica has the ability to receive the light into the Earth bringing it into effect. However, what is similar to the Earth's living element - the formation of humus - does not accept the light and does not make it effective in the Earth. Humus carries out its role without light. Steiner adds: "*... these are things that need to be known and understood.*"

Meanwhile, it is good to point out here that we are not talking about physical light but the Light ether. Try to imagine what would happen if we brought light into humus: humus would still not accept it, and the siliceous pole would not be strengthened.

We already know that excess silica tends to restrain the cosmic forces below resulting in cosmic root forms and leaving the top of the plant, especially the flower, more exposed to the earthly forces. The root then loses part of its impulse to ramify, and the flower will lose colour and tend to fall apart.

If we think about it clearly, we are now acquainted with the process through which Zarathustra was able to domesticate plants: if we look at *couch grass* (*Agropyron repens* – aka: Twitch-grass, Scotch Quelch, Quack-grass etc) we see a root with incredible wealth while the upper part of the plant is so insignificant that it is not useful - even for reproduction which must be achieved vegetatively by means of its underground roots. So if we strengthen the processes of light in the couch grass we lose some terrestrial forces in the roots and these tend to go to the flower. In other words it tends to turn into wheat. It may sound absurdly trivial, but to domesticate a wild plant it is sufficient in practice to make holes in the soil around the stem of a plant - or rather this was sufficient at the time of Zarathustra. It is now more involved to bring plants into a connection with the forces of the cosmos. Nevertheless, it is important that our consciousness has grasped this concept because this will enable us to find simple and natural solutions to this type of problem. Otherwise we will be forced to leave this field to the techniques which science is compelled to use, without having made inroads into the laws of life and, therefore, only acting on the physical aspect. However, a simple gesture, such as making holes around a plant, becomes powerful when we have the consciousness that through this gesture we can call upon the forces corresponding to the specific spiritual entities. Otherwise it remains a futile and sterile gesture that will not activate any process in nature.

The greater our consciousness, the greater our ability to solve the problems that afflict nature through simple and moral methods. Furthermore, the knowledge that we can gain will not need to be protected by patents or jealously guarded within exclusive groups, because it can never be used to produce adverse effects or be exploited for purely economic interests. Such intentions are incompatible with the level of awareness necessary to establish a relationship with the living beings and with the cosmos in which we make our humble work.

At this point Steiner begins, unexpectedly, to talk about the animals ……

Seventh meeting

Again, we are reading and remarking on the Koberwitz course from the thirty-first paragraph of lecture 2 ("*Now, the plants growing ....*"). In this paragraph Steiner speaks about animals for the first time and says that a "*cosmic qualitative analysis*" for a particular environment can be made only by understanding the factors that allow the co-existence of certain plants and animals.

The key to understanding this paragraph may be called 'the etheric geography of the Earth', which is the study of the way in which life-forces are distributed in a given region. To grasp nature's signals it is necessary that our perception has been adequately educated, so that the presence of a particular plant indicates the influence of a certain force - etheric, planetary or zodiacal - that features in a certain environment. It is an approach that often occurs in the work we do, but, as Steiner is indicating here, our analysis can only be considered complete if we are able to broaden our observation to the animals that naturally live in the area that interests us. Only an analysis of this type can enable us to establish the art of cultivating plants for quality, meaning by this expression the ability to produce conditions favourable to the evolution of the plant kingdom and, at the same time, of the people who eat these plants.

Steiner comes to the point of suggesting that the right proportion of livestock is gauged by what the farm is able to produce for the sustenance of the animals, knowing that the animals in a fair partnership will also produce the appropriate quantity and quality of manure for the fertility of the farm. We should point out that, in saying what we have just summarised, Steiner punctuated his discussion by using the word "*just*"[11] seven times. Since this cannot be random, it is worthwhile trying to reflect upon this emphasis to try to understand what may at first sight even seem exaggerated.

We know that the term 'just' indicates an attribute of the spiritual and, together with 'true' and 'good', is a yardstick to check if our ideals and aspirations correspond to the expectations of the spiritual world. But while 'good' echoes more with the soul region and 'true' with the etheric-physical sphere, what is 'just' corresponds to the will and hence the spiritual forces within us. The animal world, because it is a living world, is an expression of will because animals grow and change involving continuous movement and modification of the whole body, from blood to the bones. But if '*just*' is linked to the will and to transformation in a spiritual sense, and if we want our agricultural organism to grow and prosper, we must respect the *right* proportions of the various organs of the organism - specifically the animals in this case. Moreover the use of the term 'right proportion' refers to the animal world, because proportion is nothing more than a relationship between numbers, and we know that the logic of the soul can be summarised in the expression, 'each in relation to the others'. In other words Rudolf Steiner is actually talking about the soul-logic of the farm. Moreover, there is a sentence that should be a wake-up call for us when he says that a farm that does not have animals and so is forced to buy in the manure from outside, must be considered diseased.

In this regard we should consider what the term disease [It: *malattia*] actually means. '*Mal-attia*' derives from mal-agire (Lit: '*to act badly*'), but we can also ask: 'chi mal-agisce [lit: '*who is it that acts badly*'] in us when we are ill?' The answer is: the astral body or, more broadly, the soul. This means that if the soul goes awry, the I

[11] '*Giusto*' in Italian. 'Right, appropriate or just' as in '*justice*'.

no longer has complete control of all the components of the body. In an analogous way when we bring in manure from outside the farm, we are impeding the control of the farm's I so that it becomes unable to control everything that happens in it.

We must come to a better understanding of the role of manure on a farm, because muck-spreading is normally considered to be a contribution only to the soil's organic matter. In reality we have to know all of the forces that manure is able to carry.

First, manure comes from plants ingested by animals and thus is the bearer of the etheric forces that were present in the plant. The animal is not able to remove them completely and so a part remains in the manure. But animals - especially cows – submit the food to repeated mastication that enriches the plant substance with an important soul component. Perhaps less intuitive is the fact that since the animal also has a connection to its Group I, it needs food to sustain this relationship and one can find this support in the fact that the plants have a relationship with *their* Group I. A plant's connection with its I is normally expressed in its verticality. (Of course all these subtle 'dialogues' are scrambled if animals are fed with silage or, worse, with synthetic proteins.)

The animal, however, fails to fully extract from the plant these forces linked with the plants' I, so the manure still has a 'potential' or predisposition to the I that, when put in the soil, allows the seed to connect, in turn, with its I group.

From what we are saying one can infer why human faeces are not recommended for fertilising: the human, having an individual Ego, is capable of extracting all the etheric, all the astrality, and all the I forces from the plants, so that residues produced are a poison for the land.

Amongst all this reasoning it is important that we don't lose the understanding that manure is especially important as the bearer of the I forces. This makes it clear that it cannot belong to a different farm organism unless it serves as a remedy for an illness. Even as a remedy it should no longer be used in tonnes per hectare, but in minimal medicinal doses.

In some agricultural traditions farmers who knew nothing about phyto-hormones used to macerate manure in water to soak the roots of the plants. This may seem an operation based on a mixture of tradition and superstition, but the maceration extracts auxins that are phyto-hormones that stimulate rooting, and bathing the roots puts a spiritual process in motion, because the I of a plant is anchored below the roots.

As far as augmenting organic matter and improving the soil structure is concerned one could easily replace the manure with something else. It is well known that to improve soil structure it is sufficient to sow mustard, and to make a good amount of organic matter one only needs to use green manure. But these practices add nothing to the farm in terms of connecting with the world of the I.

This shortage is even more severe if we consider the fact that the I enables the organisation of the astral and etheric planes, and any lack in the farm results in a critical failure of organisation that shows up in the emergence of all the problems that characterise modern agriculture.

On our farms we should have animals that represent all the elements, although - depending on the etheric balance of the farm - one or another animal will be predominant.

We can now build a scheme to highlight the value of the various animals from the point of view of the forces that they bring and the processes they can stimulate.

By now we all know that the pig brings the forces of the Earth and poultry bring the forces of Water. Pig manure is colder and richer in potassium, while poultry droppings are the most fluid manure. Between Land and Water, as the *Sal* process, we can put the earthworms. With reference to the manure they produce we can put sheep, goats and rabbits as corresponding to air, while the horse is linked with the Fire element. All will recognise a *mercur* function in cattle and the correspondence with the *sulfur* process in the bee.

ANIMAL	*PROCESS & PREPARATION*	*ELEMENT*	ANIMAL
		FIRE	HORSE
BEES	*SULFUR* 501		
		AIR	RABBITS GOATS SHEEP
COWS	*MERCUR* 508		
		WATER	POULTRY
WORMS	*SAL* 500		
		EARTH	PIGS

The scheme that emerges lists animals on the right that are capable of bringing forces and, on the left, those who are able to activate processes - and processes are capable of dominating forces.

Considered in this way the scheme reveals that the bee could govern the Fire and Air elements and thus the function exercised by horse, sheep and goats; in the same way the earthworm governs functions related to the other two elements. In cows we can recognise a central role.

We could also say that the earthworm has the ability to regulate and stabilise the etheric while the bee can regulate and stabilise what is astral. The cattle can govern both and could easily be seen as the central peak of the spectrum. By *regulate* and *stabilise* we mean the ability to 'remove' in the event of excess and 'donate' in the event of shortages.

Unfortunately, now one hardly ever sees cattle on [Italian] farms with the result that to activate all the processes and to regulate all the forces we must at least ensure that there are earthworms and bees.

Let us dwell for a moment to reflect on these last two animals. The earthworm is a lunar animal, shrinking from exposure to the Sun or salt on the surface even at night. During the night of the full Moon the earthworms couple. The bee has a deep relationship with the Sun and the queen joins the drone in her nuptial flight, but when there is bad weather and the Sun is not shining in the sky the bee remains in the hive.

If we now consider the habits of cows see how they are between these others: in fact when there is full Sun they seek shade to rest and at night they ruminate. Morning and evening are the periods in which the cow eats. We could say that they are active at dawn and at sunset.

If a modern farm lacks the animals it needs to govern the natural processes we have to think of how we can remedy this. Steiner, in the fourth lecture, suggests the preparations now known as 500 and 501 to adjust the astrality and the etheric; moreover he suggests the use of a preparation capable of holding together the area in which the 500 and 501 are united. This preparation is horn clay.

In the thirty-second paragraph ("*An understanding of ...*") Steiner says that we should come to an understanding of what is terrestrial and cosmic in animal forms as he previously suggested we do to grasp the 'ABC' of plants. Sooner or later we will have to deepen our work here because it is indicated, in no uncertain terms, as being essential for the understanding of an environment. Shortly we will dip into the field of the planetary influences upon the animal world, trying to identify the normal and excessive influences somewhat like we have done for the plants.

Meanwhile we can see how Steiner indicates the distribution of planetary influences in the body of the animal and this continues in the paragraphs that complete the second lecture. Particularly in paragraph thirty-three ("*People who are interested ...*") Dr Steiner suggests the difference between the direct force of the Sun and that which is reflected from the Moon. This seems a most important indication - that embryonic development is particularly influenced by the sunlight that is reflected from the Moon and that reaches cows from their behinds. This gives a clear indication of the direction that the pregnant cow should adopt in the stable and introduces the issue of the orientation of the stables with respect to the Sun and the Moon.

If we want to have a broad indication we can say that the animals' horns should be to the east to pick up the solar stream: the tail should be located to the west to accumulate forces brought by the lunar current.

There is a beautiful study by Dr. Karl König, entitled "*Considerations on the connection between the digestive system and the organisation of the brain in humans and in animals*", in which he presents evidence that an embryo's brain and intestines are closely linked, even physically. The separation between these two organs becomes increasingly pronounced during embryonic development. However, a functional link remains that can be intuited from the fact that metabolic disorders are often accompanied by headaches. In fact human and animal nutrition has a dual aspect. There is a terrestrial nutrition stream that consists of consuming food and bringing it into the lunar sphere (the intestines), from where they are etherised and raised up and condensed to form our brain. The head is thus made of condensed terrestrial substance but in a spherical form that betrays the presence of cosmic forces.

There is also a cosmic nutrition stream taking the perceptions of our senses and leading these cosmic substances from the brain. Once these perceived cosmic substances have crossed the 'threshold of the atlas' they condense to form the substance of our body. Thus in the rest of the body we have cosmic substances woven through by terrestrial forces[12].

We could consider the intestine as an embryonic brain and the brain as a transformed intestine. Perhaps it is interesting to regard the term *intestine* as in-testine [It: *testa* = head] or an inner head.

[12] See "Course on the quality of nutrition' by the author. Not in English (yet – 2008)

Now we can continue to consider the influence of the planets upon animals. When, some time ago[13], we studied the planetary aspect of plants we separated their incarnating and excarnating activities. We also augmented these two influences of each planet with characterisations of their excessive influence. So let's proceed in the same way in consideration of the animals. Note that each planet's incarnating influence is exerted on the road down to manifestation before the birth of the material aspect in the same way as occurred for the plants. The excarnating phase begins at the moment of birth and follows the animal all the way to its death at which time it is freed from its physical connection and can return to the spiritual nexus from which it arose.

We will identify, therefore, their effect on the physical, etheric, astral, and ego levels for each of these influences (incarnating / primary, excarnating / secondary and the excess of each).

We start our journey towards manifestation by considering the normal influence of *Saturn 1* on the physical level. Saturn 1 governs the formation of the *bones*. In the physical body an excess of Saturn becomes *hardening*, but not only in the bone, even in the soft tissues. Thus we can attribute excess Saturn with the appearance of *sclerosis*.

We will also find Saturn working on the etheric level, the level of movement and rhythm, and since it is a planet with a long orbital duration, it will also induce *slow* and *heavy* rhythms and movements in animals. In this sense the animal that expresses the greatest Saturnine influence is the elephant. We can recognise incarnating Saturn in the labours of draft animals like oxen. We must also identify the effect of excess Saturn 1 on the etheric plane. We believe this can be seen in a way of being, in a temperament, which in this case is *melancholy*.

If we take this idea of slowness to the astral plane we find something that could be defined as *calm*. The idea of calm can be represented by an animal (but this also applies to humans) that remains quiet even if stimulated. The excess of calm can become *stubbornness* or *immovability*.

On the spiritual level Saturn 1 is found in the link with the individual principle of the group, either in terms of its promotion or its inhibition. This link is manifested in *instincts* of the species. Excess connection with the spirit leads to extinction of the species, or, less drastically, to *disinterest in the physical body*.

The elephant shows a strong link with Saturn 1, as does the rhinoceros and tortoise.

On the physical level Jupiter 1 reveals its influence in *musculature* and in *elasticity*. Moreover Jupiter is the planet that brings to mind the old Sun and thus the Air element. Jupiter 1 generally regulates the liver and cartilage. The physical body of an animal that bears the forces of Jupiter 1 will be *imposing* and *fleshy*. An example of the animal typically tied to this planet's influence is the beef bullock. Unbalanced Jupiter 1 on the physical level is seen in *obesity*, where the fat is in pathological excess.

---

[13] See meeting five of the present work

BODY	EXCESS	INFLUENCE	ANIMAL	CHARAC TERISTICS	PLANETS
PHYS. ***ETH.*** **AST.** ***I***	HARDENING *MELANCHOLIC* STUBBORNNESS *NO INTEREST IN PHYSICAL BODY*	BONES *SLOW, HEAVY* CALM *CONNECTION TO THE SPIRITUAL*	ELEPHANT RHINO TURTLE	SPACE SPLEEN SKELETON	**WILL** **SAT.** **I**
PHYS. ***ETH.*** **AST.** ***I***	FAT *PHLEGMATIC* DEPENDENCY *PSEUDO WISDOM*	IMPOSING, LOOSE, ROTUND, FLESHY, RED MEAT *SLOW, MUSCLE TONE, PLUMP, SYMMETRICAL* SOCIABLE, HERD / FLOCK *THOUGHTFUL INTELLECTUAL*	DRAFT HORSES BEAR OWL	FORM LIVER CARTI-LAGE	**WISDOM** **JUP** **I**
PHYS. ***ETH.*** **AST.** ***I***	GIANTISM *CHOLERIC* AGGRESSIVE *TYRANICAL, DESTRUCTIVE*	STRENGTH, GROWTH INTO SPACE SPEED, CLEAN MOVEMENTS PREDATORS, RAPTORS, LABOUR INDUSTRIOUS	CATS RIDING HORSES EAGLE	ACTIVITY EXTERIORI-SATION GALL BLADDER	**FINISHED ACTIVITY** **MARS** **I**
PHYS. ***ETH.*** **AST.** ***I***	CONCENTRATED, HARD *VANITY COMPLACENCY* PRESUMPTUOUS *SELF-SACRIFICE FOR OTHERS*	HARMONIOUS, WELL PROPORTIONED, ROBUST *HEALTHY, GOOD COLOUR* VERSATILE, GENEROUS, PRECOCIOUS, ALERT *HERD LEADER, MESSAGE FROM THE GROUP I (POLITICAL)*	DOGS LIONS	MATERIA-LISATION HEART	**RELATIO-NSHIP WITH MATTER** **SUN** **I**
PHYS. ***ET.*** **AST.** ***I***	WATER RETENTION *LATE DELIVERY* CLINGING *TAMABLE*	BODILY FLUIDS, MUCUS *PREGNANCY,* AFFECTIONATE, NURTUING, CARING *GREGARIOUS, MATERNAL*	COWS SHEEP RABBITS	INTERIOR-ISATION KIDNEYS	**CARING** **VENUS** **I**
PHYS. ***ET.*** **AST.** ***I***	DIMINUTIVE INCLUDING DEFORMATION *FRENZY* ALL'FRONT' INSUBSTANTIAL *CHAOTIC*	QUICK AGILE *VITALITY, LYMPHATIC MOVEMENT* RESTLESS, FICKLE., ADAPTABLE, IMPROVISOR, PLAYFUL *INFIDELITY*	GOATS MICE SQUIRREL POLYPS MOLLUSCS	MOVEMEN T LUNGS LYMPH	**ENDLESS WORK** **MERCU-RY** **I**
FIS. ***ET.*** **AST.** ***I***	TUMOURS *SELF REPORODUCING CELLS* CONSERVATION, TRADITION *SECRETIVE*	FLACCID *FERTILITY, FECUNDITY, HERMAPHRODITISM* INVASIVE, SELF-PROMOTION, *SELF INSINUATION*	SNAKES WORMS INSECTS NOCTURNAL RABBITS WOODPECKER CRICKET CUCKOO HORSE (SLEEPWALKER)	TIME REPRODU-CTION SEX	**VITALITY** **MOON** **I**

Etheric Jupiter brings *tone* and *elasticity* to the muscles and tissues and *symmetrical forms* in general. Etheric excess shows itself in the *phlegmatic* temperament.

We can attribute *social* tendencies to Jupiter on the astral plane, so here we can put all the animals that gather in *herds and flocks*. Excess causes the inability to take individual initiative that translates into *passivity* or *dependency.*

Jupiter brings *intelligence* to the plane of the ego. The excess of this feature in animals leads to a *quasi-capacity to think*. An animal that expresses this excess is the owl, and thus has been adopted by cartoonists as the brainy bespectacled professor.

Moving on to Mars 1 - we first recognise this in *aggression*, and in the gallbladder. In animals the physical stamp of Mars 1 is the *muscular strength* and its excess is expressed in *giantism*.

On the etheric plane it brings *speed*. An animal typical of Mars 1 is the Tiger. The excess of these forces form the *choleric* temperament.

On the astral plane we see the predisposition to aggression typical of the *predator*. We are in the world of raptors where aggression can reach the limit and betrays this excess. The aptitude for prolonged physical labour also belongs to Mars 1 on the astral level.

Mars 1 is found at the level of the I in being active: *industrious* sums it up. In people this asset is reflected in the ability for a continuous renewal of their interests, but when it becomes excessive, it leaves no room for the needs of others. Mars in excess produces egotism and selfishness that can become a kind of *tyranny*, bringing *destruction* to everything around.

After Mars we will consider the Sun's incarnating activity. On the physical level we can translate the effect of his influence with three adjectives: *harmonious*, *proportionate* and *robust*. The excess on the physical level leads to a kind of concentration. We can picture to ourselves a small sturdy self-absorbed man. The Sun is dominant in dogs and in the lion.

The animal that expresses the forces of Sun 1 on the etheric plane is full of life and *healthy*. Being so naturally healthy can lead to *over-complacency* in relation to health that can then flip over to become a kind of *vanity*, or a mania relating to the care of their personal welfare: cosmetic surgery and the like derives from such an unbalanced evaluation of their appearance.

Astrally the Sun brings a 'can do' attitude and the confidence to jump in and learn quickly. If you want to characterise this planetary force with adjectives one could say: *versatile*, *generous*, *precocious* and *alert*. These can be quickly exaggerated and easily become pathological and this could be summarised as a kind of *arrogant presumption*.

For the I the Sun 1 governs the position of *leader* who leads the flock in connection with the I. On the individual level the link with the world of spirit is typical of Saturn 1, whilst with the Sun this connection is expanded for the benefit of the whole herd. Excessive accent on this feature can lead to *self-sacrifice* for the rest of the community. This is the dog that is ready to sacrifice itself for the shepherd or the flock.

When we turn to Venus we meet the world of all kinds of metabolism in which there is a successive process of taking, developing, and then passing on. Ruminants are animals typical of Venus, an 'internal' planet which thus manifests itself primarily under the diaphragm, in the guts. Notwithstanding the bond of Venus with the digestive system, we find this influence in *mucous membranes*. We can also identify Venus with the lymphatic system, but we want to use a more comprehensive expression and say that Venus governs all bodily fluids. Continuing this reasoning, we can see Venus in water retention, which incidentally primarily affects women.

The etheric aspect of this planet is what enables *pregnancy*. Venus is not related to sexuality and conception, because these are typically lunar. It can be intuited that an excess of this influence leads to *late delivery*. This excess within relationships might be seen in an otherwise good mother who wants to keep her children attached too long, depriving them of their freedom.

Venus on the astral plane is *discrete*, *affectionate*, *nurturing* and able to take *care of people*. Clearly in excess, as we have just mentioned, this leads to be *clinging* and keeping the child to oneself so as not to lose control.

With the I, we see this putting of oneself at the service of others transformed into a *gregarious* animal. Here we recognise not only cattle or sheep, but even the rabbit; think that to keep the offspring warm the mother even pulls off her own fur. An excess of Venus brings the animal to put everything into the care of others and to give up on their own needs, and this can be very interesting because it is the basis of the animal that is *tamable*.

Quicksilver is another name for Mercury and this planet gives the physical characteristics of being *quick off the mark*, *agile* and *small*.

Excessive Mercury brings *dwarfism*. Mercury refers to the propensity to do the little odd jobs; remember that long ago there was a court jester whose characteristics correspond to those that we indicated earlier.

Mercury is always recognised as the bringer of movement and this is expressed, on the etheric plane, in *lymphatic circulation*.

One can express the astral influence of Mercury with the words: *restless*, *mobile*, *adaptable*, *humourous*. If you wanted an animal with these features you could think of the squirrel. In humans we should think about people who present themselves well, but behind an impressive appearance there is not enough real substance. We could compare such a person to a will o' the wisp, which appears to burn but when one looks there is no fuel.

In the realm of the I we can confidently call mercury *unfaithful*. Mercury is, not surprisingly, the 'patron saint' of thieves. When Mercury I is excessive an uncontrollable *chaos* arises.

Moon 1 is linked to water and proliferation. The amoeba is a good representative of the Moon. An animal's physical Moon characteristic might be *flabby* or *flaccid*. Excessive Moon is a tumor, where there is a lack of a restraining form and disorderly growth results.

On the etheric plane the Moon brings *fertility*. Here we are talking specifically about fertilisation, as we mentioned when talking about Venus. A typical lunar animal is the earthworm that we all know is *hermaphroditic*. Excess Moon on the etheric plane gives *purposeless cell reproduction* without a functional link to the rest of the body. What we are saying is clearly very much related to the reproduction of cancerous cells. Excess of this dynamic brings the possibility of cloning.

We have said that the Moon is linked to water and water penetrates everywhere. Inspired by this observation we can understand how, on the astral plane, the Moon governs *invasion*. Invading the legitimate domain of others extends to conquering their territories which will then also have to be defended and preserved. Excess intrusiveness then leads to being *conservative*.

What on the astral plane was defined as invasiveness, carried to the domain of the I, becomes *insinuating* or *intrusiveness*. Insinuation is much more subtle than invasiveness, and requires more cunning. If we were talking of people we could speak of those who are dedicated to *secret organisations*. In the animal world we could think of social animals like ants, whose society we do not perceive adequately because it is hidden underground.

We have concluded a brief sketch of the planets' incarnating influences which mainly interest us in the period preceding birth, but which are also dominant in approximately the first 35 years of life. After that age the excarnating planetary influences are dominant and accompany a person until they completely abandon the physical-material body.

If we were to stick to the temporal sequence we would proceed with the excarnating Moon after the incarnating Moon but we will continue our work beginning with Saturn 2.

Recalling that the incarnating influence of Saturn was clearest in the formation of bones, we will not be surprised that the excarnating influence causes *decalcification* and, in excess, *osteoporosis*. Moreover osteoporosis is a disease typical of an advanced age and mainly affects women – and women are typically less incarnated than men. *Rickets* is also attributable to the excess of Saturn 2. Rickets may be considered as due to a lack of an incarnating influence, but it can also be the result of too much excarnating influence. In the plant world a representative linked to Saturn 2 is the cypress that has the characteristic of reducing soil salination.

On the etheric level the function governed by Saturn 2 is the *formation of red blood cells* because in haemopoiesis one can recognise an excarnative bone process.

Astrally Saturn 2 brings *fever*, and the fact that the animal can *come on heat* is linked to the same force. (Note the connection of the word "*oestrus*" to the word '*astra*'.) Excess of the fever dynamic brings *inflammation* while the oestrus force can in excess become nymphomania.

We can see that there is a similarity between the functions stimulated by Moon 1 and those typical of Saturn 2: the Moon brings fertility and Saturn is linked to oestrus. This observation gives us the opportunity to say that on the etheric plane the communication between planets happens via a cross: the primary or incarnating function of one communicates with the secondary excarnating influence of the complementary planet (eg Saturn 1 – Moon 2). The second's incarnating influence also communicates with the first's excarnating influence – Saturn 2 with Moon 1 and so on.

Let's continue with Saturn 2. In relation to the I we see a *lack of individual awareness* or, in animals, a loss of instinct. (Instinct is the I's effect on the physical body.) For instance some animals have lost the ability to distinguish between what is edible and what is poisonous. Other animals have lost the instinct to reproduce. The excess can be called *enthusiasm* that takes us outside of ourselves towards new things. This is linked to the fact that Saturn is the exit from our system and the route into the cosmos. Representative animals of Saturn 2 forces are the birds.

Jupiter 2 on the physical level brings *overbearing form* and, in excess, *degeneration of tissue*.

On the etheric level we find *mobility of joints* and muscular movement. It is worth stressing that Jupiter 1 was related to muscle mass but this time it is the movement that is of interest, and therefore the elasticity. We could also place *tissue regeneration* here. Excess brings about sudden uncoordinated movements causing *distortion*. *Synovitis* is related to Jupiter 2 on the etheric level.

In terms of the astral or soul world Jupiter is expressed in *mimicry* consisting in moving the muscles in order to give the face certain expressions. The excess shows up in spontaneous muscle movements unmotivated by any need for action - *tics*.

On the level of the I, Jupiter 2 brings *fantasy* and *creativity* that, if excessive, brings trouble concentrating - '*spaced out'*. Let us remember that we are talking about Jupiter 2 acting fully in people who are older than 50, so we are talking about the moral imagination or the ability to solve a tangled situation in a civil way. This must be distinguished from a child's imagination seen in noisy and fantastic fantasy, and of which the moral imagination is the older sister. However, a child who is not allowed the freedom to develop its level of fantasy will be unable to develop the elder sister when adult. Thus we can understand how important it is that a good education is capable of nourishing the development of childhood imagination.

The typical Jupiter 2 animal is the monkey that is both sociable and imitates everything it sees.

If we move to Mars 2 we find its influence on the physical body in *stamina*. As a concrete example of an animal that embodies this we might think of the camel. Lets take an example to grasp the significance of an excess of this force: we all know that steel is an alloy of iron and carbon, and we also know that steel is harder the higher the proportion of carbon. But if the proportion of carbon exceeds a certain threshold the steel becomes cast iron that is extremely *fragile*. This aspect of Mars 2 is recognisable in the horse that is willing to run until its heart bursts.

On the etheric level Mars 1 brings speed, so you can easily imagine how the excarnating Mars 2 brings *slowing of vital functions*.

The influence of Mars 2 on the astral level can be understood as a reversal of Mars 1. We gave Mars 1 the characteristic of the predator, with the will strongly facing outward to manifest as a ferocious aggression. Mars 2 turns this powerful will force around towards inner transformation. Unless it becomes capable of *interiorisation of the will* it is difficult to concentrate and meditate and it is therefore impossible to follow a spiritual path. Excess of Mars 2 does not have a negative connotation because it becomes the predisposition to *learn something new*. This can mean putting everything up for discussion or going as far as evoking a metamorphosis led by an aspiration to reach higher goals. We must always bear in mind that the incarnating process always has a quantitative impact, such as the formation of bones, flesh, muscles, sociability, or is directed towards construction. The excarnating process, on the other hand, has a sublimating aspect and has a connotation linked to quality. We can now better appreciate how Mars 1 represents the warrior who moves forcefully to destroy the enemy, while Mars 2 turns this destructive capacity to burst through the inner limitations which are rooted in desires and passions.

For the I, Mars 2 causes *catharsis*. We are no longer in the animal kingdom that will only come to this after a long evolution. The excess of this force can cause *being in the doldrums* or an *excess of detachment*. It seemed important to mention the effect of Mars 2 in humans, but to return to the job in hand – the animals - we need only consider those profound changes that are seen in the animals that form a cocoon. The true cocoon is linked to Jupiter, but the impetus to hide away and transform is undoubtedly linked to Mars. The metamorphosis from caterpillar to butterfly passes through a phase of fasting and purification, which overcomes the fundamental instinct of eating to survive. To understand this catharsis from a different point of view, think that the caterpillar starts from a state based on *Water* and is reborn as a butterfly which is a clear expression of *Air*.

Let us now look at the Sun. If the primary influence was to materialise, the secondary must be *dematerialisation*. The physical characteristics linked to Sun 2 are: *thin*, *weak* almost *filiform*. Excess leads to *scoliosis*, *hunchback* and *cachexia*.

In the etheric, Sun 2 induces *retarded processes*, perceivable in clear insubstantial mucus. Excess causes *anaemia* and *aneurism*. Aneurysm is the dematerialisation of the mucosa which protects the artery, which is then susceptible to excessive expansion and bursting.

In the astral there is *delayed development*. This is understandable when we consider that development is an incarnating process that can be slowed down if it encounters an early excarnating force. Excess leads to *depression* or, if more marginal to a bit of laziness, although this can also be attributed to the Moon.

In the I Sun 1 was the herd leader, Sun 2 brings to life a *priestly* element, seen as a messenger to the Self. Excess resembles *fanaticism*, even if the fanatic is properly lunar. We could describe this aspect of fanaticism as the willingness to self-sacrifice for a cause, in other words, the *kamikaze*. The first suicide bombers gave their lives for the emperor representing God on earth, and were priests.

Venus 2 in physical governs all *excretion* and when it is in excess causes *haemorrhage*.

On the etheric level we find the *secretions* that, if excessive, bring *glandular hyperactivity*.

On the astral level the ability to *give love* is attributed to Venus 2. An excess of this giving leads to exhaustion or *love sickness*.

In the I we see complete *self-sacrifice*. If we think about the vestal virgins we have a clear example of this self-sacrifice. The vestal maids carried out an obscure task thanks to which the dedicated sacred flame was always kept alight, but when the priest arrived they withdrew with humility. The excess of this is the complete *cancellation of the self*. It was the condition of women in the Middle Ages, when she was regarded as an animal. In the animal kingdom we can recognise powerful links with Venus in the world of the bee, especially the queen bee, or in mammals with a low fertility; don't forget that the vestal girl was a virgin because she renounced motherhood for her calling.

PLANET	CHARAC- TERISTIC	ANIMAL	INFLUENCE	EXCESS	BODY
SAT. **2**	TIME MARROW	BIRDS MICE OESTRUS & PREGNANCY	DECALCIFICATION *HAEMAPOIESIS* FEVER, COMING ON HEAT *LACK OF PERSONAL AWARENESS*	OSTEOPOROSIS, RICKETS *ASTHENIA, ENERGISATION , METABOLIC EXCESS* SCHIZOPHRENIA, FEVER, INFLAMMATION, NYMPHOMANIA *ENTHUSIASM*	**PHYS.** ***ETH.*** **AST.** ***I***
JUP **2**	EXPRESSIVE MOVEMENT MUSCLES	EMBRYO MONKEY (MIMICA) HORSE	OVERBEARING FORM *MOBILTY. ARTICULATION, MUSCULAR MOVEMENT, TISSUE STRENGTH* MIMICRY *FANTASY, CREATIVITY*	TISSUE DEGENERATION, OBESITY TIC, UNCONTROLLED CONTRACTIONS, CHOREA *STRUGGLES TO CONCENTRATE*	**PHYS.** ***ETH.*** **AST.** ***I***
MARS **2**	INTERIORI SATION LARYNX	CAMEL LETHARGY COCOON CHAINED ANIMALS PHASES OF DOMESTICATION ALMOST SPEAKS	STAMINA *SLOWED VITAL FUNCTIONS* INNER WILL POWER *CATHARSIS*	FRAGILITY *LETHARGY* LEARNS NEW TRICKS *DOLDRUMS, PASSIVITY*	**PHYS.** ***ETH.*** **AST.** ***I***
SUN **2**	DEMATERI -ALISATION CIRCULAT ION	COLD BLOODED ANIMALS (LIZARD) VULTURE ABANDONED OXEN ACQURIED CHARACTERISTICS	WEAK, THIN, ELONGATED *SLOW PROCESSES, DIAPHANOUS MUCUS* LATER DEVELOPMENT *MESSANGER TO THE I (PRIEST)*	SCOLIOSIS, HUNCH, CACHEXIA *ANAEMIA, EXHAUSTED, ANEURISM* DEPRESSION, SLOTH *FANATIC SUICIDE*	**PHYS.** ***ETH.*** **AST.** ***I***
VEN-US **2**	EXTERIORI SATION VANITY	LOW FERTILITY REDUCED MILK YIELD BEE	EXCRETION *SECRETION (HOREMONES)* GIVES LOVE *SELF SACRIFICE*	HAEMORRAGE, *HYPERACTIVE GLANDS* GIVING EVERYTHING, ANIHILATE SELF *GIVING UP ONES SELF*	**PHYS.** ***ET.*** **AST.** ***I***
MER-CURY **2**	FORM GLANDS	SPONGE SEA ANEMONE MASTITIS PROSTATITIS RABBITS SHEEP MULE GOAT	HOLLOW ORGAN STRUCTURE, ASYMMETRY *HEALING* FEAR AND SADNESS *THEORISING (MODELS AND IMAGES)*	HERNIA (WITH JUP. 2), VACUOLISATION *INDURATION OF GLANDS* FEAR OF SURROUNDINGS *FIXED OR STEEOTYPED IDEAS*	**PHYS.** ***ETH.*** **AST.** ***I***
MOON **2**	SPACE DIFFEREN-TIATION SKIN BRAIN	FISH TRUFFLE ANIMALS INSECTS WITH EXOSKELETON HORNS AND HOOFS	KERATIN, SKIN, NERVES, *UPPER MEMBRANE DIFFERENTIATION* SPECIALISATION *CRITICAL DISCERNMENT*	EXOSKELETON WARTS *DEFENDING TERRITORY FROM OTHERS* *OBSESSIVE*	**PHYS.** ***ETH.*** **AST.** ***I***

Mercury 2 is responsible for *formation of the hollow organs* and *evagination*. If this influence is excessive one is prone to *herniate*.

On the etheric level Mercury *actives the healing processes*. In excess it causes *hardening*.

In the soul and bearing in mind that Mercury was the first jester, Mercury 2 brings *sadness* and, in excess, *fear*. This is also true for humans, and all *phobias* are attributable to the excarnative influence of Mercury when confronted with the outer environment.

Mercury 2 leads the I towards stubbornness, and in excess it brings *fixed ideas*. The typical Mercury 2 animal is a goat that does exactly what it wants when it wants.

The excarnating Moon is manifest in the *skin* and *nerves* and is responsible for squamous skin and membranes. The excess is reflected in *formation of an exoskeleton*. It should be easy to connect what we have just said with the world of insects.

Etherically *differentiation of cells* depends on Moon 2 and, when it is excessive, in the formation of *warts and veruccas*.

On the astral we find *cell specialisation*, which in gives *overemphasis on parts*.

In relation to the I Moon 2 brings *critical discernment* but when unbalanced this can become *obsession*.

The work we have done should not be considered finished, but as a basis on which to establish a system of diagnosis and treatment for the animal world. For each planet we can find the remedy for 'its' disease, or suggest food to support certain features.

Third Koberwitz Lecture

Eighth meeting

The third lecture of the Koberwitz Course was held on 11th June 1924. Remember what was said during our first meeting about the structure of the course and note that according to that schema this lecture echoes the cultural content of Greek times, as the previous two were linked to the Persian - the first epoch to be considered as Agriculture was initiated by Zarathustra - and to the Egyptian. We will remember that Zarathustra's vision of nature was focused on the struggle between light and darkness and therefore in the first lecture silica, a representative of light, was contrasted with limestone, a representative of darkness. Similarly the second lecture led us to the threefold vision of nature, typical of the thought of Hermes Tresmegistos, with the polarity between silica and limestone no longer considered as the representatives of light and darkness but as bearers of the processes of expansion (*Sulfur*) and contraction (*Sal*), with clay as a mediator and representative of the *Mercur* role. In our studies of the second lecture we met with the 13 'holy nights', and the clay preparations with which we could address the issues of directing the growth of plants. Our last effort was directed to consider the influence of the planets upon the animal world.

The Greek era marks a further separation of consciousness from its divine origin. The remaining spiritual connection allowed penetration only as far as the angelic sphere. Relations between humans and the Gods of Olympus - the struggles and agreements between them - is the image of what weaves in this angelic sphere. The gradual loss of direct contact with the spiritual world was replaced by a continuous and gradual accretion of knowledge and awareness. Note that the Greeks developed the system that we call philosophy and philosophy means 'love of wisdom'. Actually, the term '*philo*' means follower or lover, while the term '*sophia*' means knowledge, wisdom, or knowledge of the divine.

Thus the Greek philosopher connects with the spiritual world through logic rather than through direct inner experience, and observation of the external world resolves into four characterisations that are called the 'elements'. The elements - Earth, Water, Air and Fire - in the plant world can be recognised in the four parts of the plant: root, leaf, flower and fruit respectively. In the animal world we can still recognise the four elements in bone, in the lymphatic system, the muscles and the warmth system respectively, while in humans they become the physical body, etheric body, astral body, and I.

The reason Greek thought missed the reality beneath these four different aspects is connected to the fact that the era of Greek civilisation is the repetition of the fourth stage of incarnation of the Earth and, therefore, could only be characterised with the number four.

This way of perceiving the world remained quite intact until the 1500's when it began to lose meaning and become only the echo of past memories. After 1700 - with the dawning of the age of Enlightenment – this key has been almost completely abandoned, and replaced by a concept of a completely material world in which we recognise as real only what can be touched or measured in some way.

We must always keep in mind that the terms *Earth, Water, Air* and *Fire* do not indicate any substance, but were ideas to which the observer referred when perceiving the surrounding world. For instance the Earth element did not indicate a piece of land, but anything that could be recognised or characterised as three-dimensional and was unable to occupy the same space as other Earth[14]. This 'Earth' referred to Mother Earth, but also a pillar, a blackboard, a lamp and other things, or what we identify today as the solid state of aggregation. Ice is 'Earth', although chemically we can say that it is frozen water, while 'Water' corresponds to milk, petrol, blood and any flowing liquid, and it governs surfaces and movement. The same water (H_20) that is of the Earth element when it becomes frozen is 'Water' when it is recognised as a liquid, and is of the Air element when it evaporates.

If we consider the element 'Air' we mean those things that can contract and expand freely. The expansion following the contraction can be slow or sudden, in the latter case, the Air acquires an explosive characteristic. This principle is found in Nitrogen and is the principle chemical that allows the construction of bombs, which can be seen simply as the ability of Air to be greatly compressed and then quickly released.

The elements are all found everywhere, though from time to time one or another predominates. For example, if we consider a sponge, we must recognise that their ability to contract and expand is in this respect due to the air, but because their inside is made up of countless surfaces they can absorb liquids; in this instance we can also recognise the Water aspect of the sponge. Also, if we think that a sponge has its own corporeality we can see that here the sponge is also linked to Earth.

The fourth element is Fire that is certainly the most difficult to conceptualise because in our culture warmth is not considered a state in itself, but only an attribute of other states. We conceive the idea that water, air, or a solid body can have a higher or lower temperature but not that the heat has its own independent reality.

Spiritual science teaches us that the fire is linked specifically to man's I, manifesting in the warmth of blood.

Fire has many nuances of meaning. We all know that there is a fire that matures the fruit, and in this case we talk about warmth in its physical aspect. However, there is also a more subtle heat that comes with surges of enthusiasm. Enthusiasm is what enables the renewal of life, the adoption of new activities and new interests, but also to pursue solutions to new problems. Fire is, generally speaking, the world of quality that, in the modern way of farming, is not really pertinent. Now only the yield counts - the purely quantitative aspect of production. Wishing to clarify the concept Fire with a word borrowed from a parable of the Gospels we could say that Fire is represented by *talents*. It is often said that a person who is always ready to solve the problems of life has "*presence of mind*" and this expression means that the person's I (spirit) is always present in every situation.

Among the realms of nature the human is the one that most embodies the Fire element, and we have a manifestation of this in body temperature. At this point it may seem curious that warm-blooded animals have a higher temperature than man. This fact seems to belie the assertion just made.

Recently we said, speaking of Greek civilisation, that the acquisition of consciousness has seen a weakening of the direct relationship between the spiritual world and Man, and as Fire is the door to the spiritual world, our relationship with it undergoes an attenuation and this leads to a cooling of the body. Moreover, it is known to all that when the body temperature undergoes a strong increase because of

[14] To investigate the elements further see E. Marti "The Etheric" or E. Nastati "Study Group on healing." [Neither available in English in 2008.]

some illness, you see the phenomenon of delirium, a form of diminished consciousness.

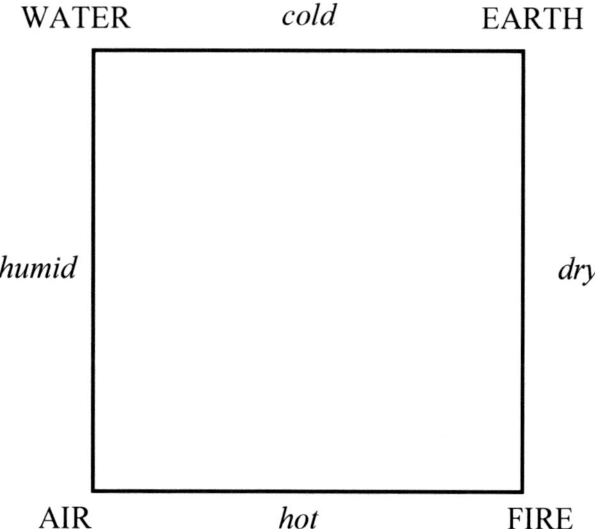

We have sketched the world of the four elements, from which we can glean many things such as the four stages of evolution of the Earth, the four realms of nature, but also the four zodiacal directions (Leo, Scorpio, Taurus and Aquarius) from which the formative forces act for the manifestation of our system.

In respect of agriculture it is particularly interesting that at the physical material level the four zodiac forces each have a representative on Earth: the constellation of Leo (Fire) is represented by hydrogen, Scorpio (Air) by nitrogen, Aquarius by oxygen (water), and Taurus (Earth) by carbon. These four substances are the four pillars of life on Earth – and of protein. Protein, as a substance that contains within itself the principle of life, is composed of these four substances, and a fifth, sulphur, which regulates dynamic relations, as if it were an inner potentiser that allows the circulation of those forces that we call *life*.

The Greeks knew the fifth substance that is put at the centre of the square of the elements, and called this the *fifth essence* (*quinta essenza* or quintessence) and in it they recognised the source of life.

It should be remembered that we cannot directly perceive the four Elements but only what is due to them, as a sort of categorisation of what we see. In contrast, we <u>can</u> perceive the qualities that are determined by the combined action of the elements. So between the Fire and Air we meet the quality of *heat*. Between Air and Water the quality of *humidity*, between water and earth we find *cold*, and between Earth and Fire we have the quality *dry*.

A proof of the fact that knowledge of Greek civilisation survived relatively intact until 1500 is that until those years there were still examples of practical implications of this understanding. An example we often suggest is the cloister, a ubiquitous architectural element within convents and monasteries, which was inspired by a representation of the four elements with the *quintessence* at the centre.

For this meeting we will pass silently over the whole aspect of the construction of the colonnade that was meant to cause a strong purifying interior experience in the monks through the use of specific forms of columns and capitals, and the dynamic vibration caused by the singing of psalms. Once purified the monks could leave the square of the elements represented by the walkway, and go right into the garden to access the quintessence, represented by the pool - the source of life.

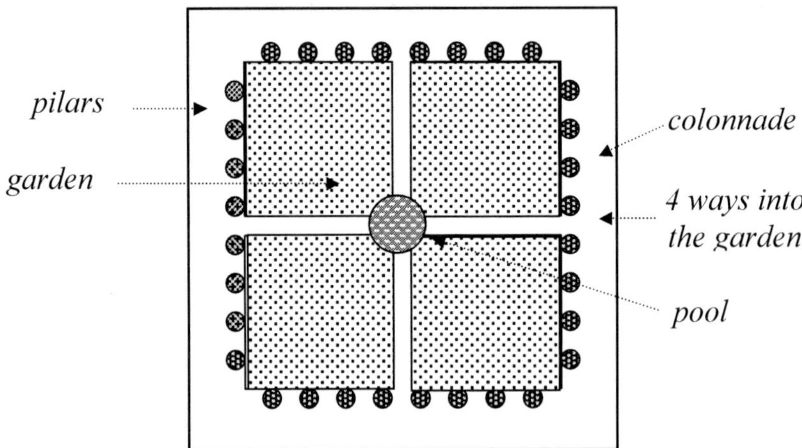

In lecture three Steiner spoke to the gathered farmers about the Philosopher's Stone, which may surprise us because maybe we would have chosen a different occasion for such a profound subject. But we must not forget that Zarathustra taught agriculture only to the best of his students. Also what Steiner has given in this cycle of lectures goes far beyond the boundaries of agriculture and up to the highest levels. As we have already said the Agriculture Course was the last Pentecostal course held by Rudolf Steiner and can be regarded as his spiritual testament, the work with which he wanted to leave the most profound knowledge with men.

The Philosopher's Stone is that which is called in the Gospels "*The stone which the builders rejected, the same is become the cornerstone.*" In a more subtle sense we could think of a part of us that is divided and placed in two specific places, that is inside the skull and another where the sacral bones end. At the centre of the skull is the pituitary gland, which has a pyramidal shape and that is contained in the *sella turcica* that is shaped like a cup (that is the image of the Holy Grail within us).

In his own fall it is said that Lucifer lost the gemstone that adorned his crown and that this stone has been received by the cup that is at the centre of our head. Then the fall has continued along the spine until the sphere of Kundalini, where now the sleeping snake of Fire is coiled, causing the formation of the second structure of carbon. It is no coincidence that the area is called the *perineum* that comes from two Greek words "peri" and "noas" which means close to the temple. Neither is it coincidental that the bone located nearby is the *sacrum*. The term *sacred* is actually used to indicate other parts of the body, for example what is called the *via sacra* is the path that the kidneys make to move from the position they occupy in the embryo (near the ears) to their final place in the fully formed body. Moreover between the embryo and the form of the grown human we know there are many other differences due to the Fall. These are found mainly in the position of the organs. The heart and liver should be positioned in the centreline of the body, but the former is moved to the left testifying to an influence from Lucifer (acting on us from the left), and the latter to the right due to Ahriman. Ahriman has also insinuated himself into our lives, otherwise we would not be subject to death. Do not forget that Adam and Eve, when they were immortal, could not eat of the tree of knowledge, but regularly used to eat of the tree of life. After their sin, having been expelled from the earthly Paradise, they are no longer able to eat of the tree of life and therefore became subject to death.

During gestation the embryo retraces the path of humanity, so at the beginning of gestation we are still beings of paradise and during the nine months we follow the steps of the Fall to the extent that we are even born with our heads down. It is still very interesting to note that the first pulsing centre in the embryo, which then become the heart of man, is originally found at the centre of the forehead.

In humans we can find three temples: the first we already encountered when we said that perineum means "close to the temple". The second is located near the point where the kidneys originate (ears have a shape that echos the kidney and these spots are still called the *templ*es). The latter temple is the *temple of the I*, which stands opposed to the first that is the *temple of Life*. We might understand at this point that the tree of life and the tree of knowledge are represented in the spine. The tree of knowledge has its roots in the brain and goes down the spine: the tree of life arises instead from the reproductive sphere and rises to the head. These two cross in the third temple, which is the heart.

We could even say that what happens in the heart is the *reunion of waters* - those same waters that were separated in the story of Genesis. It is said that the eternal Father separated the waters above and called those "sky", and those under he called "sea". If we consider what was said earlier, the water that irrigates the tree of knowledge is the cerebro-spinal fluid, while the water for the tree of life is semen. Through the exercise of chastity the undispersed semen undergoes a process of vaporisation in their ducts. Once vapourised this *water* begins to rise and purify, until it joins with the other *water* thereby overcoming the division following the fall.

However, since the waters are two, there are two ways to arrive at the reunion: one uses knowledge to enter ever more deeply into the secrets of Man and the other springs from life, changing its quality. But it is not enough to take only one point of departure because in the first case there is a risk of becoming 'cold hearted', and in the latter case one becomes a closed container of '*soul food*' that fills ones own interior but is not fruitful for anyone else. Anyone who manages to combine this inner transformation with sufficient knowledge can begin to be helpful for the development of all the kingdoms of nature.

Before we dive back into reading the lecture we can still add that the alchemists, as the previous guardians of this 'secret' knowledge, indicated the stages of this inner work of purification as the *black opera*, the *white opera* and the *red opera*, and although they did not explicitly speak of the three temples and the merger of two waters actually they were referring to their transformations. It is interesting to note how the French flag (where black has been transformed into blue) is inspired by these three phases of this alchemical Opera.

After this long introduction we can start reading. The title of the third lecture is "*Parenthesis on the activity of nature: the influence of the spirit in nature.*" The first sentence states that the duality between the Earthly and cosmic forces acts through certain substances and these characterise the Zarathustran view of agriculture that we discussed in our comments on the first two lectures. We have also repeatedly stated that any force on Earth has a reference substance and this idea is clear in the sentence suggested by Rudolf Steiner: "*there is no spirit without matter and no matter without spirit.*"

In the second paragraph ("*One of the most important ...*") Rudolf Steiner starts talking of nitrogen, and then he says that the world normally sees only the final effects of nitrogen's action but cannot penetrate into the natural network of relationships in which it operates, and he calls into consideration the importance attributed to nitrogen as a direct agent on the growth of plants. Nitrogen is actually important because it is the activator of certain processes in which other substances come into play.

Interestingly, Rudolf Steiner has addressed the issue of the four elements beginning with nitrogen just because of its connection with agriculture; this is because nitrogen, among the substances to consider, is closely tied to the spiritual world. Starting from the spirit nitrogen comes to manifestation and then from the manifestation it wants to come to life. We can imagine the course of Steiner's exposition by comparing it to a vortex.

In a vortex we perceive the movement of water down to its base, but we can well understand that the water must not only descend, but it must also be able to go come around again. In fact, the vortex that we perceive hides another polar and outer flow that rises up.

If the first vortex can be identified as the course of manifestation, which in the plant proceeds from the Earth / root towards the Fire / fruit, the second upwelling part of the vortex seizes this descending life and proceeds from hydrogen towards carbon.

Between these two vortices we can place sulphur, introduced in the fourth paragraph ("*To understand the full significance…* ") as a pivot between them. If we return to the example of the cloister now perhaps we will understand better the image of the monks who walk around the square of the elements reciting the psalms, praying and singing, in order to 'activate' the cloisters so that their consciousness can rise and grasp the truth coming down from higher worlds.

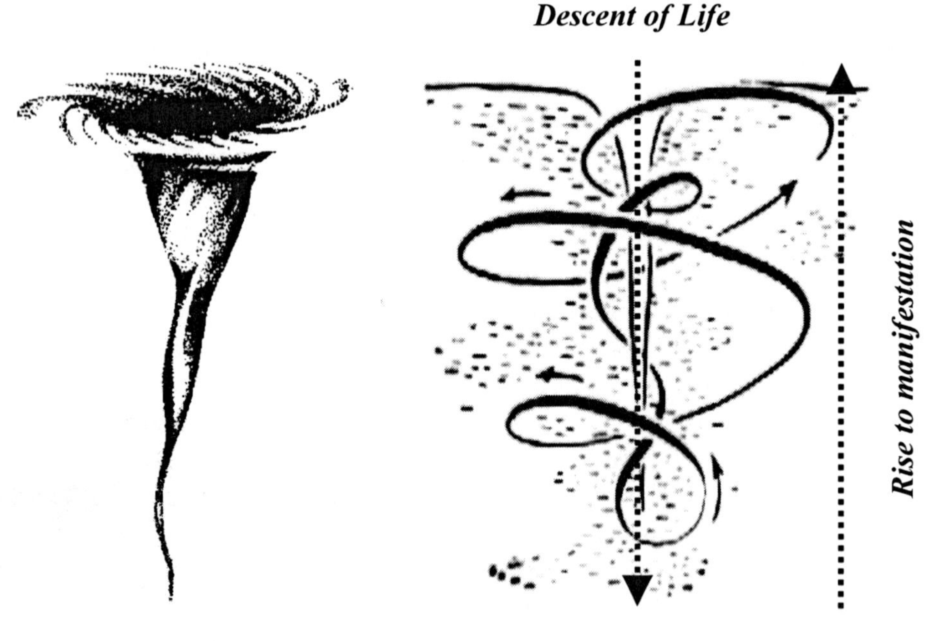

The visible vortex *Return currents of the vortex*

In the fifth paragraph ("*Since Sulphur's work …*") after a simple reference to sulphur's very delicate role in nature, Steiner introduces the three other substances: carbon, hydrogen and oxygen that - together with nitrogen - constitute the protein.

Finally in the sixth paragraph ("*How all this relates to plants ...*") we get down to the first substance: carbon. Having said that today, despite its immense significance in the universe, we know very little about carbon, and after explaining the ways in which it can present itself (as coal, graphite, diamond) he concludes by saying that when its importance was recognised it was bestowed with the noble name, "The Philosophers' Stone".

Just as sulphur is the quintessence, carbon represents the Philosophers' Stone. While the quintessence is the origin of life, the Philosophers' Stone is the seed of the new man within us. The chemical element carbon has four valent bonds and this presupposes a square structure that we know is related to the characteristic form of the Earth element. The three manifestations of carbon that we mentioned mark the route of our fall: diamond can be considered as the carbon of the Old Sun, graphite, that of the Old Moon and finally coal which is the terrestrial manifestation of carbon. In this descent carbon has completed a path that can be shown on three sides of the square of the elements from Fire up the Earth through three stages from life to manifestation of life, the descent of life, and finally death.

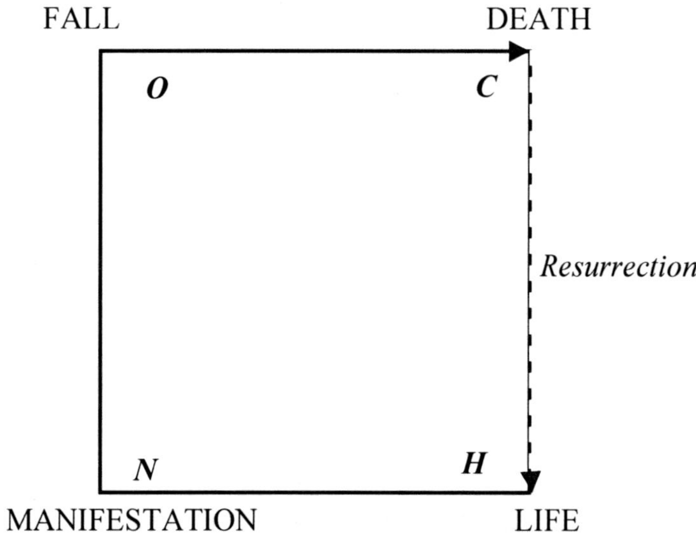

Having arrived at this point, carbon can rise back to Fire (see dashed line) and, transforming itself, go directly to the spiritual world represented by Hydrogen. The alchemists called this leap the '*philosophical death*' that involved the death of the part of Man which contains desires, passions, possessiveness, jealousy, anger, and pride that could not follow the resurrection of the spirit.

Thus the "resurrected" man would no longer emit CO_2 but oxygen, which is life. At that epoch plants would no longer exist because Mankind will have reintegrated them, thus completing the redemption of the vegetable kingdom.

This is a new way to conceive of farming compared to that of the Greeks. It is no longer based on the four elements (Earth, Water, Air and Fire), but on the transition from Earth to Fire, with the overcoming of the Fall through evolutionary resurrection, to a realm in which mankind works consciously with the forces of Christ. Clearly we are talking about agriculture because this is the object of the course, but we can bring these thoughts into the fields of education, architecture, and medicine and indeed to all areas in which people are active.

Moreover it is clear that the conditions on the Earth have become so compromised that it is no longer sufficient only to work with the otherwise laudable intention of not adding to the worlds' pollution or bringing in wheelbarrows of manure to the fields. We must now activate a process of resurrection that - thanks to the indications of Rudolf Steiner – we are in a position to pursue with greater safety.

Thus, we arrive at the seventh paragraph (*"And why was it carbon? .."*) in which Steiner explains why the Philosophers' Stone of the alchemists is carbon. We know that the alchemists were labeled pejoratively as '*blowers*' because they very often used bellows to pump air to keep the fire roaring. But if we take our point of departure from this cliché and treat it as an allegory that uses a physical image which actually proposes something much more subtle like an interior process, this '*blowers*' can be likened to our breathing and the bellows is the pulmonary apparatus. Everything in Alchemy is described as a process taking place on the physical level but actually they occur in the human body that is the true retort (*atanorr*). The function of the lungs is to send air throughout the body. In ancient times *pranayama* was practiced which was a yogic technique of breath control that served to fuel the 'inner bellows'. We can easily understand how the changing phases of breathing cause simultaneous movements at the level of the diaphragm and of the cerebrospinal fluid. This liquid is what we have just called, in the words of Genesis "the waters above". Each breath produces a current in the liquid mentioned (which runs between the two meninges), and gives a gentle caress to the base of the brain, like a kind of massage. But the breath does not act only here; its impetus moves to the last cell of the body thus linking the two trees of which we have spoken.

Respiration therefore involves not only the pulmonary apparatus: throughout our body we are Air, just as throughout the whole body there is Fire, Water, or Earth, except that in the individual parts from time to time there predominates one or another of the elements. From this point of view we can say that people have a body of Air. Remember that Air has the characteristic of expansion and contraction and that it moves very easily, and therefore the union between the lower and upper temples first occurs through air and only later through water and Earth. Moreover, the rhythm of our breath is connected to the heartbeat and to the Platonic year: our heart beats 72 times a minute and 72 is the number of years that the Earth's axis takes to move a degree compared to the Zodiac - the phenomenon known as the precession of the equinoxes. Our breath has a rate four times slower than the heartbeat, so in a minute we breathe 18 times. This means that we breathe 25,920 times a day - a number that corresponds to the 'cosmic year', or 72 x 360°. This means that there is a profound and direct relationship between our breathing and the macrocosm.

Steiner starts this part of the course speaking of carbon and we know that carbon represents the world of death in that it is related to the Earth and fixed form. Likewise hydrogen, which is linked to the Fire, recalls the *sacrifice* of Thrones, nitrogen recalls the *gift* of Dominions and oxygen recalls the *rejection* that occurred on the old Moon and brought about the birth of evil. (If carbon is not reduced to a rigid form it is for a particular interaction with other substances that we will review shortly.) What we are presenting is the *moral* aspect of creation. On this path the last step is death, but Mankind has the opportunity to change his fate and, approaching death, we can trigger a process of resurrection involving the transformation of carbon within us. This would actually be the last step of the re-ascent because we will have had to cleanse our soul, overcoming passions and desires, and this will lead to the re-absorption of the animal kingdom. Then we must purify our etheric body, or the body of our habits, temperament and attachments, allowing reabsorption of the plant world. Finally, the victory over our instincts will allow the absorption of the mineral kingdom. At this point the man would be liberated from the physical body.

Mid-way through the seventh paragraph we find an important statement: "*In fact, Carbon is the carrier of all of nature's formative processes.*" So carbon is the physical element that carries within itself the form that comes from cosmic distances in the form of images. We will try to clarify exactly what we have just asserted: we can say that the *image* of the plant is in continuous change and manifests itself in different *forms*. So the *form* tends to fixity while the *image* is in perpetual motion. If we wish to understand life we must learn to see pictures in their metamorphoric processes since the form is only a snapshot of a moment of life. It would be as if we thought we had grasped the mind of a director by seeing a single frame of a movie. This work is the *black opera*. But we should learn to do even more, namely to grasp the forces acting behind every image to be able to characterise them in all their manifestations. This is the training to get to the door of the Sun, to the temple of the heart, thanks to the *white opera*.

Steiner says in paragraph eight: "*A hidden sculptor is at work in carbon and this hidden artist makes use of sulphur when it builds up the many different forms that have to be built up in nature.*" Sulphur is the mediator and we could picture sulphur as the water with which the sculptor wets the hands to shape the clay. To quote Rudolf Steiner: "*To look at carbon in nature in the right way, we must see how the spirit-activity of the universe works as a sculptor, moistening its fingers with sulphur so to speak.*"

Also in the eighth paragraph we are presented with how the spirit, '*moistened*' with sulphur, weaves in the wake of carbon, building and dissolving forms. When we read this it should fill us with devotion and reverence for the spirit that immerses itself into things to form and configure them. Now we can better understand what we said about the study of form, namely that through the understanding of forms we can reach up to the forces that determined these forms. The form is immaterial and is the boundary between us and the spirit and through it we can know the divine forces, the entire universe with planets, the zodiac, beings and entities that act from time to time. The approach to the spiritual world through reading forms also has the great benefit of being conceptualisable and performable in a 'scientific' way, so that this approach can be the subject of communication.

Thus we come to the ninth paragraph ("*In earlier epochs ...*"). The Luciferic fall led to the condensation of carbon, but Lucifer has 'opened the door' to Ahrimanic forces that have resulted in the entry of limestone into us that - unlike carbon - is completely inelastic. Limestone therefore is the power of death and hatred, but, as we have now learned, when it comes to the extreme site of death life arises again because the bone is also the site of the production of red blood cells, which is a process representing the action of the influence of love and of new life.

It is interesting to note that the only bones that remain elastic even in adults, are those that form the chest, which is located in the central area of the body, the part of *mercur*, which contains the heart.

But we must always remember that because both Lucifer and Ahriman are in our hearts our ransom must be twofold. Besides, the manifestation of Christ has been through two birth events: the first at Christmas in which Jesus was born from the heart of Mary, and the second symbolised by the baptism in the Jordan when Christ began to descend upon Jesus.

In future we must first have a change of heart, so that what has been transformed can be taken by Christ as 'substance' begotten in us. We used the term '*begotten*' because, when referring to Christ it is not possible to use terms like *birth*, *create*, but only 'to beget'. The *Credo*, the fundamental prayer of Catholicism, refers to Christ as "*begotten, not made, being of one substance with the Father* ". It remains our task to provide the raw material - our transformed and purified heart free of the luciferic element. Then the second resurrectional birth can occur that overcomes the Ahrimanic element.

Ninth Meeting

At our previous meeting we talked about carbon which is a chemical element or, better said, a substance commonly considered only in its chemical aspect. We should start thinking of substances from another point of view according to which the substances are viewed as 'ambassadors' on the Earth of specific planetary or zodiacal forces. Generalising we could say they are ambassadors of supersensible forces and, in the final analysis, of spiritual entities.

We could also say that the whole spiritual world uses reference substances through which it can act on Earth. What we are saying can be translated into the following familiar sentence: "*No matter is without Spirit and no Spirit is without matter*", or in other words we can say that in our reference system, the Earth, there is no spirit that does not act through a substance and no substance that is not the bearer of spiritual forces. Following a point of view we will find that the Ram, for example, needs silica to act, while the Scorpion uses carbon and the Sun makes use of gold. Enlarging and generalising the concept we can argue that Father Sky (this includes all spiritual beings), in order to influence Mother Earth, uses specific substances as ambassadors. Approaching these same thoughts in a different way we can say that any material is the bearer of specific spiritual forces. This should stimulate us to ever-greater caution when using substances in agriculture and indeed in many other areas of life, and to ask what forces we are activating.

In the section we read at the end of the last meeting Steiner gave us a key to understand carbon, linking it to the physical. We know that carbon chemistry is organic chemistry and every organism has a form. Steiner here shows us that carbon is the bearer of the principle of form. In other words carbon is the substance that allows the Group I to mould organisms into forms typical of it.

As mentioned, carbon is the earthly ambassador of the constellation of the Scorpion. We know that scorpions have venom and are therefore the principle of death in the Zodiac in the same way that our physical body is our corpse. The connection lies in the fact that our physical body is based on physical carbon that is attached to the Scorpion and thus to the celestial analogue of death. The sketch below shows the relationship between the various substances, elements and forces that work in plants.

It should be noted that the life 'descends' from Hydrogen-Fire through the four elements (which represents an aspect of the 'fall' and therefore 'death' for Zoe) using the world of silica to reach the their goal in Carbon-Earth-Form linked to the calcareous.

In its life-cycle the plant follows the same route but in the opposite direction: from carbon-Earth, to Hydrogen-Fire all with the assistance of sulphur. The life of the plant takes it course between this descent from the spirit (its Group I) and the re-ascent to it. This is a process of a Raphaelic type[15].

Very different is the direct relationship that exists between carbon-Earth and hydrogen-Fire: this path - from death to resurrection - is the typically Michaelic path that allows the stimulation of new plants. One can travel this path in proportion to one's consciousness.

[15] See 'From the healing of Raphael to the healing of Michael" by the author.

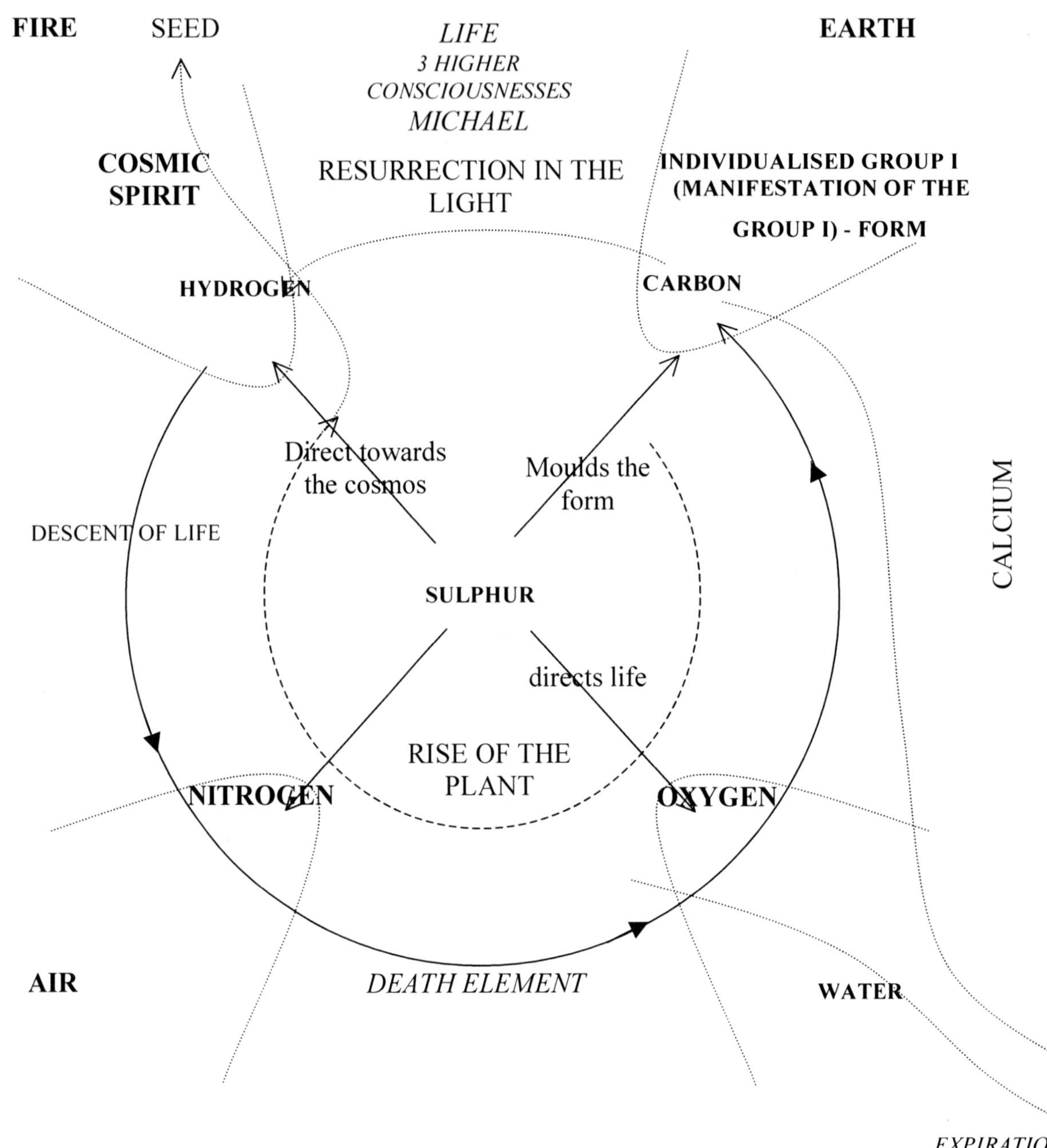

If we look instead to the etheric body it becomes interesting to connect to the influence of oxygen which bears the forces of Aquarius.

Anticipating what Steiner will show us in a few paragraphs, we can draw an imaginary cross in the sky with carbon at the top linked to the Earth, oxygen, which is linked with water, nitrogen (Air), and finally the last part linked to hydrogen (Fire). At the centre is sulphur. What we are tracing here is the image of protein, where sulphur - the fifth constituent - binds itself to the other four and puts them in motion. Sulphur is linked to the twins.

The tenth ("*You can conceive of this ...*") and eleventh paragraphs ("*Now you see, ...*") of the third lecture from Koberwitz tell us that where there is a structure of carbon there is always an etheric body, because carbon chemistry is organic chemistry and wherever there is an organism there is life.

The substance that enables the etheric to connect to matter is oxygen and oxygen does so with the help of sulphur. Sulphur, as the quintessence, governs the plasticity of carbon and the connection of oxygen with the etheric.

In other words we can say that sulphur, acting with carbon, moulds the form while sulphur with oxygen brings life. This seems a strange thing to assert even though in conventional agriculture sulphur's capacity to activate life processes is known. There is a practice that is called dusting the fields [It: *gessatura*] which is generally practiced in the last year of a farmers lease, with the aim of saving on fertilisation and exploiting the last residual life of the land that must be handed back. The *gesso* or chalk that is used is calcium sulphate. This practice is best avoided because sulphur brings life to oxygen, giving a false stimulus to the ground leaving it completely impoverished.

In the fourteenth paragraph ("*Seen in this light ...*") Rudolf Steiner makes another statement that deserves consideration. Steiner says that excessive ethericity makes us lose consciousness and this is not immediately understandable. But it begins to make sense if we think that in the human lifecycle the peak of ethericity follows our birth and this is also the time when our consciousness is virtually non-existent. By contrast, during the period of old age, vitality has been eroded by years of attrition, while the processes of consciousness are at their highest level - at least they should be.

We must slowly familiarise ourselves with the thought that not only the physical food must be properly digested, but also the more subtle substances. As the physical nourishment must be digested by our gastric juices, in the same way its etheric component must be digested by our etheric body and its warmth must be digested by our I. The heat-regulating diseases that can also be caused by heat stroke, are due to our inability to digest the change in temperature because of a weakness in our I.

We can find other examples of the polarity between vitality and consciousness, for instance, if we think of an area of the world where the oxygen is particularly rich in life, as in the tropics where plants can also develop aerial roots. We can see that indigenous peoples do not have a high level of consciousness, without denying their human dignity. It is not a matter of race but a condition linked to the environment, so that if *anyone* were to live there for an extended period of time, they would experience this dimming of consciousness. Moreover, the luxuriant nature and the deep green plants testify to a superabundant vitality.

The last sentences of this paragraph could be developed into the subject of a whole conference on working the land. Actually, cultivating the soil should be done with the aim of bringing oxygen into the soil. When we work on the soil, oxygen cannot but come in which - as we have already read in the second lecture - on entering into any living organism works to slightly raise its vitality. It is obvious that the life we are talking about is not the life of Mother Earth, but it is that of Father Sky since it is brought in from outside, which we can see as a fertilising with cosmic Life. It is important to know and understand what forces are active in the sky (constellations and planets) when we do cultivate, because they are the forces that penetrate to fertilise the Earth at that moment.

One could begin a whole new chapter of study that will be the subject of much effort in the coming years. We will have to clarify how the different types of cultivation should best be performed according to the way they want to get oxygen into the soil.

Of course when Steiner speaks of "*becoming alive*" on the part of oxygen, he does not refer to the physical world but to the world of the ethers. From this point of view, when oxygen enters into us or into the soil, it becomes the bearer of the alchemical ether.

We shall not comment on the fifteenth paragraph ("*Naturally, it's hard ...*") because it seems easy to understand. We will just repeat the last lines in which it is revealed that: "*... oxygen is the carrier of the living ether, and this living ether uses sulphur to gain control over oxygen*".

To simplify the first few lines of the seventeenth paragraph ("*I have just juxtaposed ...*") we can say that through carbon the Group I of the plant works on the single plants, as the I of a person acts upon his physical body.

Steiner then mentions the human to say that the etheric plane, which is linked to oxygen, must be able to find a way to bind with the spiritual or the form carried by carbon, and wonders how this can happen.

The problem may be conceived as follows: between the physical and etheric planes there must be a link, otherwise the two planes would remain detached and we would not have any incarnated life-forms. This link cannot be made by sulphur, which we saw as active in the link between the etheric plane and oxygen and between the Group I and the principle of form via carbon. This time something else serves to bind the physical and the etheric, to bridge between matter and life.

The answer is provided in the next paragraph: "*The mediator is none other than nitrogen. Nitrogen guides the life into the form or structure embodied in carbon.*" In all nature's kingdoms the bridge between oxygen and carbon is made by nitrogen. (We note that sulphur, a true champion of the spirit, is once again active with nitrogen.) This is commonly referred to as the interweaving of the astral element. We know that the astral is linked to the planets, so what we have just said can be seen from another point of view: namely that the nitrogen - through the seven planetary currents - can cross the etheric plane and flow into plants, giving rise to the seven life processes.

The resulting seven life processes emerge from the influence of the astral world (the seven planets) upon the etheric-vital plane. They are:

PLANET	PROCESS	EXCESS	
Saturn	respiration	dissipation	Ahrimanic
Jupiter	generation of heat	burning	
Mars	nutrition	sedimentation	
Sun	*separation*		
Mercury	growth	maturation	Luciferic
Venus	preservation	hardening	
Moon	reproduction	generation	

These seven life processes are linked to the light reflected from the planets and constitute the seven formative forces of the etheric body of the Earth and all beings. As can be seen from the scheme three of them may fall into pathological excess as Ahrimanic diseases, and three as luciferic pathologies. Only the process linked to the Sun, separation (or individualisation), cannot be corrupted.

In the expression used by Rudolf Steiner nitrogen simply allows the etheric (oxygen) to enter or to flow into the physical (carbon). But we have to understand that for the etheric it is not natural to connect and interpenetrate the physical body since the etheric and physical planes obey opposite laws.

We have repeatedly emphasised this: while the body is linked to materiality, and therefore is centripetal, gravitational and ephemeral in duration, the etheric is tangential, peripheral, sucking and lasting. Physical laws impel towards the ground, etheric laws pull us away. Because of the physical laws the plant develops geocentric roots. In harmony with the etheric laws plants develop their heliocentric airy parts.

Without an effective mediator the etheric would prefer to remain freely wafting outside the physical being. That is why the influence of nitrogen is essential in agriculture. Actually this element should not be considered a fertiliser, but the means by which life can take a firm grip on the soil and plants. Do not forget that the etheric world is manifested in the living mainly through metabolism, growth and reproduction. The real substance of manuring is oxygen as it is the bearer of life, but oxygen would not be able to lead life into the plants without nitrogen.

To develop another consideration we can say that to be an effective mediator between life and the physical world, nitrogen should have a strong connection with the world of life and so should be of organic origin. A chemical and therefore dead source of nitrogen evidently cannot be considered in the same way. In other words only a good nitrogen - naturally occurring organic and therefore 'alive' - can communicate with Life (with organic living oxygen) and ensure that life binds into the carbon that is the basis of organic chemistry.

Seen from this point of view we can understand why the air we breathe is overwhelmingly composed of nitrogen and a relatively small proportion of oxygen - the element which is essential for our survival.

Perhaps it is worth making a digression here to explain a little known aspect that concerns the relationship of the plant with the astral world. We have always said that the plant does not have an incarnated astral body and that the bloom is the effect of the free astrality upon the peripheral part of the plant. But if we only deepen this statement a bit further we realise that is not entirely satisfactory because it is not clear why a plant is so interested in this free astrality. In fact, this interest is the result of a friendly relationship between two sympathetic forces that come into contact – one from within the plant and one without. But if the plant does not have an internal astral body, what part is it that is sympathetic towards the free astrality?

The plant is also the field of activity of forces that come from under the ground and that completely pervade the plant. These forces are due to spiritual entities living underground, and through the plant they establish a relationship with cosmic spiritual entities, thereby bringing about the activity. These little-known entities of darkness are called the *Chthonic* beings. We can imagine that the cosmic hierarchy of spiritual beings has a corresponding spectrum within the Earth, which is intended to act within nature to prepare for the influence of cosmic beings. This second series of spiritual entities is like a mirror reflection of the series of nine higher spiritual hierarchies. These entities of darkness cause negative consequences in the natural kingdoms. For example they are responsible for all forms of pests[16]. But when their existence is recognised with the eyes of wisdom and love it becomes possible to realise that there is a collaborative relationship with the beings which are commonly called 'higher'.

[16] A development of this assertion is found in "Understanding and dealing with pests," by the author – not yet translated to English (2008).

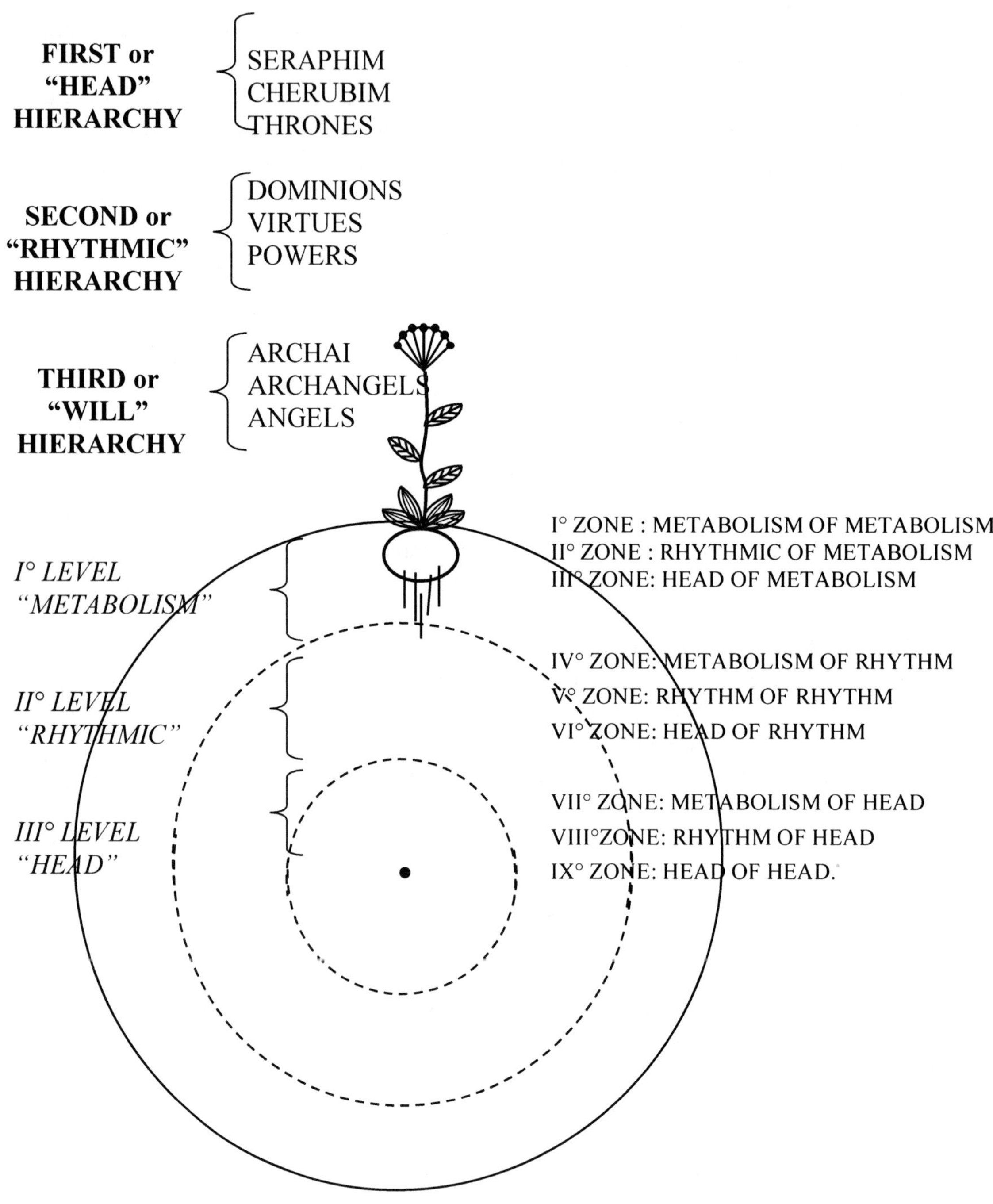

It is interesting to note that we define plants that are typical of or native to a certain area with the term *autochthonous*, giving the term a purely geographic slant. Actually the term is of ancient origin and refers to a certain system of forces in the area and the geographical location is only one of its characterisations. In fact, looked at in a more subtle way, the forces acting in a given geographical area are none other than the result of the spiritual forces that act there, and that actually are the imprint of the Chthonic gods.

For a better understanding of what we are saying it can be useful to take an image from the world of mythology. We all know that the Greeks recognised an underground deity called Vulcan who was in the blacksmiths trade and, with the help of an underground fire, built all the weapons of gods and heroes from Jupiter's lightning to the shield of Achilles. In these mythological descriptions the spirit of cooperation is clear between the lower and the upper beings. It is also clear that in ancient times these relations were generally known, though not in such rational terms. They have now completely disappeared from our culture.

Accordingly, it is clear that all agricultural work is none other than a continuous conversation with the spiritual world. This truth is a source of joy to our hearts, but also brings a substantial responsibility because it makes us realise that we are the instruments of spiritual beings that inhabit the cosmos and the interior of the Earth. The sense of responsibility increases if we think that the level of consciousness that we have developed also marks the limit of the beings that can be appealed to in our work, and this raises the need to strive to ensure that our level of consciousness continues to rise.

We are discussing agricultural work and not meditation or prayer because without taking away from these practices and even though it sounds strange, the greatest connection with the world of the spirit is obtained through manual activity. Humans began to change nature when we gained the ability to stand upright and free our hands. The hand has become the specifically human way of expressing ourselves - after the word. At the time of Atlantis the greatest spiritual evolution was expressed by the word *Manu*, which was the oracle solar.

Remember that when Atlantis disappeared the native American Indians called their god by the name of Mani-tù. Furthermore we call u-mani those deprived of Manu who have fallen and must return to the level from which we originated. Note how the U is just the sound whose sign represents this fall and then rise. It is clear that humans express ourselves with our hands [*It: mano*], and it is not just a coincidence that we are unique in the world in having the ability to oppose the thumb to the other fingers. Do not forget that human karma began when we began using our hands, because beforehand we were not responsible for our actions. From the time we gained the ability to act on nature we could choose whether to act as helper or as a destroyer.

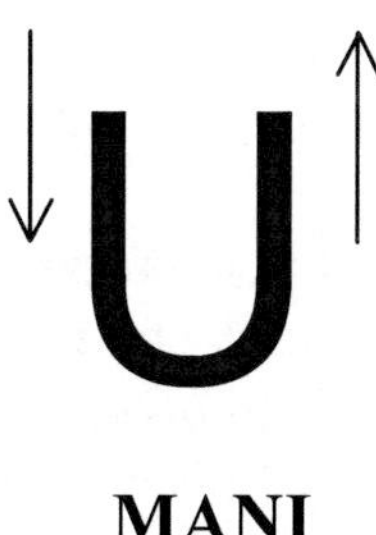

MANI

The Gospel offers us the real name of Jesus as *E-Manu-El*, which means *Divine Manu*. So when we move our hands we can resonate with the Earth's Manu and the Divine Manu, thus becoming instruments of the Spirit. Moreover it is also true that normally the hand bestows the blessing. Even now it is by shaking hands that we symbolise the meeting of the two I's that connect mutually when concluding a contract.

It is no coincidence that in English the words man and woman are used. One can establish with the "V" a strengthening of the function of the fruitful female and recognition of the woman's vital role of carrying forward life and the whole world.

The farmer therefore moves the spirit when he uses his hands - should he have adequate awareness. Agriculture is perhaps the most effective of prayers because it is more important to act to affirm the laws of the spirit than merely to pray for the same ends. Moreover, the task before us is the conquest of free will [*It: libera-azione*], namely the ability to act freely, by which we mean to be active only for love, without any kind of personal implications.

Think for a moment about what the word "love" [*It: amore*] means. It means '*a-more*' or '*without law*', or - better - without the rules and regulations of the world. This implies a 'presence of spirit', that is being present in every decision, recognising what is right from what is not, in order to act properly.

In conclusion, the hand is the instrument through which man can realise his destiny, which is to become a being who is free to collaborate with the world.

The real dimension of man is not earthly or planetary, but the Seraphimic and Cherubimic Zodiacs and where the Spirits of Love and of Harmony weave. In fact, the rules that we set are in the world of planets, because they are based on rigid mathematical rules that govern their rotation, revolution and mutual relations, which depend on eclipses, trigons, squares etc.. The stars have no rules. They seem immobile because they are not concerned with issues of space and time. The stars, as we have said many times, represent the world of the spirit while the planets represent the soul world. The ten commandments (rules) were given to bring the soul into order to regain the world of the spirit - our cosmic matrix. Obviously we cannot get to the Zodiac without first having ordered the world of planets, or having put order into the world of karma, which is linked to the hand of '*u-mano*' beings.

We will leave our digression now to resume reading the lecture of 11 June 1924 at the eighteenth paragraph ("*Spiritually speaking ...*"). This paragraph does not require special comments but we will emphasise a couple of the sentences: "*The etheric life principle would float around like a cloud, would not take the carbon framework into account at all, if the nitrogen did not have such a strong attraction to the carbon framework.*"

In the nineteenth paragraph ("*This nitrogen is really ...*") Steiner emphasises that nitrogen is important in human breathing. It allows the meeting between oxygen and carbon enabling the elimination of CO_2 continually dissolving that rigid form typical of a carbon structure. The next three paragraphs are also self-explanatory.

Particularly interesting is what is proposed in the twenty-third paragraph ("*Astrality is everywhere ..*") which can be summed up by the last sentence: "*In short, nitrogen pours out over everything a kind of sensitive life.*"

Talking about '*sensitive life*', however, means talking about sensory organs. In other words nitrogen in the soil enables perception of conditions of sympathy or antipathy, and brings the elemental beings into action to bring balance to all situations. Steiner gave the example of water (in paragraph 23) which, for instance, is particularly fitting because nitrogen needs water to bind with the physical.

This aspect of nitrogen's sensitivity allows us to explore another consideration; faced with a drought living nitrogen may intervene to 'recall' the distant missing forces. Conversely chemical nitrogen, which cannot link to the etheric, cannot be helpful, and it is also well known that chemical agriculture requires large quantities of water.

One might ask what we need to do to bring living nitrogen to our farms. In the first instance we can consider the introduction of an animal that can bring astrality or, better said, those that can stimulate the capacity for plants to connect with the distant source of the etheric. The best animal is undoubtedly the goat. Be well aware that we are not saying that you should introduce a herd of goats, but to introduce a few perhaps, depending on the sise of the farm, which only serve as a stimulus for astrality. It's like a homeopathic intervention. The goat, in fact, precisely because of the manure that it produces, is not adapted to areas lacking in the etheric, but should only serve as a stimulus for the astral.

Another way to bring nitrogen if there is a lack of the etheric consists in introducing bees, as regulators of the astrality on the farm. It is really worthwhile to deepen our familiarity with bees. First, we can say that the bee acts upon the astrality of a property through the forces of love. We must bear in mind that the 'being' of the colony has an individualised I almost like that of a human. It also has a more advanced evolutionary level than we do - by two 'stages' - because having already conquered love it is already at the level at which humans will attain only in the future Venus. We could also say that the bee has already won the Philosopher's Stone. So, as the higher self, it has already mastered the ability to direct astrality. That is why in very dry conditions, when the etheric forces are scant and the poor quality astral is in excess, the bee releases the excessive astrality and leaves the right amount of nitrogen to make the farm more responsive to the needs of plants. In this situation the plant can intensify its communication with the distant etheric. Therefore the bee, which apparently has nothing to do with the etheric, purifies and restores the sentient capacity of nitrogen, which can then lead the etheric to the physical plane.

A third way to bring living nitrogen onto ones property is to grow legumes. The *leguminosae*, with their ability to fix atmospheric nitrogen as terrestrial nitrogen (thanks to the astral work of nitrogen-fixing bacteria that act around its roots), becomes the organ of the living plant world capable of strengthening the functions of nitrogen in the soil. We maintain that living nitrogen acts as the sentient organ of the farm and can act as described not only in relation to water, but to any substance that may be out of balance.

Paragraph twenty-four ("*Ordinarily we are not aware ...*") highlights that nitrogen, as a sensitive intermediary, does not function within narrow limits, but can sense what emanates from the farthest stars and acts in the life of the Earth and of the plants.

In paragraph twenty-five ("*We have seen now ...*") we stress only that life looks on the one hand towards carbon and on the other to nitrogen.

In paragraph twenty-six ("*This is the situation ...*") Steiner is preparing to introduce hydrogen into his scheme. We can further deepen our understanding of what works and lives in nitrogen. We have already said that nitrogen is the bearer of planetary forces and the zodiac, but now we can specify that images of the cosmos live in nitrogen.

We know that during the 13 Holy Nights the forms and ideas of future plants descend from the cosmos, but these descend in the form of images which are taken in by the Earth's elemental beings who then give them to the plants so they can manifest upon the physical plane. These descending forces live as images in nitrogen. When we breathe in, as well as bringing oxygen and other vital gases into our body, the images of the cosmos also come into us. These images will then permeate right into the cells of our organs to reproduce over time with the same forms and under the same laws. In fact our body is built in the image of the cosmos.

So when we inhale nitrogen we bring the great cosmic sculptor within us. It should be noted that the cosmic images that penetrate us through the nitrogenous component of air, pass through our astral body (nitrogen is linked to the astral body) and imprint themselves upon our etheric component, thus shaping our organs.

If man, with appropriate inner discipline, manages to connect his centre of self-knowing, namely the I, with the etheric in his organs, then a new consciousness arises in him: imaginative consciousness. This is the first stage of clairvoyance.

Be that as it may and returning to agriculture, if we have plenty of living nitrogen on our farms and in our gardens during the 13 nights this nitrogen can gather in these images of the cosmos. When we go to spray the preparations we help the Earth to grasp these images of the cosmos since even the Earth, having been mistreated in every way and subjected to all kinds of pollution, is no longer able to perform this function alone.

Tenth Meeting

We are going to continue reading and remarking upon the third lecture from Koberwitz - based on Greek culture, at which time life was understood as an articulation of the four elements: Earth, Water, Air and Fire. Steiner did not, however, use the language typical of Aristotelian philosophy, but has translated the concepts of that philosophy - as he has done for the philosophy of Egyptian and Persian times - into agricultural terms. Zarathustran agriculture was represented as the agriculture of silica and limestone instead of as the agriculture of light and darkness; Egyptian agriculture was not presented in terms of *Sal, Mercur and Sulfur*, but in terms of limestone, clay and silica. Finally the Greek agriculture with its four elements, is illuminated through the chemical elements that represent these elements. From this point of view we can talk about the *Earth* element as carbon, the substance bringing form, because it is the whole physical basis of plants, animals and humans. We can talk about the *Water* element as oxygen, bearer of the forces of life; about the *Air* element as nitrogen which, through inspiration, brings the cosmic images into humans, plants and animals; and finally the *Fire* element as hydrogen, to which we will shortly be introduced.

But before addressing hydrogen with Rudolf Steiner, perhaps it is worth going over some aspects of nitrogen because, for many of us, it presents new aspects of being linked to the world of astrality – also known as the world of feelings – which certainly is less known than the physical and ethereal facets of life. Let us first note that the terms inspiration and expiration mean bringing the spirit inside us and releasing the spirit from us. In this way nitrogen becomes the bearer of the hidden sculptor that conforms our body to an archetypal image flowing from the cosmos with every breath. In the light of this phrase perhaps the expression of Steiner, which we have often cited, becomes more comprehensible - that the plant is an image of the cosmos.

Normally this power of breathing escapes us because it is so familiar as a continuously repeated unconscious process. It assumes great importance and brings immense effects that only our obtuseness in grasping the messages contained in the development of life prevent us from grasping. We need only think for a second about what's happening in the heart when a child completes its first breath. We all know that during the entire gestation the auricles of the heart are communicating with each other, but at the first breath of the child after birth the septum closes almost instantly. This should make us reflect on the power of breath. From another point of view the closure of the septum is also the repetition of the exit from the earthly paradise. The fact that we exit from the mother, who represents the spiritual gestation immersed in a world of spiritual hierarchies, and begin our independent life on earth with the first act of breathing, immediately brings about a different conformation of the heart. Also the heart is the centre of the circulatory system and we know that blood is the physical basis for our contact with the spirit world.

We are therefore constantly immersed in nitrogen and through it we have constant connection with the spiritual world, understood here as a flow of formative images. However, nitrogen, as we saw in our last meeting, also has the ability to induce the etheric to connect with the physical. So on the one hand we find nitrogen sculpting the form and on the other it allows the etheric body to bring life to the physical body.

Nitrogen is also the substance that allows the physical body to communicate with the astral body, making us capable of having sensations and sensing our primary needs such as hunger, thirst and so on.

Nitrogen is therefore extremely important in a farm, but it is crucial that this nitrogen is alive, otherwise it will not be possible to forge an effective link between the etheric and physical levels, as it reduces the flow of cosmic images to our plants.

Only if nitrogen is alive, that is only if it is connected with the being of nitrogen, will our plants be able to associate with the surrounding plants, with the clouds, the Sun and the stars. These connections are fundamental for the plant: do not forget that the plants do not contain an astral body and have not developed their own vital organs, but have their organs in the planetary world with which they must be linked. Our human organs are actually nothing more than internalised carriers of the forces of the planets. In this way, for example, we can see that our lungs might be compared to our internal condensers for the forces emanating from Mercury. Since plants do not have lungs they depend upon constant connection to this planet, and are similarly dependent on other planets for other vital functions.

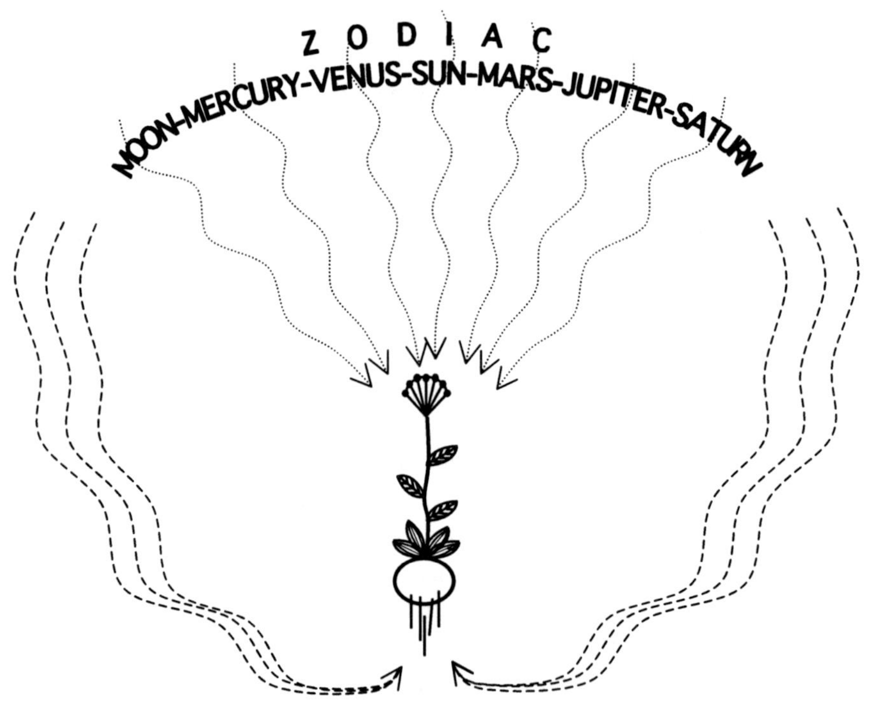

All these connections are possible thanks to nitrogen. Our task is to encourage these links by bringing living nitrogen to our farm, namely nitrogen derived from natural processes, or through the action of legumes or from bringing in manure, etc ...

Nitrogen can thus play its role as a bridge between the planetary world and the plant. The following table shows the main correspondences:

PLANET	*ORGAN*	*VITAL FUNCTION*
MOON	BRAIN - GONADS	REPRODUCTION
MERCURY	LUNGS	GROWTH
VENUS	KIDNEYS	PRESERVATION
SUN	HEART	INDIVIDUALISATION
MARS	GALL BLADDER	NUTRITION
JUPITER	LIVER	GENERATION OF WARMTH
SATURN	SPLEEN	RESPIRATION

This vision requires that we expanded our usual concept of the plant beyond that which ends with its physical material manifestation. The plant needs to be seen as an accumulation of activities that originate from the edge of the universe (and from the centre of the Earth) a plant without borders. Actually the plant in its configuration as a set of root, leaf, flower and fruit reaches beyond the planetary limits and finds reference in the Zodiac. The Zodiac is also the origin of the special arrangement of the plant, beginning with the birth / manifestation and continuing with the rooting, the rising and so on, until you get to the maturation of the fruit and the death of the plant. The following table summarises these links:

CONSTELLATION	*FORMATIVE PROCESS*
RAM	COVERING
BULL	CLINGING
TWINS	FLOWERING
CRAB	EFFUSING
LION	RECEIVING
VIRGIN	FRUIT BEARING
SCALES	REDUCING
SCORPION	STIFFENING
ARCHER	GERMINATING
GOAT	SPROUTING
WATERMAN	VERTICALITY
FISHES	BUDDING

In theory a farmer's only purpose would be to help the cosmic images shape our plants and one can do this by making enough organic nitrogen available. However, in a world where the ecological balance has changed due to all forms of pollution and thus where the conversation between Earth and the cosmos is skewed by 'filters' of every kind, the task of agriculture actually becomes one of finding a way to restore the heavily compromised communications system with the cosmos. The role of biodynamic preparations is mainly to this end. Making use of biblical terminology we could say that we wish to establish a new covenant and to transform the tragedy of our time – ever more distant from God – into an opportunity for rebirth or resurrection, which is to glimpse a new paradise.

One hundred or two hundred years ago, old traditions based on an instinctive adherence to the laws of the spirit were sufficient to make agriculture work. Today we are obliged to regain a relationship with the laws of the spirit through consciousness, and we must make the effort to understand - through observation and study - how the spirit acts upon matter. Since this depends on our choices and upon us, it is also an opportunity to engage in the struggle for freedom, and this is the real task of humanity. We can be the new point of contact between the physical world and the spiritual world: the farmer is no longer seen as the slave of the soil, but assumes the dignity of becoming a minister of this new connection.

In the past copious spiritual forces poured down upon nature, dispensed like rain and bringing Life. Now mankind must reopen these channels from below and these will gradually enlarge so that the forces of the spiritual world can return to fertilise the Earth. The people of the past could be considered as children to whom everything was given: now the children have come of age and must be active to ensure the means of subsistence. On the other side are the people who believed they can do without God and it will be up to these people to lay the foundations for a new alliance without which nothing more can be achieved.

What we have tried to characterise is nothing less than the transition from the old to the new initiation. The pre-Christian initiation was actually established to try to optimise the reception of what came from God; after Golgotha the initiatory journey has changed and requires that people make the first step towards God to try to restore a link through the verticality of our I. Mankind, in our resurrection, can also bring with us the plants, animals and things that have been entrusted to us to husband, and therefore we can be the redemption of all nature. Clearly there is a nadir in the transition between the old and new ways and this is the time in which we are living. Unfortunately this time is characterised by environmental disaster, but we have the possibility of reversing this and establishing a new bridge. But we should not waste time because there are many who are trying to escape from the nadir mentioned by seeking an alliance with other beings, implementing practices that have the precise purpose of bringing down forms of life that are no longer able to participate in this new approach to the love of God. We say this because biodynamic agriculture (and homeodynamic that is following it) is a direct daughter of the science of initiation, and if this is 'forgotten' the mother is inexorably destined to fade.

In today's meeting we will encounter hydrogen, the fourth component of protein, which corresponds to Fire and thus to the spirit. Through hydrogen we will try to understand how the spirit acts in the plant world. Do not forget that hydrogen corresponds to the spirit as does the soul to nitrogen, oxygen to the etheric or to life, and carbon to the physical.

We know that the plant itself has no internal spirit, as it has no soul, but we will see how the spirit nurtures the plants.

We are at paragraphs twenty-six ("*This is the situation ...*") and twenty-seven ("*We could say that ...*") where Steiner concludes his discussion of nitrogen and presents in a few quick phrases his thoughts on hydrogen, perhaps disappointing those who were expecting a long exposé in view of the fact that hydrogen is considered to be the messenger of the spirit. We must realise that Steiner has gone into detail on the first component of protein in order to establish his approach, since his intention is that the student should already possess the means to achieve a sufficient understanding of the last element from these scant indications.

We know that hydrogen, in chemistry, is the lightest element and we also know that in the lower atmosphere close to the earth's surface it is present in very low concentrations - about 0.5%. Rising to the higher atmospheric strata the concentration of hydrogen increases continuously until at approximately 150 km up it is virtually the only constituent. One could glean something of hydrogen's links with the spiritual world from this. Being relatively insignificant close to the ground and almost unique as it approaches the cosmic distances could be a confirmation of the fact that the forces of hydrogen are greatest in the spiritual world and minimal in the physical world.

Precisely because it is so strongly linked to the spiritual world hydrogen is the element that requires the least amount of energy to penetrate into the material world. To do so it needs the assistance of nitrogen, to which it entrusts its own images, so that the nitrogen carries them to the physical level, thanks to the mediation of the etheric plane. The spirit thus directs the material without entering it. Indeed it merely skims by working through a mediator.

We all know that in a plant's different growth stages we can witness that its climb toward the sky represents ever closer ties to the hydrogen forces. Moreover, in an annual plant, this ascent coincides with a loss of embodied vitality after which the hydrogen can interact directly with the plant and make an impression affecting its processes of quality.

The highest expression of hydrogen in a plant is in the formation of essential oils, which represent the transformation of Water in the sap into Fire. For those that observe nature thoughtlessly this phenomenon presents no significant food for thought. For others, entering into the continuously repeated phenomena of nature, these transformations are almost miracles and should fill us with wonder. We normally think that Water and Fire are antithetical, since Water extinguishes Fire and Fire evaporates Water. Despite this with the formation of oils Water is transformed into Fire, not only because it transforms its quality but because the oil is combustible.

What is absolutely clear is that the transformation of which we are talking cannot be reproduced in the chemistry laboratory but the plant can do this repeatedly and with apparent ease. If we were to borrow a phrase usually used for a similar transformation we might speak of the *spiritualisation* of sap. In other words we can say that the sap, linked to the etheric, is transformed into fire by the work of hydrogen.

We are accustomed to thinking of water as a compound of two hydrogen atoms with one of oxygen, but if we think on we also know that hydrogen and oxygen are two gases, and it is not so immediately understandable how two gases combine and can result in a liquid rather than a gaseous mixture. This combination becomes possible only in the presence of a catalyst - and this is Fire. Clearly not physical fire, but philosophical Fire, Fire understood as a virtue. Remember that oxygen in this compound is the etheric plane, and therefore represents life within biology, and hydrogen is the spiritual and thus the life of the I. These two lives are joined in water which, on this basis, is considered the foundation for life on our planet. This is why, in many religions, immersion in water is considered an act of purification, not only an act of physical hygiene but also considered spiritually. Moreover many sacraments of Catholicism use water.

We can now understand better why homeopaths use water because, besides being the bearer of memory, the water can communicate with organic beings and with spiritual beings simultaneously - and healing processes require both these two planes.

Besides St. Francis described water as a sister and qualified this with the adjectives "chaste and pure." In this way he encompassed water's biological (sister), soul (chaste), and spiritual (pure) aspects, demonstrating that he had fully grasped the being of water.

But back to hydrogen. We can see that the first time hydrogen takes hold of the plant is when the pollen enters the ovary. This brings about a total transformation of the plant. At this point the plant ceases to be 'selfish' and begins to build, through its own self-sacrifice, gifts for other kingdoms of nature (fruit) and for future life (seed).

Wanting to be more precise we could say that hydrogen, when it takes hold of the plant, disorganises that which the nitrogen built as an image of the astral body of the plant, melts the ethereal which is beginning to become labile and also dematerialises or crushes the carbon. It may be interesting to note at this point, that the action of hydrogen does not always lead to the formation of oil, but sometimes the process is a little more tied to water and will lead to the formation of latex. The formation of resin is also an incomplete transformation of Water. To better understand the relationship between resin and oil we can use an example and observe what happens when a flame consumes a candle. We all know that the contemporaneous products are light and smoke and we can all intuitively understand that smoke is a result of a lower level of transformation than the light. Well, resin can be considered as analogous to smoke, which is the shadow of light that is represented here by oil.

If we wanted to ask the theoretical question of how it might be possible to increase a plant's ability to produce resin, we should think about how to connect with the forces of living oxygen. Although the resin, compared with oil, comes from a lesser transformation and if the water is composed of hydrogen and oxygen, this formation will be more related to water while the hydrogen will better promote the transformation of water into oil.

We could, for example, limit ourselves to a herbal intervention and consider making an extract from tropical plants with aerial roots and spray this extract on the plants that we wish to produce more resin. Plants that have aerial roots demonstrate their ability to connect to the living oxygen that is so clearly present in the rainforests. If we wanted to make a more effective intervention we could think of emphasising the forces of oxygen by linking these plants with the constellation of Aquarius, through for example, the choice of an appropriate time for sowing.

We said that hydrogen's influence on the physical plane consists in the fragmentation of what carbon has built. This effect can be encapsulated in the sentence, "*dust thou art and to dust shalt thou return.*" If we consider life as a unity that has several modes of expression, the destruction of the physical body can be seen as a necessary moment in which to check upon the degree of transformation that we have failed to achieve within our bodies. In the same way, during each step of creation, God stopped to see if what he had done was good.

Clearly these breaks are necessary to move from one phase of creation to the next, just as successive incarnations are needed to complete the various experiences that we must undergo. From our point of view the influence of hydrogen brings death, but from the point of view of God it is a way of moving to another stage of manifestation in the cause of our evolution.

Plainly hydrogen needs sulphur to integrate its activity with the other substances.

Thus, in paragraph twenty-seven ("*We could say ...*") hydrogen is again presented as the preferred vehicle of the spirit through which the spirit laps against the most material aspect of existence. However, it cannot remain in this enclosed form for long and needs to be free to return "*into the indeterminate chaos of the universe.*" Hydrogen therefore dissolves the casings the spirit had accepted with the cooperation of sulphur - which is the intermediary between all interactions with the spiritual world and protein's component substances and the interactions between them. We maintain, however, that without this succession of steps from the spirit down into matter and its subsequent dissolution, reincarnation would not be possible because even the most vital condition of matter, if not dissolved, invariably falls into mineralisation which is death on the physical level (even if in the opposite direction compared to the dissolution mentioned above).

It may be interesting to note that Steiner also brings the moistening action of sulphur into this paragraph. Indeed, the transition from the spiritual to the physical world, which we could also see as the transition from the diffuse Life to the incarnate life, needs to pass through *water*, in both its fall and resurrection. As we have shown earlier, water represents the synthesis or the union between the two meanings of life as revealed in its make up of hydrogen and oxygen. Of course the conversation between the plane of biology and the spiritual life needs sulphur's brokerage. If we recreate our image of the *hidden sculptor* that sculpts the forms inspired by the cosmic images we can compare this to a sculptor who needs moistened fingers to shape the clay.

Within us the sulphur is especially present in the area of the coccyx and is the famous rejected stone that becomes the cornerstone on which the entire structure of the (new) house is built.

The twenty-eighth paragraph ("*So here we have ...*") may surprise us with an assertion that at first seems to contradict all that has been said so far about hydrogen. The last line says that hydrogen is the least spiritual among the substances that "*represents what's working and weaving in everything that is alive, and also in everything that temporarily appears dead*". If we reflect, however, we notice that the contradiction is only apparent because hydrogen can imprint itself upon the plant only when the plant has lost its vital momentum and is on the way to becoming dead organic matter – abandoned to the physical or lowest level. Moreover carbon requires a particularly strong spiritual force to act at the level that is farthest from the spirit world.

The next paragraph, the twenty-ninth ("*At this point ...*"), contains an explanation of what Rudolf Steiner means by meditation and he describes some of its effects on human physiology. First, Steiner shows us how meditating begins to change the process of breathing, but not as a result of a purposeful technique focused on this modification, but as an indirect result of exercises within the soul. In particular meditation modifies the amount of carbon dioxide that we normally exhale into the nitrogen-rich atmosphere, retaining a greater quantity within us. During the normal breathing process we all emit a certain amount of poison - as carbon dioxide must be considered. The plants, which are themselves pure because they do not have an astral body, absorb the carbon dioxide thus purifying what we have poisoned. It is clear that if we start to produce less carbon dioxide the plant loses part of its function and then we will have completed the work of inner transformation, or in other words will have won the Philosopher's Stone. We will exhale only oxygen and no more carbon dioxide. At that time the Earth will no longer need the plant world and it can be reabsorbed. Mankind will then have completely transformed its carbon. There is a terrestrial carbon that finds expression in coal, then there is a carbon of the Moon that is graphite and there is a carbon of the Sun that is diamond. When we breathe out oxygen we will have become *adamantine* like our ancestor who was called Adam for this reason.

The reabsorption of the plant kingdom by humanity must not be seen as a lowering of conscience but, on the contrary, is the result of a broadening of our awareness which should by then have fully understood all the processes that occur in the plant kingdom - for example, the formation of resin or sugar within plants, what happens during fruit set, how cosmic images enter the plant, and so forth. When this has been achieved we will be fully joined with the plant world in full waking consciousness and it will unite with us.

In this way the normal view we have of agriculture has changed completely, because it has shown itself to be the best way to develop this consciousness of the world of plants and to let it grow within us. Agriculture becomes a true path of inner transformation like other ways hitherto followed by humanity. It is a grace that we have been given the opportunity to combine the means to maintain our earthly lives with the possibility of stimulating the growth of our spiritual life.

In the thirtieth paragraph ("*You see, if you bump ...*") Steiner continues to speak of meditation, explaining that through it one acquires knowledge of what lives in nitrogen. He says again that if the farmer ponders he can acquire knowledge of the forces acting upon his land and know many deep secrets simply by becoming receptive to the revelations of nitrogen. This is because nitrogen is the bearer of cosmic messages.

Paragraph thirty-four ("*And in fact ...*") presents us with the farmer as one who learns by walking on his fields and meditating on what has been perceived. Meditation is particularly fruitful during the winter nights when the farmer is less involved in the hurly-burly of cultivating the fields and in a period when all of nature is directed inwards in order to grasp what descends from cosmic distances in these winter months.

In the following paragraph ("*These are the kinds of* .."), Steiner emphasises the crucial role of an approach to nature which is not based upon intellectualism but on observation and close perception. The coarse but defining concepts that may result from intellectual appraisal cannot grasp the living and subtle weaving of nature.

Once the peasant-farmers used to walk in the fields alone, especially on Sundays before or after mass, and they carefully observed the situation on their farm. These walks enabled the gradual acquisition of sufficient sensitivity to understand what was in harmony from what was not. They began to understand the needs of individual plants without having read a page of a treatise upon agronomy or hearing a lecture at the university. This could also be done with animals - and doctors should learn to do this with their patients.

We should bear in mind that Steiner wanted to offer three courses in agriculture and what we are studying is only the first of this hypothetical trilogy and we do not know what he would have said in subsequent cycles. However, as a good engineer can appraise a set of foundations and ground floor, we might guess how high the completed building will be and have some idea of the harmonious design of the upper floors. Perhaps with a little effort we can try to guess what Steiner could not teach directly.

Clearly what we are now looking at is a model of agriculture based upon the elements. From our point of view we believe it is necessary now, in the fifth cultural era, to bring a healing impulse to agriculture (and more) based on the forces arising from the Milky Way, because only at the Milky Way is there *Perfection*. In Christian terms, at the dimension of the elements where the lunar influence is paramount, the achievement of *salvation* is possible. The second step, which is *Bliss*, can be reached by crossing the gateway of the Sun. The third step is reached at the Zodiac and is called *Sanctity*. This is the ability to be linked with Adam Kadmon or with the Universal Being (depending on the culture). Finally, as we said, there is *Perfection* at the level of the Milky Way. The level of agriculture addressed by Rudolf Steiner in the third Koberwitz lecture is that of *salvation*, which is aimed at saving nature, to clean her up and restore her vitality. In the upcoming lecture Steiner will lead us to the next higher sphere because when the preparations are buried they are christianised, and this lets one pass through the door of the Sun.

So we arrive at paragraph thirty-three ("*So, all of these elements ...*") in which we are told that the substances described above are related to other substances in the plants upon which they depend. Hydrogen can intervene in the release of these substances in two ways: by encouraging the total dematerialisation of the substance or by bringing the plant to seed. In this second way the other protein substances become receptive to the cosmos. Moreover, the emergence of a new plant takes place thanks to the communication that is established between the chaos formed in the seed and the chaos of the cosmos.

The seed wakes up when in contact with water which is the basis of biological life. However, since water is also the basis of spiritual Life the seed is also freed to connect to this Life so that the forces capable of renewing the biological life will descend from the cosmos. Into this capacity for union we can apply our own efforts to induce certain forces to manifest in a particularly strong way, according to what we want to stimulate in the plant or based upon what was lacking in the mother plant. However, this connection between seed and cosmos has now become difficult because of various forms of pollution caused by humans, and so it becomes necessary for other interventions to restore this interrupted dialogue.

We could also say that the seed can follow two types of impulse to start building a new plant. The first concerns heredity and following this momentum the new plant will express what has come down from plants that make up its 'family tree' in the same way that a physical-etheric person is the result of a combination of the characteristics of the four preceding generations. We can imagine this hereditary line as communication on the 'horizontal' plane. But the plant also has the ability to receive impulses from a different line, the individualising one, that makes direct 'vertical' contact with forces from the cosmos. Both these transmission lines are now polluted, and both must be reorganised. However, the most delicate is the vertical line that suffers mainly from a type of pollution that is very special - pollution of awareness. People no longer remember that up in the sky there is a Father, and even those who still believe this think there are no means for making a connection. Even those within whom spiritual development is alive feel compelled to use the means that scientific orthodoxy provides when they are in the countryside and use the methods that they have been taught at school. This is a real and unsought for tragedy because a lacerating inconsistency is cultured in the soul which is the source of many health problems. The most common of these is the state of anxiety resulting from the feeling of failing to adopt behavior consistent with the ideals that inspire us. When one of these people manages to adopt farming and gardening techniques in line with their own spiritual evolution and that of nature, they experience a deep joy. This is a consequence of the liberation arising from finally establishing a proper relationship between their inwardly cultivated ideals and what is being carried out on the ground.

This is a specific attribute of spiritual science. It allows us to establish relations with the spiritual world based on the rigor of knowledge typical of the scientific approach. At the same time we can devise techniques to operate in the world inspired by a morality that is respectful of the laws of the spirit.

Returning to the Koberwitz course we said that what should manifest from seed follows two lines: the first biological and 'horizontal' is based on water as its life support, and the second 'vertical' line sees water from the point of view of its capacity to purify and to prepare for the descent of the new impetus that descends from the cosmos. Perhaps it would be more correct to speak of liquid instead of water.

At this point Steiner completely changes the formulation given to the agriculture course. After having given a theoretical framework imbued with profound knowledge, the author goes on to the practical aspects and talks about the seed in order to stress the importance of seed work in agriculture. By now we will understand that only through sexual reproduction can one really direct the growth of plants, because only through the seeds can one institute a dialogue with the beings of the universe that can bring about their influence through the vertical current we mentioned earlier. With vegetative reproduction one can only transmit messages of a hereditary nature and thus it will not be possible to regenerate plants and allow them to reveal anything new. Working this way can only repeat what has come from the past and that is no longer suitable to support the evolution of modern people. Among other things, we cannot overlook the fact that vegetative multiplication inevitably leads to a weakening of the vital forces and thus to increasingly weak plants. Some of our plants, such as potatoes, have now been vegetatively multiplied for centuries and this cannot fail to make us reflect on the quality of those forces that potatoes are able to bring. Only through the seed do we have the possibility to renew the lives of plants and thus, for us, attention to seeds assumes increasing importance. We have expended great energy to create seed baths and we continue to stress their importance. Self-grown seed is the most significant asset that any farm could have

We stress once more the importance of seed as a synthesis of the plant world and the chaos that together direct the revitalisation process. Bearing in mind the shared etymology of the words *chaos*, *crisis*, and *Christ*. Chaos is the basis for any real change. Moreover, chaos is disorganised only when seen from our viewpoint. However, chaos is the order that governs the cosmic distances. It is no coincidence that chaos theoreticians have been studying the laws of chaos and fractals for the last fifteen years. We can observe that if chaos is governed by laws it is far from random and disorganised, even if it seems not to be reflected in the laws of physics. The most chaotic movement we know is that of the vortex and we know that the vortex is an opportunity for the descent of new Life from the cosmos to Earth. It is no coincidence that the biodynamic preparations are activated with a dynamisation that consists in the formation, for a specific period, of a succession of alternating vortices. Equally the internal shape of the *Diffusor*[17] is not arbitrary. It exactly replicates the vortex and draws inspiration from the proportions of the orbits of the planets and their average distance from Earth. The same proportions are found in the maternal uterus which

17 A device for spreading the homeodynamic preparations without having to spray.

undoubtedly has a strong connection with Life! In other words we can say that although the vortex appears chaotic, it is a tool for communication with the cosmos following very strict laws.

Clearly this discussion about chaos in the seed could expand to include all the issues relating to the choice of the time of sowing to determine what forces should be encouraged to form the new plant, or to those relating to adequate preparation of the seed always with an eye to the improvement and regeneration of the plant etc ... but this would lead us too far from the issue that we are addressing now.

We should always keep in mind that any real change must pass through a period of chaos. To try to grasp this idea we can refer to a common experience such as climbing a flight of stairs. There is no doubt that the possibility of going upwards is linked to the existence of the steps, but in the construction of staircases the builders also provide landings and the function of landings is irreplaceable: without them it would be an impractical succession of various flights thanks to which the climb is achieved. For if after the first flight of stairs there was no opportunity to reverse the orientation of the climb we could not even build the staircase, since space inside a building is limited. But the landing is the place of chaos which enables the change of direction so one can change ones point of view, to prepare for a new stage of ascent.

Before moving to the next paragraph try to grasp another image of the seed in which we can picture the seed as a single geometrically dimensionless point which allows connection with the infinite. This allows us to emphasise once more the difference between the vastness of possible links when working on the seeds compared to the limited options of vegetative reproduction.

With the thirty-fourth paragraph ("*Let's take a look now ...*") we are practically back to the first and second lectures. Steiner puts carbon as a centre and says that carbon can team up with limestone or silica giving rise to the manifestation either of siliceous or limestone types of plants. In practice what has been proposed is a summary of what has already been brought to light but with the additional wealth of knowledge acquired in the meantime.

We can then develop a final reflection on this lecture, especially regarding the protein process and its four main constituents.

In its evolution, the Earth has passed through a phase in which it was completely protein. It was a living pulsating organism just like one immense protein. The surrounding atmosphere was also of the nature of protein like a kind of thin albumin. The living forms 'breathed' this airey living albumin. The plants also lived in this milieu and reached the planet Earth only after the Fall of Man from the earthly Paradise. The first plants that lived in this protein atmosphere were essentially all 'flower'. Of course, all this has nothing to do with their current forms.

In its 'descent' to the Earth the flower-plants gradually developed their leaves, stem and finally their roots. The plant is a being descended from heaven to Earth and brings with it, as we know, the four aspects of protein: Fire-hydrogen-fruit, Air-nitrogen-flower, Water-oxygen-leaf, and Earth-carbon-root. In its actual expression of life, the plant 'goes back' from Earth to heaven and this gives it its virginal character.

What we have offered should help us to understand why Steiner looked at the plant through the process of protein: he captures the event through the sphere of the elements. In the previous lecture the plant was more 'cosmic' and in a certain sense the plant was still living in the cosmos. It is no coincidence that Steiner talked for so long about the planets.

Eleventh Meeting

We concluded the previous meeting discussing protein made of carbon, oxygen, nitrogen, hydrogen and sulphur. We have already seen how these first 4 chemical elements of protein are connected with the four elements of Aristotelian philosophy. Steiner does not make an explicit link with the old Greek civilisation but it can be inferred from a careful examination of the structure of the Koberwitz course. Nor is it clear, in a still broader vision, that there is a connection between this lecture and a particular phase within the incarnation process of our planet. Actually, in the current fourth stage of manifestation of the Earth, there was a period during which all the living world seemed like one enormous protein that then divided and multiplied to give rise to all the forms of life we still see. In this context we are mainly interested in the plant life that still bears within itself, as a record, the result of carbon in its form, the action of oxygen in its life, the influence of nitrogen as bearer of the cosmic images and, finally, the role of hydrogen in its maturation seen as a process of drawing back to its origin and to death.

Certainly the idea of the Earth as a single matrix of protein - and therefore of all life-forms - is rather unusual, but in some of his early lectures Steiner referred to some creatures on Earth - in particular he speaks of shellfish - big enough to cover the entire surface of France. Evidently Steiner refers to the period in which the single protein matrix we mentioned had recently begun to differentiate and multiply. The Moon had only recently detached from Earth which still had the forces that tended to create these gigantic forms of life. These beings progressively diminish in size until becoming the size with which we are familiar. These animals, however, have not left any fossil record for us because fossils are the remnants of much firmer animals with a limestone structure whose consistency is much closer to those of current creatures. It is also pertinent that our whole planet has gone through phases of different consistencies from today. If we now see dinosaur prints in some rocks it suggests that these animals were moving on soft surfaces that could receive such imprints.

Modern plants are daughters of that primordial protein and reproduce the four vital functions. Remember that for an individualised life to exist there must be a structure formed in harmony with a cosmic image and this presupposes assistance from a carbon framework. In this sense we can say that the plant is the image of a planet or that the man is the image of the zodiac.

Note also that the term *form* can be properly understood as the result of the influence of cosmic forces capable of shaping material. In this sense form should not be understood as the impression of a rigid mould on any material, but is the result of a dynamic and active influence that shapes the matter from the outside like a hidden sculptor. Into this sculpted matter we can think that life has been blown in - to paraphrase the Bible. In terms more appropriate to our way of looking, the breath going into the material carries the life of oxygen. It is certainly not a coincidence that the plant breathes in CO_2 that is carbon and oxygen.

All this must be continuously transformed in the course of life and therefore needs continuous feeding by the cosmic images, via nitrogen. Life needs to be fed continuously or it is destined to degenerate. If we put a plant under a powerful electromagnetic field or subject it to strong nuclear radiation, in order to inhibit a proper conversation with the cosmic distances, we can observe how the plant becomes deformed. We believe that everyone can remember the deformed plants after the

Chernobyl disaster.

Finally, thanks to hydrogen, the mature plant-being can disappear only to return to manifestation dressed in new material.

In this lecture the plant is presented as linked to the cosmic Life in a way that could become very practical if we were only able to penetrate it with sufficient judgment.

Besides, the title of this lecture is "*Parenthesis on nature: the influence of the spirit in nature*," and its contents can be a bridge between the life of the plant and the diffuse Life we have called *Zoe*, which is none other than the primordial Life. In other words, this lecture should give us the means to achieve a form of agriculture able to reconnect the plant to the sources of its life.

We are now at the thirty-fifth paragraph ("*Lime and silica, however ...*") of the third Koberwitz lecture. The previous paragraph ended with the statement that: "*In human beings and in animals, carbon is not the sole determining factor; it builds on the formative activities of lime and sulphur.*"

After the first brief sentence of this paragraph comes a second sentence that seems insignificant at first sight but that hides a huge amount of information. First the [Italian] sentence begins with the word *dobbiamo* or '*we must*', a verb very rarely used by Steiner, and this stresses the importance Rudolf Steiner puts on this work. He speaks of the three systems of the human organism - the digestive, respiratory and circulatory arising from three embryonic layers, ie from the influence of the Sun and the external planets upon us.

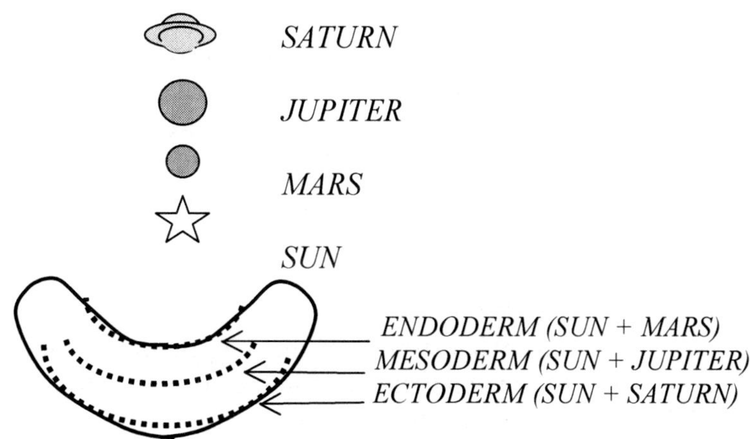

In other words Steiner is inviting us to study how the Sun works with Mars, Jupiter and Saturn, warning that we cannot do without developing such knowledge. Remember that the planets that we have mentioned bring the influence of spiritual beings like the *Powers, Virtues, Dominions* and *Thrones*. Therefore, what is being proposed is a very elevated model of agriculture, much more so than that to which organic farming aspires - which essentially limits itself to the lunar sphere.

Steiner then recalls the role of carbon and we have often stated that carbon can appear in several guises: as coal typical of the Earth, or as graphite as present on the old Moon, or as diamond as it appeared on the old Sun. This triple manifestation leads us to the threshold of the Sun, but we are invited to go further on the path of inner transformation and to consider the relationship between the Sun and the outer planets. From spiritual science we also learn that the combined strength of the Sun and the outer planets, in addition to forming the three embryonic layers, governs the formation of the larynx when combined with the forces of Mars; combined with the force of Jupiter it finds expression in the forehead as localisation of thought, and finally combined with Saturn is expressed in the top of the head where is to be found the chakra with a thousand petals, a symbol of higher knowledge. In practice Rudolf Steiner is pointing to a way for the completion of inner transformation and to connect with Zoe. At this point we can become practical, because the connection with Zoe enables us to operate with extraordinary effectiveness on the lower planes of life.

This sentence from Rudolf Steiner initiated all the work conducted in our February conference that was entitled: "*From the Healing of Raphael to the Healing of Michael*"[18]. The way of Michael may also become practical and can be applied.

Do not forget that Michael's way was defined as the 'dry path' (*via secca*) as opposed to 'wet path' of Raphael (*via humida*) that was followed by alchemists. The Michaelic route connects directly to the cosmic forces.

In the third short sentence of this paragraph Steiner hints at a way to observe nature when he speaks of so-called *insight.* He says that before observing the details we must grasp the context and only afterwards go into the details. Moreover, observation of the world of plants and of the soil on which it grows can enable understanding of aspects of life that we can hardly glean from the other kingdoms. Using the example from a later sentence, it is not possible in humans to see how "*oxygen is caught up by nitrogen and carried down into the carbon – that is, carbon insofar as it is supported by lime and silica*". We could also say that Steiner wants to guide our observation so that we might grasp the way in which life is integrated into its physical support, thus making the material substance flow.

The last sentence of the paragraph reiterates that what lives in the space surrounding the earth and is thus connected to oxygen, must be able to penetrate the soil with the help of nitrogen, to rely on silica, but taking the form from limestone. Actually, the contractive action of limestone drags everything into death if it is not balanced by silica with its ability to maintain links with the periphery.

In the thirty-sixth paragraph ("*If we have any ...*") there are a few stimuli to new ideas. For example, we may ask what stimulates the oxygen and the nitrogen when they are breathed into us. The oxygen that enters into us becomes living and strengthens our biological life, while nitrogen brings in the cosmic images which are then held back from the bones. Through nitrogen we are continually remodeled, because every cell that is born within us is built in accordance with the cosmic 'images' brought within us with every breath. If our cells were modeled according to the laws brought by oxygen instead of the images brought by nitrogen, they would be cancerous.

At this point it is interesting to recall that the healthy respiratory rhythm is 18 cycles per minute. This allows us to calculate that an average breath lasts 3.33 seconds. We all know that this is the number linked to the life of Christ and gives us the opportunity to confirm a link with cosmic Life - Zoe.

In this regard Steiner, in his "Astronomy Course ", stresses the importance of the

---

[18] Available in English from 2009.

link between the individualised life-forms and this cosmic Life, thanks to which each of us can say that we are an image of the cosmos. This teaches that the best way to study life is not just to look at the cell but to study the cosmos.

A similar instruction is contained in the 'principle of correspondence' formulated by Hermes Trismegistos in ancient Egypt and expressed in the sentence: "*As in the heights, so in the depths; as above, so below.*"

In the second part of the paragraph Steiner proposes knowledge of the *leguminosae* as organs of respiration and compares them to our lungs during inspiration. The expiration phase, in this view, is all the other plants found in nature. The Earth interacts with the universe through other channels: we should not forget what we said about metals and minerals. There is also an important connection with the cosmos through the animals. The insects are organs of perception towards the cosmos in particular, while the other animals can be seen as microcosms on Earth. In this understanding humans are very important because we are free to choose to arrange for the interactions to happen in the best way possible, perhaps by restoring interrupted contact ... or we can ignore the whole issue. It is only a matter of awareness and conscience. Unfortunately modern farmers usually establish their activities on material assumptions and utilitarianism, even forgetting that there is a Father in heaven. But precisely for this reason those who have the gift of being able to recognise the spiritual reality behind the appearance of the physical world must engage with faith and with all their forces, so that at least their piece of land can flow with the abundant forces of the universe to the benefit of the whole Earth and humanity.

Proper understanding of the way through which plants and animals can communicate with the cosmos is necessary because it allows one to establish the full agricultural organism which consists of all the organs needed to establish an effective relationship with the planets and constellations.

In this way we arrive at the thirty-seventh paragraph ("*Our task then is to ...*") in which it is pointed out that in order to set up a farm organism properly one needs to learn to recognise the being of every plant. Then each can be placed in the right place in the same way that every human organ must be in place to form the human organism. Note how Steiner does not speak of the need to know plants botanically, but speaks of the being of the plants, ie their spiritual matrix that we have sometimes characterised as the idea that embodies the plant.

This paragraph also stressed that knowledge really has to include a 'spiritual substratum', otherwise we will inevitably slip towards a gradual loss of knowledge. Criteria that are used to gauge individual problems without being embedded in an overall vision lead agriculture on a false path. Unfortunately this has already happened. It very often happens that we see properties that are totally missing some vital organs and this is sometimes the case even when the farmer shows a certain sensitivity to the ecology and the natural balance. Of course, as foreseen by Rudolf Steiner, we have lost the right knowledge and we have taken a wrong path. It is clear that despite the good intentions such an organism can no longer exercise its true function which, remember, does not equal the sum of the functions of the various organs, but includes a higher function which is to 'commune' with the cosmos. To further clarify the concept we can make an example of a human being, which we know is composed of a multiplicity of organs such as the lungs, stomach, kidneys, liver and so forth. But nobody would dream of considering a person as the sum of the functions of their organs. In fact a person expresses their uniqueness in the ability to think and to be conscious of themself. Whilst this higher level may be fully exercised only if all the organs of the body work reasonably, it is not attributable to any organ in particular.

Regarding our work we can broaden the concept that we are trying to clarify to all branches of our activities and we can then try to establish a link with Zoe whether we are building a home, preparing food or manufacturing fabrics for making our clothes. (It is not always possible to achieve all this and therefore we insist on trying to create at least one possible contact with Zoe every day. The easiest way to do this is to try to swallow at least one mouthful of adequately prepared and blessed food, or to drink water that has connected with Zoe through a particular form[19].)

In the thirty-eighth paragraph ("*You can see how* ...") Steiner describes the action of the *leguminosae* by highlighting the fact that these plants have the ability to channel the forces linked to nitrogen to the ground far more than any other plant species. This observation deserves to be filled out with the assistance of a drawing.

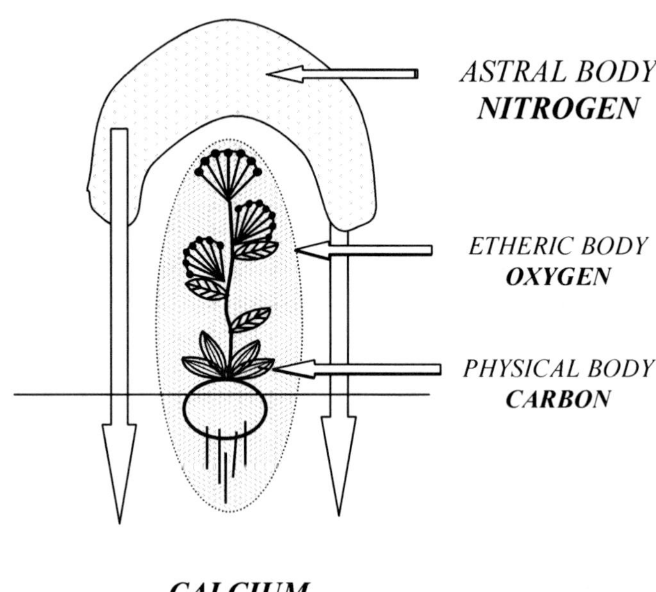

We sketch our plant in the usual way and above it we can identify the zone of action of the astral, of nitrogen. Lower down we can identify the zone where the etheric works, which we know is related to oxygen. In the lower part of the drawing we can represent the ground with a horizontal line which we can identify with the physical world linked to carbon. The *leguminosae*, like all plants is penetrated by the ethereal, but has the characteristic of bringing down the astral plane, which we identify with nitrogen. We know the fact that the *leguminosae* grow very well on 'physical' ground and thus on limestone and this leads us to write 'calcium' on the ground. Finally, with two arrows we can indicate the soil influence mentioned above, bringing the astral to the ground which - being calcareous - exerts a significant downwards pull, accentuating the *leguminosae's* characteristic properties.

---

[19] See the vitalisor from l'Albero della Vita

The astral power that penetrates through the plant right into the soil is an anticipation of the process of formation of the flower and then the fruit which appears in several places below the apical area of the plant. In *leguminosae* we can find further leaf production after the first flowering, followed by yet another flowering and leaf, and so on. This phenomenon is typical of all the plants that produce alkaloids. Legumes also contain alkaloids. For this reason Pythagoras actually forbade his disciples to eat beans to avoid blocking their route to initiation. Legumes are related to limestone and thus bring low quality astrality. Therefore they are contraindicated for anyone who is working on inner purification of their astral nature with a view to achieving their higher spiritual goals.

There is an interesting tradition of well-wishing according to which one eats lentils on New Year's Eve. Unlike other legumes that have a shape similar to that of a human kidney - the typical form of the astral - lentils have a circular shape betraying the presence of solar forces. It is also noteworthy that lentils are soaked before being cooked and consumed. On the physical level this soaking only facilitates cooking, but at a different level it has a much deeper meaning because immersion in water symbolises purification and allows one to understand that the preparation of lentils for the New Year symbolises the hope of transformation of the astral body.

In the second paragraph we can read that the legumes "*have a tendency to wait for winter; they would actually like to wait for winter with what they develop.* " In other words we can say that the plants in question tend to wait when the cosmic images descend from heaven because at that time their work is facilitated. From this point of view the function of the *leguminosae* takes on a truly significant importance on our farms because their role consists in bringing cosmic images into the ground. This translates into opportunities for other plants like nettle, cherry or peach, to continue to be nourished each year because the forces of their spiritual matrix can find its reflected image on Earth due to this living nitrogen. Unfortunately in our times the descent of the cosmic images is extremely hampered by manmade pollution - particularly electromagnetic – and therefore it needs our commitment to enable this descent to a sufficient degree to animate nature.

Of course legumes capture images up to the cosmic planetary level, and so that the Earth can also receive cosmic images from the zodiac it is necessary that our farm also uses yarrow which has the ability to establish a bridge between the Earth and the more distant cosmos. The yarrow is an *umbelliferae* and thus, even in terms of its form, is the image of the starry sky. It is very rich in potassium and sulphur that have opposed functions: the first is centrifugal, the second expansive. It is as if potassium leads to the soil those things that sulphur brings from the cosmos. (We will not now mention the critical function of metals and minerals because it would lead us too far from our current subject.)

It can be interesting at this point to consider why some legumes fix nitrogen in a short time, such as clover, and others including alfalfa need a much greater period of time. We can help answer this question by observation of the roots of the various species that we have mentioned. Clover has a fibrous root (terrestrial), while alfalfa has a tap root (cosmic). Now if a legume's ability to fix nitrogen depends in large part on the contractive influence of the ground, a plant with an earthly root is in "sympathy" with limestone and can quickly begin its operations. On the other hand a plant with cosmic roots will need more time and should also wait for the time when the forces descending from heaven are abundant until it starts the process of nitrogen-fixation.

From what we have said the importance becomes clear of the pulses being able to discern the forces that descend during the Holy Nights. Of course, the simplest way to do this consists in having the *leguminosae* in the ground at that period. It is relatively simple for alfalfa which remains on the ground for many years, but it is much more difficult for other legumes such as beans or peas. In these cases we would like to suggest inserting the legume seeds in a horn and burying them on the first holy night as when creating a biodynamic preparation. Do not forget that we have a preparation which is called 'Pro-nitrogen fixing' which we have made for the function we are describing, which could also be used as a seed bath.[20]

The thirty-ninth paragraph ("*That is an example ...*") states that the greedy limestone has an extraordinary affinity with the world of human desires and therefore tends to attract itself first to oxygen, and with it life, and then the nitrogen with the astral. Steiner uses suggestive terms: *satiated, restless, longing, instinctual craving*, etc. The greedy limestone mainly attracts the low-quality astrality both in the soil and within us. In fact our bones attract the lowest desires and are the seat of hatred and that can't really be considered to be a very positive sentiment. The idea is of a selfishness that expresses itself in greed, which seeks only inner pleasure so that self-satisfaction becomes an end in itself.

What we have said is also important for buildings because we might consider that the limestone with which we make our buildings first 'sucks' the vitality and then the astrality, creating an environment of low quality astrality. This information is crucial because once we know it we can find ways to deal with the adverse situations in which we find ourselves.[21]

The remainder of the paragraph says that the limestone is not content just to absorb life but "*wants to combine with any kind of metallic acid, even bitumen, which is not even mineral anymore*". We believe that with the words " *which is not even mineral anymore,*" Steiner was alluding to *sub-nature* - the forces opposed to life - in the sense that limestone accumulates the forces of sub-nature. This sentence may seem a dire sentence for those who live on a calcareous soil, but if the limestone is enlivened and transformed into humus, the limestone is ennobled and may become a preserver of life.

Do not forget that the tomb of Jesus, from whence he began his resurrection, was carved into limestone. We have often said that limestone is a fallen Sun that may rise again. Naturally because there is the possibility of resurrection it is not enough to leave such things alone, but we must activate the process by which the Sun can shine in the darkness again. To do this we were given the preparations, linked with the forces of the risen Christ by their interment.

The discussion continues in the fortieth paragraph ("*This noble substance - ...*") in which we are presented with silica as the noble element that has no more desire and is simply resting within itself. We are told that silica is not only the familiar mineral that we call by that name, but is also ubiquitous in a homeopathic state. Silica is the external sensory organ of the Earth, while calcium is its longings. We have finally revived the use of clay as a link between the limestone and silica poles.

To put it more simply, we can say that the activities of silica can be summarised in the word '*give*', while limestone is encapsulated in the word '*take*'. But if we go back

---

[20] A separate seed bath has been created since this meeting and is available to l'Albero della Vita members.

[21] Archtectural works are available – see appendix

to what we said, paraphrasing Steiner, that the Earth was once a single large protein, we can understand that with the gradual hardening of its constituent substances it began to materialise along the lines proposed by Steiner. Some parts moved towards silica with its expansive force, the other towards limestone with its contractive strength, and in the middle clay emerged.

The Earth is built in this way. To the east is silica in the 5,000 km Himalayan chain, and to the west there is the 8,000 kilometer limestone chains called the Andes and the Rocky Mountains. In between is the clay, Europe. The whole Earth is organised in this way both in general and in its particulars. We can imagine the Earth as a big plant whose roots are in the Americas and whose flowers are in Indonesia.

But we also find this theme in the individual continents; in the USA for example there are three peninsulas: California, Florida and Mexico, whose form is a repetition of the above-mentioned model. California has a rectangular form, Florida has a more open gesture (out to the Caribbean where the expansive form is even more obvious), and Mexico has an intermediate triangular shape. Even in Asia the same model is apparent with the rectangular Arabia, Indonesia formed by an explosion of islands and the triangular India in the middle.

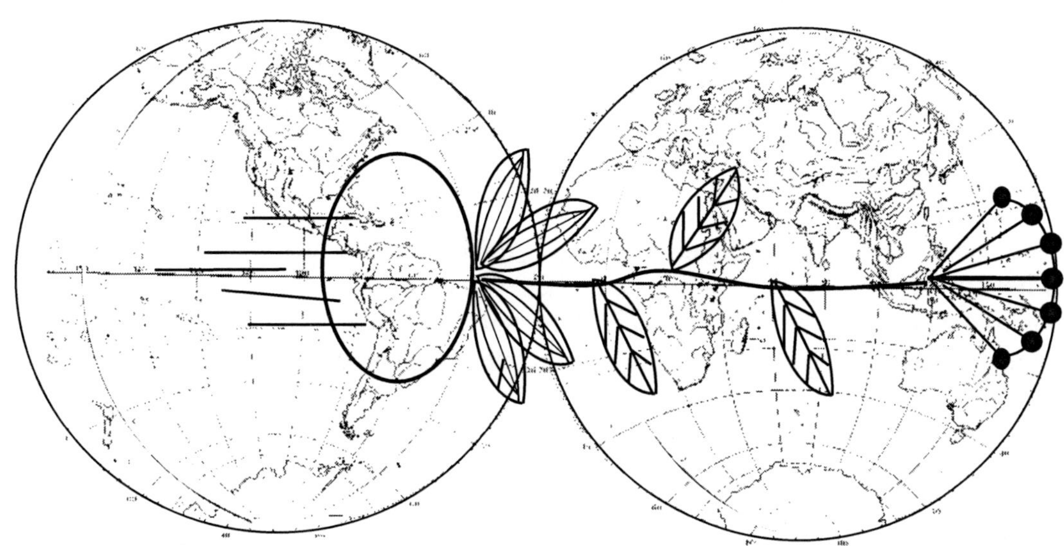

If we move our focus to Europe we find the square Spain, Greece with its expansive gesture (open hand) and Italy with its triangular Sicily in the middle. We can imagine that the expansive forces of Europe (silica) tend to radiate westwards, as the contractive Western forces (limestone) tend to draw things to themselves, making an impulse from east to west that is manifest in all the seas in the world, the formation of many islands on the west coasts, and smooth well delineated contours on the east side. But all this has already been discussed several times and we should not pursue this sidetrack further.

We shall move on to the forty-first paragraph ("*We need to come ...*"). In the first sentence Steiner speaks of "*sensitive knowing*" or of a knowledge connected to the laws of the universe. The sentient soul is the part of the soul that connects to the spirit in a way that is normally unconscious. Once again Steiner says '*we should*' to underline the fundamental nature of this kind of knowledge.

Then Steiner speaks again of limestone and silica, but this time uses the verb *to experience*: "*We should be able to experience lime as a creature of many desires .... Silica as a noble aristocrat ...* " He brings to mind the inner soul elements of the alchemists who experienced the processes of nature within themselves and from these profound experiences they forged the capacity of transmuting matter. Steiner is inviting us to undergo similar experiences.

In the second part of this paragraph silica is presented as a force capable of liberating what the limestone has captured - life. Clearly, the purpose of this liberation is to allow life to flow into the plant world.

The forty-second paragraph ("*Carbon serves as the ...*") describes the difficulties encountered by carbon in the course of terrestrial evolution. The second sentence is one way of describing the fall from the earthly paradise. Steiner says that carbon could form all the plants if there was water under them. This is because the plant has a living form and without water there can be no life. We have already seen how the water is composed of hydrogen and oxygen and is, therefore, the basis of biological life because it contains oxygen, and spiritual life because it contains hydrogen. Water is used as a drink, but also as the stuff of the sacrament of baptism. However, biological life would not be possible if life did not flow from the spiritual level, so water becomes an essential bridge without which it would not be possible to find any life forms. All of this would work without problems if there weren't any limestone to disturb the vital processes with its insatiable greed that leads it to absorb water. This is why carbon must work with silica and even clay to overcome the resistance of limestone and bring the forms to nature.

In the forty-third paragraph ("*So what is it like ...*") nitrogen is inserted in this whole process as the bearer of cosmic images that performs the function of regulating the interaction between the spiritual world and the ethereal. The fourth lecture of Steiner's agriculture course shows how this nitrogen can enter into the plant world in the proper way.

At the very end of this lecture Rudolf Steiner mentions cosmic and terrestrial forms again. We should be able, at this point, to find the opportunity to work practically.

We can sketch the square of protein again that we will supplement with the label of limestone above the side between carbon and oxygen – between the physical world and the etheric. Of course, on the opposite side we will put silica and clay in the middle.

Between the limestone that wants to grab the plant and restrain it and the silica that tends to make it slender and fibrous, we find the action of carbon mediating between the opposing forces and giving our plants their final form. It is important to note that limestone and silica represent processes, while the carbon - in as much as it is a materialising substance - must be seen as the representative of an element, in particular of the Earth element. Now every element brings to the Earth the force of a Zodiacal substance that is then materialised as a static form, and to express themselves in a living way they must be inserted into a process.

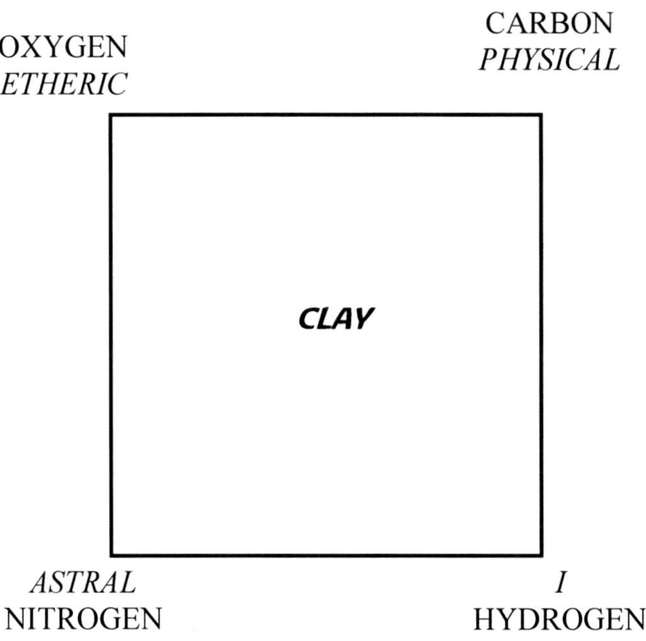

At many other times we have clarified the difference between matter and substance. We said that substance is the essence of the matter which, in order to manifest itself on the physical plane, has to materialise as a chemical element or compound. Therefore, when we speak of carbon or oxygen we refer to the physical manifestation of substances that find their origin respectively in the Scorpion and in the Water-bearer just as - to mention the other components of protein - the substance nitrogen is of the Bull, hydrogen is tied to the Lion and sulphur to the Twins.

At this point Steiner says that just as the astral body brings order between our I and our etheric body, in the same way nitrogen intervenes between limestone and silica. It seems as if the two references to carbon and nitrogen overlap. In fact we consider that it is useful to point out that if we consider only the appearance of the chemical elements, the cosmic images flowing in nitrogen are taken up by carbon and locked in a rigid form. But if the world of elements is inserted in the resulting processes and enlivened, the flow of cosmic images allows continuous re-sculpting of physical forms, because the forms are pervaded with life.

The journey of the conquest of the Philosopher's Stone, on which Steiner insists in his agriculture course, can be achieved in another way than through the sketch described above (following the square of the elements or of Raphael). This other way allows us to reach our goal directly by the *dry way* as described in the February conference as the way of Michael.

At that time we described the processes of life that takes place in a clockwise direction from *Fire* to the *Earth* element, or from hydrogen to carbon. Having come to the end (carbon) it can lead to death, or to rebirth directly to hydrogen by passing through a higher level of existence. This step is the Michaelic initiation that in

Anthroposophical terminology is described as coming to the door of the Sun.

All this can easily be represented in our sketch. If the limestone governs the elements Earth and Water and silica governs the elements Air and Fire, the two sides between Water and Air and between Earth and Fire are governed by clay that we have defined repeatedly as a solar force in our meetings.

On our square of the elements, the Michaelic route is located to the east, close to the Earth side-Fire. In this context we are most interested in the Sun in its role as a source of strength, as the bearer of Christic forces of which Michael is the countenance.

In the square of Michael we have positioned nine substances and the first vertical line (on the left) corresponds to the plant world. The practical aspect of this reasoning is the conclusion that the domestication of a plant can be very fast (direct) by spraying the seed with the three substances on the line relating to the plant world.

All can potentially make this qualitative leap and the first task before us is to remake the path of domestication of plants first laid 7,000 years ago by Zarathustra. We can now use the new knowledge we have. The technical work is extremely simple, but whoever wishes to perform this work must have made a profound transformation of their own soul. Without this, results are unlikely.

Fourth Koberwitz Lecture

Twelfth Meeting

In our meeting today we will begin to comment upon the central part of the Koberwitz course. Of Rudolf Steiner's eight lectures that make up the agriculture course the fourth and fifth can be regarded as its heart.

We have often asserted that the first Koberwitz lecture corresponds to the ancient Persian period, the second to Egyptian agriculture, and the third to the Greco-Roman period. It follows that these two central lectures relate to the modern era and so we could think therefore that in them Steiner will describe the agriculture appropriate to the present. In these lectures Steiner presents us with the preparations, although in this meeting we will only have time to comment on the initial part of the fourth lecture, and it is therefore not appropriate to speak about the preparations today.

The fourth lecture is entitled, "*Forces and substances that penetrate from the spiritual sphere: the problem of Manuring*". For those who do not know the preparations and Rudolf Steiner's vision of the world, this title is an enigma that we will try to solve as we read this lecture.

In opening the lecture, in the very first sentence ("*We have seen that ...*"), Steiner gives a practical lesson in methodology by suggesting a system for dealing with whatever problems life presents to us - whether or not they are linked to agriculture. This method consists in lifting our focus from around the problematic situation until we are able to grasp the context in which it is embedded. This makes it possible to understand the complete system of forces that can act and the way in which they are actually acting to produce the unbalanced situation that we want to bring into harmony. If we consider an agricultural property in this way it can be regarded as a cell within a larger body from which it cannot be considered as separate; the conditions of the wider area - such as climate, soil type, orography and others - are key components upon which to base our rebalancing activities. Whoever wants to consider a farm without considering the context in which it is placed would commit the same mistake as those who wanted to investigate the health of a cell or group of cells without any relationship to the state of health of the organ of which they are a part, and to the health of the body to which they belong.

Unfortunately modern science increasingly addresses issues in a materialistic fashion, and so tends to commit the error of evaluating its problems via extremes of analysis, and runs the risk of losing the sense of the overall problem. The example given by Steiner is of those who hope to understand the nature of the human being by minute examination of a little finger or earlobe.

It is true that when one is dealing with a farm we are dealing with a much larger reality, but even this is framed within a broader context. It is therefore imperative to have clarified all the elements necessary to make a diagnosis before focusing all ones attention, for example, upon a single sick leaf. One must observe the broader environment surrounding the farm, then observe the farm itself, and only afterwards evaluate the individual problem: the cause of the leaf disease is never only in that leaf. Our investigation should not even neglect items that may seem irrelevant such as the name of the place where the farm is located. Often that specific name indicates a general characteristic of the surroundings that can be crucial. For example, the place where we are is called Monselice and obviously we are at the foot of a mountain composed of sandstone. But if the whole area was formed by siliceous rocks there would have been no need to characterise this particular place in this way, then we can infer that the Euganei hills are probably mainly of limestone and this mountain differs markedly from its neighbours. We could thus consider three different environments: the mountains predominantly of silica and therefore an area dominated by cosmic forces, the plain characterised by limestone and thus predominantly subject to earthly-lunar forces, and the foothills of the mountain which is the area of dialogue between the mountains and the plain, the area where we have the greatest wealth of vital forces. Following what we have said we can identify environments we might demark as those of *sulfur*, *sal* and *mercur*. This is a view that we could not neglect if we had come to make a diagnosis of a farm.

Steiner concludes the paragraph by noting that using the method he has described, we need to build up a real science that deals with the universal relationships.

In the second paragraph ("*Just think of how often ...*") Steiner shows how a science that bases its inquiries upon the particular details, often gives partial answers that it is forced to retract or reappraise following the emergence of further detail. The example that is used belongs to the field of nutrition and refers to a theory, supported by the 'secure' contemporary scientific evidence, according to which a man weighing seventy or seventy five kilograms would need to eat one hundred and twenty grams of protein each day. These statements were not open to objections because they were a scientifically proven fact. Too bad that after only a few years the scientific facts were disproved and the amount of protein needed for the same person became 50 grams per day. Scholars today are arguing that 30 grams of daily protein keep a person in good health. Evidently the problem had been tackled without all the necessary knowledge. Anyone following the directions of science from the beginning of the century would suffer a stretch of negative consequences because it is known that excessive intake of protein causes the formation of toxic intermediates within the intestines.

We note that Steiner used an example that relates to the field of human nutrition but, as is usually the case, he simultaneously brings some important indications for nourishing the soil - for fertilisation. We are saying that excess protein in the diet is the main source of arteriosclerosis in the mature human and that something analogous is true for plants. Over-feeding - which can be translated as excessive artificial nitrogen fertilisation - carries similar consequences. Atherosclerosis in humans is the daughter of an excess activity of the current of forces that starts from the head and works down. If we bear in mind that the plant is like a man flipped over, these 'nerve currents' translate as the impulse of ascending sap. Being *Water,* which normally brings vigour but carrying minerals, it also brings the I-activity of the plant. Artificial nitrogen fertilisation enhances the astral sphere at the expense of the etheric and the relationship with the group I.

We already know that whenever the etheric or I forces are reduced the astral is dominant and is the origin of all diseases. The age of maturity for a human - that is the age at which sclerosis appears – has its equivalent in the plant in the transition from the vegetative to the flowering phase when vitality is reduced. It is at this stage that diseases appear resulting from this imbalance.

The third paragraph ("*Spiritual science is ...*") reaffirms that spiritual science is not limited to the observation of physical forces and gross material substances, but also considers forces and substances that resonate with and are penetrated by the spiritual world, and investigates these broader connections of life.

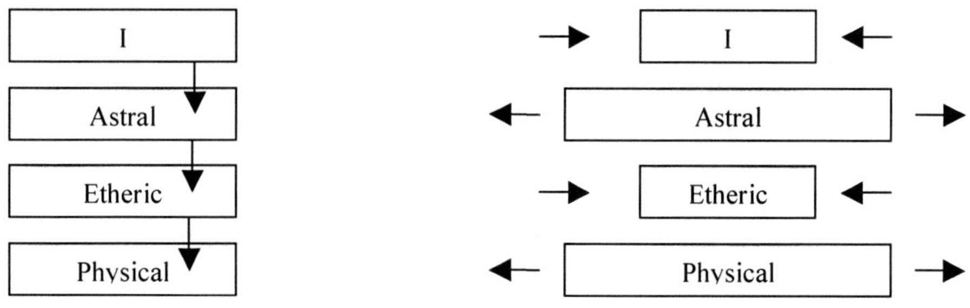

Normal equilibrium in a person

Excess astrality

In paragraphs four ("*Even the way the* ...") and five ("*People have believed that* ...") Steiner continues to develop the parallel between fertilisation and food and highlights the fundamental mistake that science makes when addressing itself to nutrition. The error is always in considering only the material aspects. The same is the case when considering feeding the soil. The scientific model considers fertiliser only as a source of nutrients for plants completely forgetting the forces of which the substances are the vehicle. Moreover, the food that is eaten every day is regarded as the sole basis of human nutrition.

We know that our nutrition takes two different paths that we could call gross (or terrestrial) and subtle (or cosmic). Science not only completely ignores the existence of the second, it also does not even give the first adequate consideration. Whatever we ingest, if only considered on the physical-chemical plane, is mostly re-expelled and plays no role in building up our organism. Food serves mainly as a vehicle for life-forces to get into us. The physical component assumes a different significance. Elsewhere we have already had occasion to study how our vital body can be nourished only by the vital component of foods, in a process that may seem strange at first. When we eat food of high quality it actually has its own strong etheric body and our etheric body is forced to engage deeply to overcome it. It is this great effort that strengthens our etheric body. It should be remembered that the laws governing the etheric plane are opposite to those of the physical plane: demanding physical work consumes forces whilst the exact opposite happens for vital forces. So the more vital the food we eat the more must our etheric body struggle to destroy it and the etheric body is reinforced in proportion to that struggle. A strong etheric body is able to support all life processes and the capacity (will) to perform physical activities and internal movements (such as circulation).

So we come to the sixth paragraph ("*On the other hand* ...") in which Steiner speaks about the cosmic nutrition stream that enters us through our sense organs and our respiration. The fact is that what is introduced through the mouth - the gross nutrition - is eliminated through the intestine, and what enters through the senses is eliminated by the growth of hair, nail, skin loss and breathing. It is also pointed out that what goes into our stomachs, if it has sufficient vitality, functions as a fuel to develop the will to act in the body.

Regarding this cosmic nutrition we can say, even if it seems curious, that our body is built with the forces and substances that the senses capture from outside and which are then distributed in our body, together with the overcoming of the vital forces of our physical food, through a descendant current.

In manuring the problem is similar, in that it is not so important to make large quantities of organic matter, but to make organic matter that is rich in vital forces. If the fertiliser is vital, it is also likely to put all the metabolic processes related to the will in motion. We know that any agricultural soils have a certain amount of humus that represents, in its formation, the initial phase of a process of vegetalisation affecting the whole Earth. Currently the Earth can realise her desire to vegetalise only through the seed, which can then create a single living plant. We can intuit that if the soil is rich in vital forces it can effectively transfer its willingness to vegetalise to be individualised in plants. But if these vital forces are insufficient, the plants will not only be weak but devitalised.

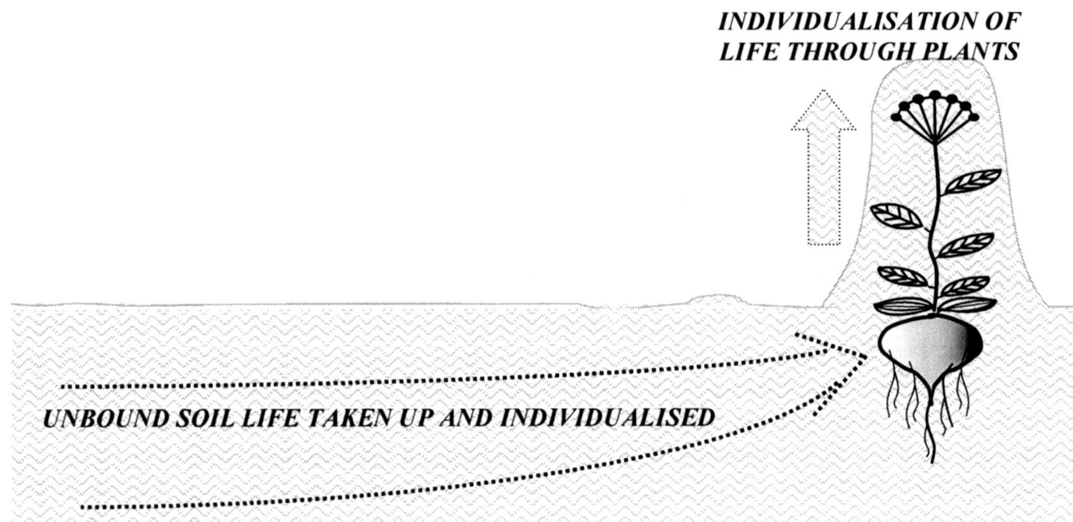

It may be interesting to investigate another nuance in the way Rudolf Steiner has chosen examples in this lecture. He has suggested a parallel between human nutrition and manuring, but as nutrition is destined for a living man complete with all his organs, for parity he should also have compared it to a soil that is part of a functioning farm organism. Otherwise the vital elements could not be properly exploited. Besides the title of this lecture says '*forces and substances*', and we must not forget that the substances are linked to the world of the environment and only a living environment can properly receive and use the forces that the substances can bring in the soil.

Continuing to work upon the examples to find themes that interest us, the upper parts of the plant do not create their substance from the gross nutritional fraction (and thus out of what the plant draws from manure), but from what the plant takes as subtle nutrition - from the cosmos. Returning instead to cosmic nutrition with regard to farming, the soil can receive subtle nutrition through plants, minerals, and also the horns put underground for the preparations. Do not forget that at the end of the second lecture and after speaking of the forces of the internal and external planets, Steiner compared the farm to an animal with its head buried in the soil, apparently without further substantiation. The horns, which are obviously right on the head of the animal, are organs of perception for the forces of the cosmos and when buried - which is to say placed in the part corresponding to the head of the animal - can become organs of perception for the soil and thus involve the land in this subtle nutrition.

Paragraphs seven ("*You see I am telling ...*") and eight ("*We must recognise...*") do not clarify this any further, but only suggest that somehow we must bridge the divide to discus this with orthodox science, even though this is often extremely difficult. This is mainly to stem the continuing flow of totally inadequate practical guidance in various areas of life like agriculture.

In these paragraphs we feel directly called to instigate involvement because there are no signs that the orthodoxy is moving closer to our principles to verify their validity. However, we must try to show the validity of our ideas in terms of science, on the common ground of perceptible results. We must therefore strengthen our efforts towards experimental verification, following the protocols recognised as valid by the scientific world, to make the effectiveness of our preparations and the approach that we are taking clear to all.

If you want to summarise the thoughts expressed by Steiner in the first paragraphs of this fourth lecture we could say that there is a life that depends on our food and there is a life that depends on cosmic perceptions. So there must be life in the food that we take in from the environment in which we are immersed. In other words Steiner is talking of embodied life (food) and unbound life (environment), and we are saying that we must have a proper relationship with both.

At this point we should ask ourselves what is the true nature of this widespread and unbound Life. We think that there are aspects connected to the Father, the Son and also to the Holy Spirit.

The widespread Life of the Father can be recognised in all that we received 'free' from nature and therefore fundamentally from humus. The Life of the Son - an active life that renews itself - can be recognised in biodynamic preparations, in the 'holy nights' and from the four archangels in the course of the year. Life connected to the Holy Spirit is found in people's sense of the sacredness of life, or in other words in the morality which acts within nature.

This threefoldness of life can in its turn also be threefolded for a total of nine aspects. We will leave further exploration of these aspects below to scholars.

ASPECTS OF UNBOUND LIFE			
	FATHER	The giver of life	**F.** *Humus* (Unbound Life of the Earth) **S.** Water (The basis of Life) **H.S.** Air/Light (The formative forces of Life)
	SON	The renewer of life	**F.** 13 Nights (The annual gift of Life) **S.** BD preparations (The Christianising of Life) **H.S.** Spiritual scientific attitude in their use (The knowledge of Life)
	HOLY SPIRIT	The sacredness of life	**F.** Sacredness (Participation in Evolution) **S.** Love (Dedication to Life) **H.S.** Morality (Sacrifical veneration)

It is interesting to note that a person's sense of the sacredness of life can even compensate for a lack in the other two aspects (Father and Son) and can substitute for imperfectly made preparations or a lack of humus.

The sense of the sacred, the act inspired by this impulse, becomes a true 'spiritual manuring'. People are the strongest aspect of our method, but if farmers are not sufficiently aware they may turn out to be the weakest link.

This is certainly not the first time that we have pointed out that a healthy awareness can run a farm without using preparations. The preparations are really only a support or crutch to raise awareness through our activity, a 'ritual', but one which we can increasingly omit as we mature.

Unfortunately, we often see that farmers have an extremely selfish attitude towards the Earth, so the preparations are used only with the aim of getting greater yields - and this can lead to poor responses from the same preparations.

We believe that preparations should be used primarily for the good of the Earth, to help it on its path of development. Through preparations we have the ability to regenerate the land and to restore to it what our selfish exploitative attitude has taken over the millennia.

At the time of Zarathustra it was expected that farmers would leave the soil to rest every fifth year. At the time of the Greeks this period had already reduced to once every two years. The Romans still practiced 'sidereal' fertilisation ie every seven years the ground was left uncultivated so that it could replenish itself with cosmic or sidereal forces. This practice has now been all but lost. At our latitudes [N Italy] it is probably enough to rest the land for a year in every 21 years of cultivation. The rhythm we just mentioned was appropriate for a hot climate. The adoption of this practice is not such a great sacrifice and should especially be respected by those who practice organic farming. For those using the homeodynamic method properly such rest for the land should not be necessary as long as one has the right frame of mind when using the preparations ie, the precise intention of returning to the Earth what has been removed for human greed.

The living element that we previously named 'Mother Earth' is due to the Holy Spirit in the table above, and the attitude of sacredness that we should develop is the ideal fertilisation for Mother Earth. Humus, in the same sense, is the ideal manure for 'Father Earth'.

In the ninth paragraph ("*To understand what I mean ...*") Steiner begins to speak of embodied life. He immediately talks of the tree declaring that it is profoundly different from any annual herbaceous plant. It is interesting to note that Steiner said this difference is characterised by the fact that "*A tree surrounds itself with bark.*" Clearly the tree is not identified with its physical appearance but by being surrounded with bark, the same way we dress in clothes. This is as much as to say that this being is essentially the periphery of the trunk. To help us better understand the being of a tree Steiner compares it with a mound of soil which is rich in humus and organic matter.

The comparison continues in the next paragraph ("*Here on the left ...*"). Steiner makes two drawings, the first of which is a heap of earth with a crater in the centre. The second is a tree.

In the tree sketch the bark is particularly clearly marked - something which he had previously described as something which is around the tree. Steiner has offered the two sketches for comparison.

First, the bark of the tree is compared to the crust of the mound. Furthermore the basis of the heap of earth has the same shape as the lower part of the tree: both have a form that recalls that of a vortex. The pile of earth has a crater on the top that can be seen as an invagination, which is the form through which unbound Life can become incarnated. Let us not forget that when we put a seed in the ground we create, in some way, an invagination.

We can see that Steiner's sketch of the heap of soil is like an M, a consonant that is repeated three times in the word 'mamma'. The eurhythmy gesture of M is represented with reciprocating left and right hands pressing forward and back. This step can also be performed vertically to represent the seed that sends up a call to the life of the universe and the answer from the Life of the cosmos itself. Actually embodied life (*Bios*) comes from the unincarnated Life and our common task is to create favorable conditions so that this transition can take place with the utmost ease and completeness.

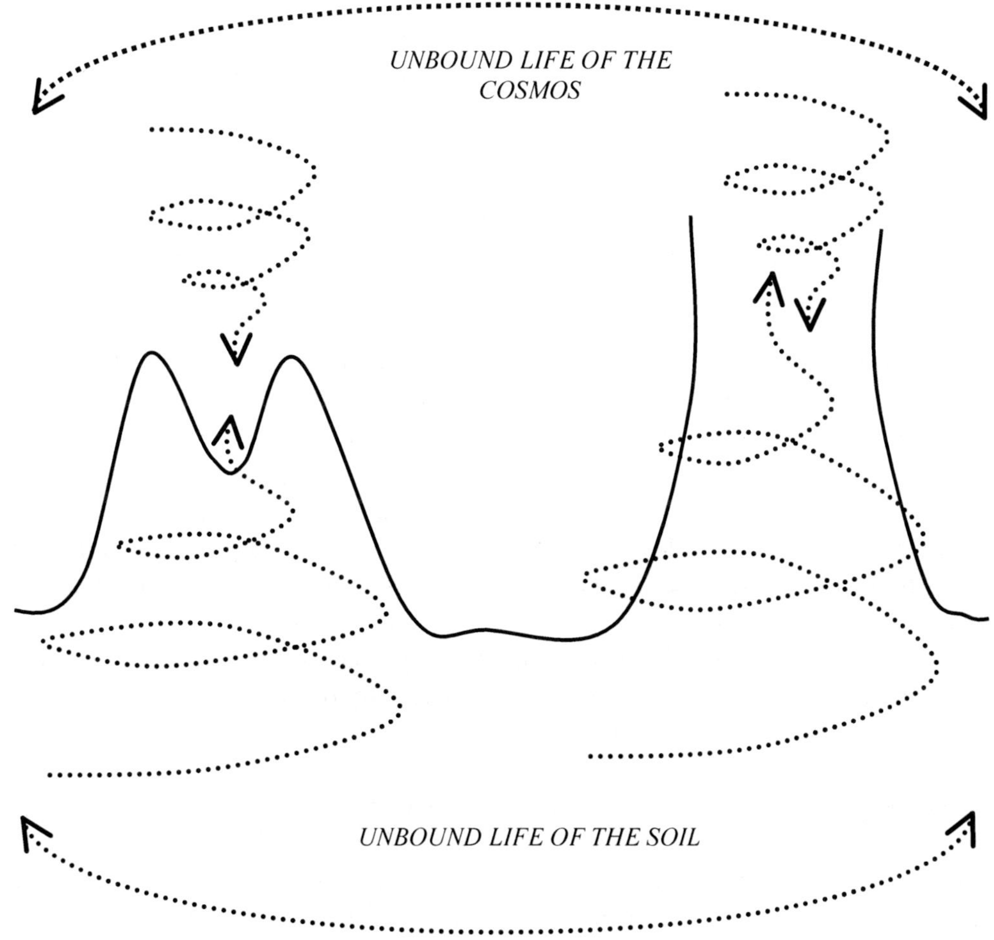

When we have understood this passage thoroughly enough we will be able to create 'etheric machines'. It seems incredible but in the past there were people who had the capacity to build etheric machines. In the second issue of Albios[22] we reported that an exhibition held in Trieste on 'Scientific Imagination' displayed many inventions. One was a machine consisting essentially of a two-handled bronze basin full of water. When these two handles are stroked vibrating sound is created, like 'om', which is audible even to those who are in distant rooms of the building and that raises many jets from the mass of water. The vibration is clearly also transmitted to the body of the operator and it is so strong that one cannot continue to generate the sound for long.

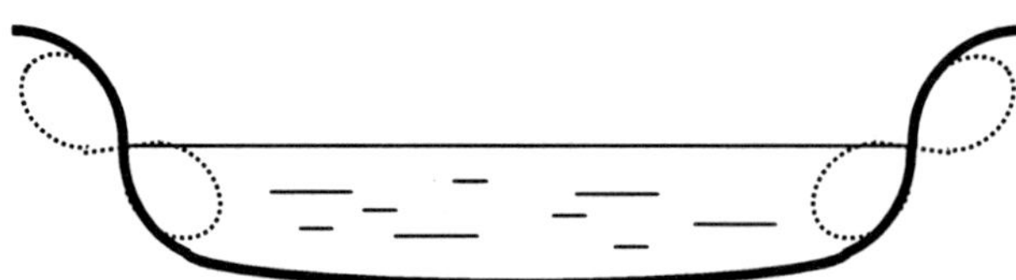

Thus we come to the eleventh paragraph ("*For any given locality ...*"), where Steiner clarifies that whenever you form a soil level which is sufficiently raised above the general ground level in the area it will have a particular tendency to life - to be permeated by etheric-vitality. It is not by chance that monks and hermits often built their monasteries on top of a mountain. Indeed it was because the silica in the mountains eases the ascent to spiritual heights. Furthermore, if the area of the hermitage was generally limestone they built a base of siliceous stone a few meters high at the base of the building, with the specific intent of reducing the downward pull of the limestone that would hinder the elevation of the spirit. You can find many examples of this way of working; Attica in Greece is a peninsula that stretches from northwest to southeast. At the end of the peninsula is a place called Sunio where the temple of Poseidon (Neptune) once stood. The temple represented the marriage between the waters of the land and the waters of the sea. Unsurprisingly very little of the temple of Poseidon remains but you can see what once supported it. The area is a limestone region so the temple was built upon about three meters of huge siliceous boulders to enable the spirituality that was cultivated there to connect to the cosmic spheres.

Remember that the place that hosts this meeting today lies at the foot of the so-called "Montericco" ('Mount of riches). This obviously does not mean to imply that the inhabitants of the area are all wealthy, but that the siliceous mountain is extremely rich in cosmic forces.

When we prepare a compost heap we take advantage of this ability of any mound that is raised above the local *niveau* to be permeated with vitality.

The tree can be regarded as no more than a mound that is a few meters above the surrounding soil and therefore offers an opportunity for life to incarnate and spread within it. The tree can then bring to completion the will of the Earth to permeate with life, to make plant-like, or to vegetalise. From this we can understand that the tree is a precursor of all forms of plant life and its importance against desertification.

At the end of the paragraph Steiner wisely concludes his comparison between a

---

[22] The quarterly magazine of l'Albero della Vita

mound of soil and the plant as follows: "*The same process occurs in the formation of a tree. The soil mounds itself up and surrounds the plant – it envelopes the tree with its etheric vitality*".

The word *tree* [It: *albero*] expresses exactly the concept that we have just discussed. The first letter is an 'A', a sound that is represented in Eurythmy with a gesture of openness - an opening towards the cosmos beginning with the shape of the base of the tree that follows the shape of the vortex. The second letter is an 'L', which is expressed in the rising sap. Then comes the 'B', the sound that connects to the constellation of the Virgin and that is found in the words 'daddy' [It: *babbo*] and 'embrace' [It: *abbracio*]. It indicates an attitude of support and protection, in this case for life flowing inside. The following tone, which is a 'E' recalls the planet Mars and is an enclosure. Life is contained within the tree for its protection. Having examined the word '*albero*' as far as the fourth letter, we have 'albe' which is clearly similar to 'alba' (Italian for '*dawn*'). Even *alba* has the meaning of a life that is becoming or being born: the new day. But it is also no coincidence that the word 'alba' ends with the letter 'A' (opening gesture) because the dawn opens to unbound life, while in the tree the life is embodied, enclosed in a casing, a skin. Then comes the letter 'R', the consonant of Taurus that puts everything in motion. Words that roll or slide have a number of 'R's. Movement is essential for the maintenance of life in the tree: just think of the sap flowing. Finally, the 'O' that embraces everything is the uniting foliage of the tree.

In the twelfth paragraph ("*You see, I am telling ...*") Steiner helps us understand that in fact a discontinuity does not exist between the life of the soil and the life of the plant. There is a zone of interchange that does not end where the root ends or where it is physically possible to find the tangle of capillaries of the plant, but it extends all around, sometimes for great distances (remember what we said about nitrogen) and this zone is the transition between embodied and free life.

This statement leaves no doubt about how some modern agricultural practices such as hydroponics or aeroponics should be regarded. The root must be able to communicate with the soil which must itself also be alive, therefore manured and worked as Steiner points out at the end of the paragraph. Of course if all this is not fully realised by the farmer, or if what is missing is what we have called - in the key that we built - the aspect of the Holy Spirit, it will be difficult to make a fully satisfactory application. Satisfactory, that is, from the point of view of consciousness and, consequently, from the point of view of production.

Thus we arrive at the thirteenth paragraph ("*We must recognise ...*") where Steiner speaks of manuring in a very interesting way. Indeed he does not present fertilisation as a practical imperative, but more like an exceptional intervention to be undertaken in places where the soil can not develop and maintain itself naturally. This seems absurd from the point of view from which agriculture is commonly conceived today. I am not now even referring to those who use chemical methods, but to those who practice organic farming where manuring and composting is assumed to be the basis of fertility. If we return to the schema which we drew to understand the three aspects of unbound Life we can understand how, in the absence of adequate awareness on the part of the farmer (Holy Spirit) and the link with forces from the cosmos (the Son), the farmer was forced to operate with only the humus (the Father) and thus with manuring which, furthermore, often occurs through poorly made heaps or even with fresh manure. The fertiliser should be able to vivify the ground and this must be achieved through a series of processes that allow unbound Life to incarnate in order to then multiply in the soil. Inadequately treated manure will not bring large amounts of Life and therefore must be used in massive doses to make up for the deficient quality. By now many people here have the experience that the use of the Holy Nights preparations, accompanied with the appropriate moral attitude, makes possible the cultivation of the land without resorting to physical manuring.

Anticipating what Steiner will say very soon, we can stress that the farmer should "*establish a kind of personal relationship to everything in farming, especially to the various manures, to the methods of working with them.*" This personal connection is only possible if the farmer has developed a strong sense of morality (Holy Spirit), combined with an adequate knowledge of the laws that govern the world of the living and of the forces that interweave there (Son).

In the fourteenth paragraph ("*So you see that ...*") Steiner twice refers to the personal connection that the farmer has to develop with the manure. Later he repeats this two more times to highlight its fundamental importance for those who want to use the approach that he is advocating to engage in agriculture today.

When we prepare a compost heap we appeal to the forces of the Father. If we understand this deeply we will certainly create a special relationship with that heap of muck, but if we only come to know the coarse material of that manure then we will see just smelly cow-shit unworthy of respect. If we despise it or minimise its value, how can we expect this organic substance that we take out to the fields to bring in something as precious as life?

With that particular nexus into which Steiner invites us, we will imbue the compost with the aspect of the Holy Spirit and - due to the preparations - even with aspects of the Son. The manure and compost heaps are then particularly suited to support what we have described; we only need to bear in mind that the Holy Spirit in us manifests as heat and compost warms up in its first phase of the maturing process. But what develops spontaneously in the compost heap is physical warmth and when left to itself this can only disperse from the Trinitarian vision. However, aided by the etheric warmth of the farmer it is instead suitable to becoming rich in new forces of life and thus embodying the unbound Life in its threefold aspect.

We can conclude our reading with satisfaction that we have acquired two basic concepts: the three aspects of free Life, and how this unbound Life is embodied.

In our next meeting we will talk about the two preparations 500 (horn-manure) and 501 (horn-silica), which have the function of promoting the incarnation of the Son's aspect of the free Life.

The other preparations, which we will discuss later, are the means for promoting the incarnation of free Life in its Fatherly aspect.

People thus become the instrument for opening the crater (500) and enabling the descent of fertilisation with Light (501).

We can now better understand the architecture of this fourth lecture: Steiner first spoke about cosmic and earthly nutrition, stressing the importance of food permeated with the free life of the cosmos. He then spoke of the transition from free Life to embodied life. He then introduced a discussion about compost (Father), and concludes by introducing the preparations (Son). People (the Holy Spirit) are emphasised as bringing the crucial personal connection of which we have spoken.

Thirteenth Meeting

Let's continue reading the fourth lecture of the agriculture course. As we saw in the first part of this lecture Steiner spoke about unbound and incarnated life. He addressed the topic of nourishment, distinguishing between 'gross' and 'subtle' nourishment to get us to understand the principle that all living organisms are sustained from two streams. We all know about the food we take in during meals, while the subtle nutrition stream comes through our sense organs. This subject has been widely covered elsewhere[23] so we will not dwell upon it here.

This two-fold nutrition, and in particular the ability to be nourished by forces from the cosmos, does not apply only to humans but is common to all the kingdoms of nature. Indeed, it has a greater significance for the other kingdoms than for people. If you think of the plant world what I have just said is clear. All the wild plants, with their ability to live on poor soils and sometimes in the almost complete absence of water, demonstrate how it is possible to draw nourishment from those subtle forces. The intervention of humans in the life of plants, including what led to their domestication, has caused a very strong reduction of this capacity compared to the original plants. It is true that their domestication has enabled the plants' group I to have evolutionary experiences, but in terms of nourishment it has been limited. The current orthodox practice of agriculture is obviously moving in other directions because it has removed vitality from the ground so the plant no longer finds adequate subtle or gross nourishment. It is obvious that the way out lies in reinvigourating the plants' ability to draw life from the forces of the cosmos. Precisely for this reason Steiner gave the world the biodynamic preparations.

The thought that we have just formulated can be carried one step further and enable us to understand how the Earth is nourished by the cosmos. Then we can consider, besides the usual fertilisation practices such as green manuring and irrigation for example, that there may be other more subtle ways to feed the soil. Indeed we have become aware that the Earth and its mineral realm feeds primarily from the forces of the cosmos, much more than happens to men or animals. Moreover, many of the farms that use our method now fertilise without the input of ponderable organic matter and this has been the case for many years.

At the end of the fourteenth paragraph ("*So you see that ...*") Steiner says that every living being has two aspects: one within the skin and one without. This is an assertion that may seem trivial but actually Steiner is talking about the dual influence of the Moon on the etheric body. The internal lunar influence stimulates the processes of the etheric body that act mainly on the glands that produce secretions. The Moon's influence on the outside stimulates processes that produce excretions. This polarisation testifies to our link with the Moon that, as we have often said, is expressed in the number *2*. We must therefore expect that in the following paragraphs Steiner will talk about the etheric body in relation to the Moon and begin with the internal aspect.

The internal aspect relates to the full Moon, because it is more closely linked to water. Do not forget that it rains more at full Moon than new Moon. Our body is predominantly water - around 72%. This also establishes a link with the physical aspect of the Earth where water covers around 72% of its surface.

---

[23] See "Course on Nutrition" and "Fertilisation" by the author

The fifteenth paragraph ("*The inside of something ...*") begins by saying that there are currents of forces within us that flow to our skin and then bounce back as if the skin was an impermeable barrier. But there is another set of forces - free forces – that must be able to act inwards from the outside. This field of forces exploits receptors. For the human body our hairs are such receptors. In animals the horns and hooves perform this function.

The third sentence of the paragraph says something very interesting: "*Now there is something that expresses very exactly, if rather personally, how something living establishes a balance between its insides and its outsides*". First it seems self-evident that there must be a stable relation between what is inside and what is outside and that this must be absolutely personal. As we have already said, this applies not only to humans but even to the mineral world. So if we learn to heal nature or an agricultural organism, we must first understand that there are no rules that apply to all terrains, but for every situation we must find the link that can restore the conversation between the forces of the farm and those present in the cosmos. The characteristics of different farms are always unique for a number of reasons, sometimes because of climate, at other times linked to a special aspect of a farm, or the composition of the soil, or even the vicissitudes through which it has passed to get to the present. It is unthinkable that you will always face the same variables that resulted in the imbalanced conditions and which are therefore amenable to the same remedial action.

Everything which is inside the skin needs to develop its odour and the fact that the smell is retained within the skin is even presented as "*what it means to be alive*". We think now that smell is a characteristic of the astral body which is linked to nitrogen. In practice Steiner is saying that if we want life processes to take place in the best conditions it is necessary that nitrogen is not lost but is retained within the body. He is therefore clearly speaking of manuring and compost. A good compost heap, which is the basis of good organic fertilisation, must be able to retain nitrogen since nitrogen - when spread on the ground - is able to bring oxygen to the soil, which is the element of the etheric body and thus of life. The soil must maintain living nitrogen within it as humus, and in the light of this we should encouraged those practices that minimise the loss of nitrogen. In other words we can say that the astral acts upon the etheric resulting in the seven life processes, and without the astral there cannot be life: therefore, without nitrogen Life cannot be maintained in the body. (Obviously this nitrogen must be alive.)

All that we have described must be kept in balance, according to the famous "*personal relationship*" repeatedly cited by Steiner in the lecture. Actually the life processes - which are essential for life - must be kept within certain optimum ranges, failing which diseases will emerge. However, the etheric, the realm of vital processes, is moderated by the astral and that is why the astral is considered the true cause of diseases. For the astral to act in a balancing way it must in turn be guided by an I that is inspired by the values of the spiritual world. Now, the mineral and vegetable kingdoms do not have the ability to choose and decide how to relate to spiritual laws, because their ego or I, which is a group I, is not incarnated in each of them. The ability to decide for the plant world is the responsibility of humans because we are the only incarnated beings with free will. It is an arduous task that was entrusted to us, and seems even more challenging in light of the fact that the capacity to act therapeutically requires a second and higher consciousness, called *inspiration*, which is the consciousness of the life Spirit and is linked to the Son. The farmer must first be able to grasp the essence of the plant to understand what the plant needs and also the essence of the various types of compost, and then build the personal relationship between the two essences. The understanding of what is essential requires imaginative consciousness, but to create a link between two essences needs a still higher consciousness. Farming then, if this is understood, also requires efforts to develop higher consciousness. Clearly, if we are expected to have achieved inspired consciousness first this will condemn nature to a grueling wait. We must begin to work right now asking for the help of the spiritual world and also begin to work for our personal growth. In this effort the agriculture course, upon which we are struggling to comment, can be of great help because it outlines a journey of initiation within and through nature.

Midway through the paragraph there is the statement that "*a plant organism is not predestined to give off odours, but rather to absorb them ...*". Apparently Rudolf Steiner wishes to remind us that the plant organism does not have embodied astrality but does have a very strong etheric power and this brings the danger that the life processes are not sufficiently supported. We will see later that the preparation that supports astral harmony in the organism is the horn silica.

We are not accustomed to thinking of the case in which the farm organism is so rich in etheric forces that it rebuffs the regulating processes by which life is manifested and preserved, namely the harmonising astrality. Excess etheric forces may be due to excessive fertilisation, green manuring, excessive rainfall or the proximity of woodlands.

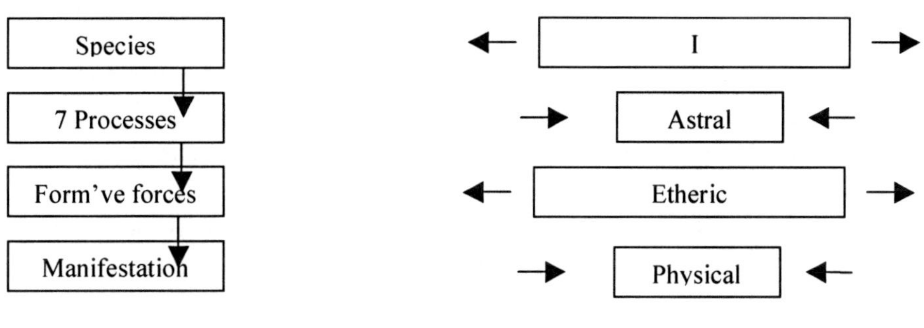

Normal equilibrium between the bodies

Excessive etheric

This sentence is followed by a different expression which seems insignificant at first glance but which is actually very significant: "*... if we comprehend* " which once again refers to the higher consciousness that can be developed with a thorough penetration of "*... the beneficial effects of an aromatic meadow full of fragrant plants*". One becomes aware of the phenomena of mutual biological cooperation between the etheric and astral components. Life already offers us the solution to our problems but we are not normally able to read these solutions. Perhaps Steiner, in this sentence, is suggesting a route to a remedy for situations of poor communication between the astral and the etheric.

The paragraph ends with the call: "*... We must have a personal relationship to all these things – then we will truly participate in nature*".

Thus, we arrive at the sixteenth paragraph ("*It is important to ...*") in which Steiner speaks again about nitrogen as a vehicle for oxygen that is, in turn, the bearer of vital forces. In this paragraph we are cautioned about the misuse of nitrogen. In fact, we understand that if nitrogen is the regulator of life processes it must be present in the soil in a balanced way. In conventional agriculture they know how to abuse nitrogen. Just enough nitrogen needs be placed in the ground to convey the living elements below the plants, in the zone where it communicates with the plant's Ego. Steiner has inserted the subject of fertilisation in a discussion about quality; if, by some misfortune, this context escapes us we would have to consider the contribution of nitrogen only in terms of quantity. Clearly the subtle function of nitrogen is possible only if the nitrogen is introduced in the plane of life; if chemically-derived nitrogen is used this communication fails and only the path to quantity remains accessible with all the consequences of which we are aware.

Viewed in the way this is presented by Rudolf Steiner, fertilisation becomes the key to the entire future life of the plant, because it acts by providing an opportunity for the incarnation of its archetype for which nitrogen should be considered as a stimulus. The rest of the life of the plant will be the consequence of how this process of incarnation comes about.

Paragraph seventeen ("*Now, this could already ...*"): with purely mineral fertilisers it will be impossible to vivify the solid element of the Earth. At the most we can reach the fluids and this will have repercussions on the plants: they will show a response to this limited nourishment by developing quantity at the expense of quality.

We can try to help understand what is said in this paragraph with our usual sketch of the plant and with reference to the elements at the root, leaf, flower and fruit. To complete the picture we will show the area of the I beneath the root towards which we will convey the vital elements mentioned in the previous paragraph.

When we use a mineral type of fertiliser we will act mainly in the area identified by the oval, or in the area of the leaf and turgor. Our efforts will not come near enough to the Earth. Do not forget that the Life ether belongs to the Earth element, while water is the basis of life. Obviously any agriculture as described above will require large quantities of water distributed on a regular basis.

By now we know enough to understand easily that a fertilisation which aims to stimulate the watery pole inhibits the action of the Fire element, and we know that the Fire element is essential for the quality of fruit. All of this leads to the common experience that the fruits grown with conventional agricultural means may have a very attractive appearance but inevitably entwined with the consequence of very poor quality, as is clear from the diminished taste and smell, and the reduced shelf life.

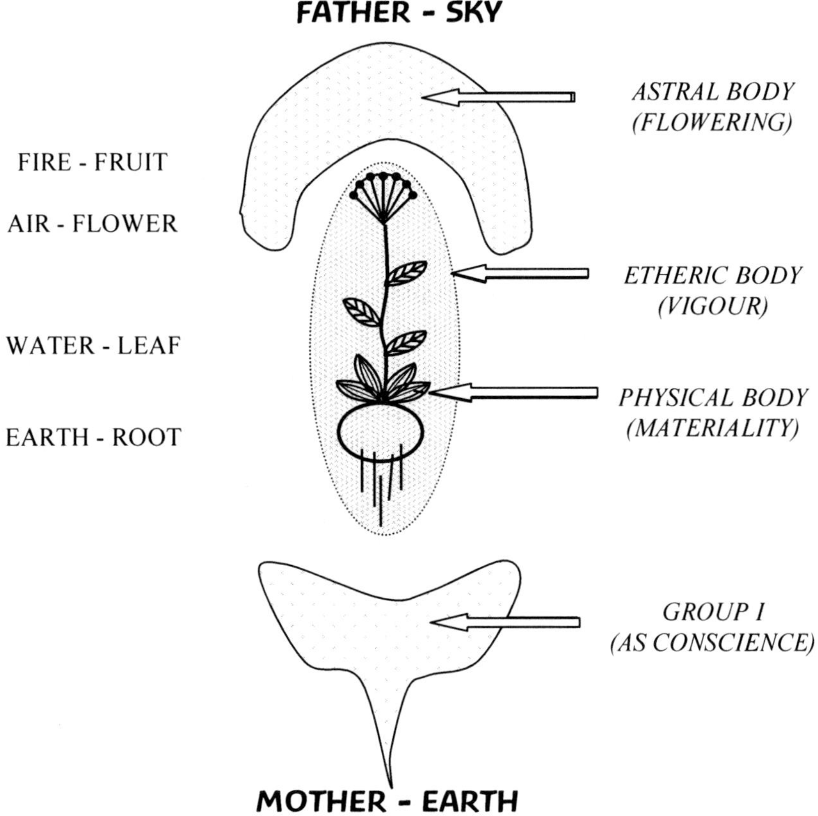

The situation is even more tragic if we also consider the condition to which we have reduced our water, namely to the level of receptacle of pollutants of all kinds. Water is for the Earth what blood is for us[24]. The ocean currents are comparable to venous and arterial blood vessels while rivers, streams, rivulets and groundwater flows are like the capillaries that carry blood to every cell of the body. Similarly water should nourish each clay-humus domain. Continuing this simile, consider that any doctor would be deeply concerned at the sight of a patient with thickened blood, or which carried such a toxic load. Yet who does care about the current level of water poisoning and the subsequent degeneration of the Earth? The current mentality is only fit for grasping the maximum short-term income without any consideration for the health of the land.

It seems a contradiction but at the time of Zarathustra, when the Earth was more vital, the water was clean and techniques of cultivation were very far from being exploitative. Growers used to allow the Earth a year of rest after every three years of cultivation. They wanted this to allow Earth time to get back into contact with the forces of the universe and the possibility of rebalancing itself by producing what it lacked. Even the Greeks and the Romans recognised the right for the soil to reconstitute its integrity, even though the time between two cultivations had become five and seven years respectively.

Well, despite the current level of exploitation, few care any more to ensure that the Earth is able to recover. The objective seems to be to take for as long as we can.

It is worth a comment on our proposed method in this regard: this involves a series of sprays during the 13 'Holy Nights' aimed at facilitating communication between

[24] See 'Nine encounters with Biodynamics' by the author. In translation (2008)

the Earth and the cosmos. In this way the universe powerfully fertilises the Earth and thus it becomes superfluous to have a rest period every seven years. Each year, in fact, the Earth has the opportunity to receive the cosmic nutrition that other cultivation techniques are unable to provide. We wish to add that the upcoming Nights (2002 - 2003) are particularly rich, since there will be a powerful Christic influence beginning in September and ending at Easter.

In the eighteenth paragraph (“*If we want to understand ...*”) Steiner invites us to observe the processes taking place in compost heaps and to learn to understand the processes that enliven the Earth. The stuff with which a compost heap is made is precious little: essentially ‘wastes’. These scraps retain something of their living experience and still carry with them the links just mentioned by Steiner: the material that we put in the compost heap brings with it the lawfulness that created harmony between the etheric and the astral. When we spread the mature compost on the soil, in addition to organic matter, nitrogen and other things, it brings such laws and the resulting nexus. All we have described here does not occur automatically. It is necessary that our awareness continues to act so that these links or these spiritual forces remain bound to the matter that we will distribute over the fields. Otherwise we will be spreading only an organic fertiliser, unable to bring life in the sense described.

This should not be really new to us, because in our meetings we have talked about how consciousness is crucial in all subtle processes, so that we can also remediate any deficiency using the preparations or with physical interventions.

The paragraph continues by reiterating that the astral finds an exuberant etheric vitality to be an impediment to its own action. This is due to the polarity between the etheric and the astral that we should already have grasped. We are mainly vulnerable to this latent danger when we build a compost heap of plant matter, potentially very rich in etheric forces, because the forces that moderate vital processes are astral forces and these could be inhibited in their action.

Steiner gives an indication that the introduction of quicklime enables a strong retention of the etheric without too much influencing the astrality to become volatile. We could suggest that in this paragraph there is a mistake, perhaps due to the transcription from stenographer or translator. We will try to back up this opinion with the help of a drawing.

In the lower part of the drawing is the world of limestone, which is partly calcium. Calcium is a very special substance although it receives little appreciation since normally we are only familiar with its dead façade. In reality lime also has a living aspect. If we go back in time to the ancient Lemurian epoch lime had not yet fallen to a material state. It was a substance that hovered above the silicaceous Earth. In that state animals and man were clearly arranged in a very different way from today and breathed lime that had the consistency of milk. It was an etheric limestone that was therefore living. Even then the limestone had the function of attracting imponderables to itself. Moreover we are what we are today because we have some limestone (the skeleton) that holds our soul-part - which is linked to our consciousness - and our spiritual component. In the compost heap we should be concerned to retain the imponderable etheric and the astral to establish the nexus we mentioned earlier and for which goal the limestone is a material crutch.

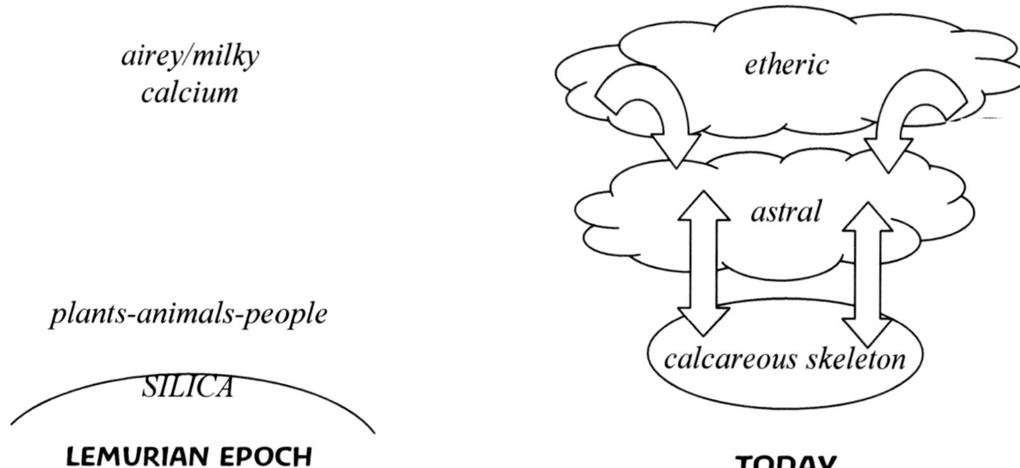

The age we are referring to was also the time of the fall of Man, after which event things began to manifest as physical-material. People, and even animals, began to have a skeleton within themselves that bound the etheric, astral and I components within their corporeality. People began to be an individuality enclosed within a skin. The astral within the physical body could be called *animality*. We know that the physical world has affinity with the astral and that both are polar to the etheric. We also know that the limestone can be characterised by its 'greed' and, by the consideration we have just made we can understand that this should be attributed, in the first instance, to the astral. Only later did limestone attract the ethereal, to quench the thirst remaining after absorption of the astral.

It follows that to attract the etheric it is necessary to have more strength than to retain only the astral. Between quicklime and slaked lime the most *eager* is the quicklime, and the latter will be the lime with the ability to contain even the etheric. Now in a manure-based heap there is a wealth of astrality and we must create a link to the etheric, while the plant-based heap is already highly etheric and we must draw in the astral. It therefore seems that the quicklime is better for manure heaps and slaked lime is more appropriate for the plant-based compost.

We might ask at this point whether it is indifferent to the compost heap if spring lime or autumn lime is spread. Well considering that lime exhales a weak ethericity in spring because of the contractive gesture that it bears, while in the autumn it inspires the etheric, we might conclude that in a manure heap in which it is necessary to retain the etheric within in the heap, we would do well to use quicklime made from limestone extracted in the spring.

In the nineteenth paragraph ("*What arises there ...*") Steiner says that if the astral acts strongly in the soil without the mediation of nitrogen it provokes a very similar process to the one within people that enables the absorption of nutrients. Steiner is talking about the process of digestion within the stomach. Indeed, the stomach is a very astralised organ as shown by its high acidity: pH around 1.5.

The stomach is also a hollow and highly asymmetric organ, both of which are characteristic imprints of the astral body. But if the stomach is such an astralised organ it seems legitimate to ask, *where is the nitrogen?* We know that astrality is linked to nitrogen but, since there is no such trace at the level of the chemical elements, we have to look for it in a supersensible form.

By now we should all be clear that nitrogen brings in the cosmic images. Well, right at the mouth of the stomach is a major port of entry for these images, which in oriental culture is called the '*chakra of the mouth of the stomach*' which we know better as the *solar plexus*, using a term most linked to the physical. The cosmic images enter through this chakra in many of the current forms of clairvoyance.

If the stomach becomes excessively astral we can experience the annoying inconvenience we call *heartburn* that is really an inflammation. If such an inflammation persists it is a disease we know as *gastritis* that can also show up as ulceration. If the process is allowed to continue it degenerates towards necrosis and cancer can occur.

A soil that has been astralised by misguided fertilising first undergoes a process of inflammation and then hardens. It is curious to note that when the land is subject to an inflammatory process it will initially increase its production, but then goes fatally necrotic and dies. What we have described is the result of chemical fertilisers, and this is usually accompanied by agronomic practices to diminish plant vigour such as the use of weak rootstocks that further accentuate the push towards astralisation.

Today it is our good fortune that we have the biodynamic and homeodynamic preparations available to rescue us from the consequences of the typical agricultural techniques of our time. So let us strive for the highest level of understanding of what is presented in the pages of the agriculture course, because they contain the secret to bring the life of the cosmos back to the Earth and the plants, and this is an appropriate impulse for our times.

It is important that we commit ourselves to what we have all understood here until it is imprinted on our consciousness because you may well leave here and have the feeling of not remembering all the points we have covered. However, in due course what we have learned will arise in the form of new faculties. We are like the 'fertile soil' for the cosmic images that nitrogen brings. On the contrary if our awareness is too limited we will not be able to discern from among the images that come to us which are the just ones from those behind which hides some being that has no real interest in the evolution of nature and Man.

Returning to the sketch of our plant that showed the polarisation between the contractive gesture of the lower part (which we could also call *egoic*) and the expansion of the upper part, we can reconsider what Steiner is suggesting to us.

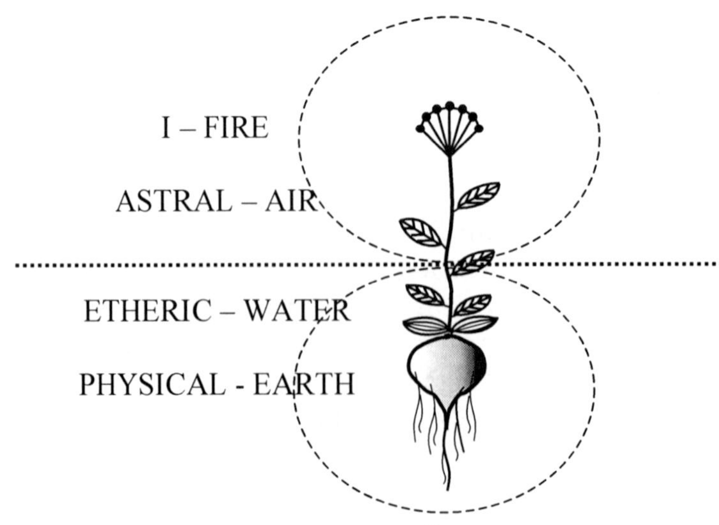

Indeed with the help of another sketch we can better understand that the real boundary between the various bodies of the plant - this also applies to animals and to humans - is between the etheric and astral bodies. Steiner invites us to create a living link so that the boundary does not lead to a separation and death. We have just now recognised that link in a substance called lime but not as the chemical element calcium (Ca), but as a *process* of limestone which we can find in quicklime or slaked lime, exhumed in spring or autumn. Indeed, in our view, the symbol Ca by which calcium is identified does not really say enough because it should always be supplemented by a qualitative aspect that will enable us to choose the right form in all circumstances. We know - we have talked about it so often - that there are seven types of limestone and only our discernment will guide us to make the right selection.

The third phrase of the nineteenth paragraph reads: "*When we treat the soil in the way I have described we stimulate it to the same energetic state.* " Once again he refers to the "*nexus*". Later when he describes the preparations he will give us the means to establish a bridge between the two poles of the plant: one is the oakbark (containing 72% calcium), the other is the nettle. In this paragraph Steiner repeatedly uses expressions such as a "*plant-like process*", "*do so conscientiously,*" or "*insight into the whole nature of the process.*"

Looking at what we have read in the fourth Koberwitz lecture as a whole, we may notice that Steiner speaks from the start of a deep, intimate rapport with the compost and then uses a variety of expressions to underline the fact that the quality of the compost does not depend solely on the ability and technical knowledge we have acquired, but by a intimate relationship with the process that is based on the awareness of the farmer. In all this talk Steiner has hidden another meaning that normally is not grasped: he is talking about Life.

ZODIAC	*MEANING OF LIFE*	SPIRIT – SACREDNESS
EXTERNAL PLANETS	*LAWS OF LIFE*	ASTRAL MORALITY
SUN	***THRESHOLD***	**INITIATION**
INTERNAL PLANETS	*FORM OF LIFE*	ETHERIC AESTHETICS
EARTH	*MANIFESTATION OF LIFE*	FREE CHOICE

If we remember the first lecture we will recall that the relative presence of silica and limestone influences the form of the plant. Understanding form can bring us from the physical manifestation up to the Sun, which in our scheme is positioned at the nexus, and therefore between the etheric and astral planes. By now, with reading and studying the first lecture, we have begun to develop imaginative awareness, the ability to understand the language of form so we can begin to approach the second higher level of awareness that is *inspirative* consciousness. Development of this consciousness leads to an understanding of life. What Steiner gives us in these lines is an exercise in super-solar logic that requires inspirative consciousness.

We can now understand why we need so much effort to make this study. The book that we are reading can be considered to be the greatest book of alchemy or the greatest esoteric book. It is a treasure that Steiner has seen fit to entrust to farmers, because the Koberwitz course was not aimed at doctors or philosophers, but at farmers.

It is not said that our task is always and only to cultivate our fields as best we can. If we can also grow internally by cultivating ourselves we can also become worthy of the highest achievements such as greening the deserts.

For the moment we are studying the forces of life that reveal their presence in the world of form. We must, however, approach the laws of life governing these forces. This second step, in the esoteric sense, consists in learning *to read the hidden book of nature*. Today we have started to '*read*' the page of limestone in its hidden, super-sensible aspect. But to become collaborators with evolution we must go further and learn to understand the '*meaning of life*'. When we have made sufficient progress then we will be able to revitalise the deserts.

The meaning of life allows us to understand why there are GMOs, and why the desert is expanding rapidly and profoundly. Perhaps we will see a different picture in which GMOs could bring a great benefit to all humanity, or where the desert can be regarded as something other than the death of the Earth, and might also be opportunity for a process of resurrection. But we must be aware that the plane we are talking about is one in which we can achieve freedom.

We have previously recognised freedom as another aspect of love so we may understand that this plane will be accessible only if our actions are no longer informed by any personal interest.

It is not enough only to develop big heartedness, because this would lead us to act generically for the good of others, and, in the absence of a proper education, there is always the risk that our willingness will be exploited by inappropriate forces. It is therefore necessary that we learn to dedicate our actions only to the laws of the Spirit, which in other words means developing an adequate sense of morality.

If you look at our drawing, the moral plane is that of laws, which as we said, must also be exceeded, because morality (It: '*more*') means more than laws. We must get to the plane where we reach Love (It: '*a-more*') that is beyond morals, or without laws. As long as we are subject to laws we cannot be completely free. We will be free when our actions are no longer inspired by morality but by sacredness, or rather guided from ourselves when we have developed a sense of the sacred in our lives.

In the twentieth paragraph ("*Now, if you simply leave ...*") Steiner gives some practical suggestions to ensure that the compost heap does not disperse the nitrogen that will gradually form. Meanwhile he reminds us that a compost heap that holds nitrogen should not be smelly. Thus he suggests building the compost heap in layers of wastes each separated by a layer of peat.

The reason peat is recommended is understandable when you consider that peat is a material of plant origin which has passed through a process of dying, but without having completely undergone the process of decomposition; it is as if *enchanted* or – better - filled by many enchanted elemental beings. Peat also has not yet completely lost its original form and thus maintains a fibrous appearance. We could say that it hovers in a region standing between complete death and the forms of life. Not having completed the cycle of death, it becomes a bridge between death and life.

Now if we consider a plant-based compost heap, we can recognise in it a process aimed at transforming the original plant towards the mineral, and we use it before this transformation is concluded. But the peat is also the result of a process of incomplete decay. Peat in the compost brings in a message that we could consider to be like *information* or an example not to fall all the way to the dead mineral, but to develop a state of suspension between life and death, or between vegetable and mineral: this is humus.

Steiner even says to construct the compost heap in layers in order to build a sort of accumulator of energy so that the whole process will be strengthened. We really want the compost heap to accumulate forces - the famous imponderables. If we move from a physical logic to a logic of forces the different layers create fields, such as the accumulation of potential difference that occurs in a condenser or a battery and which governs the interior activities. These would not be available in a homogenous heap.

In conclusion, with lime and peat we retain the imponderables and thus create interior alchemic movements. In the compost heap we want to reproduce a process in a few months that occurs in a peat bog in many years. The compost heap is that by which we wish to achieve a real alchemy of nature, because the merely organic heap created to bring organic matter into the fields now has little meaning. We aspire to create an inoculation of processes that then continues to enliven the soil. This provision of multi-layered compost heaps is also an attempt to develop the mass so it can bring *the alchemical ether* and so help all transformations and transmutations.

Peat also has the characteristic of resisting electromagnetic fields that did not constitute such a big problem in Rudolf Steiner's time. Certainly in our day, however, such a property is important.

The keystone is always our consciousness. The appropriate consciousness enables us to bypass much of what we have said. The homeodynamic method offers a dynamised product for the manure pile or compost heap that allows us to proceed without any amendments – not even lime and peat.

Fourteenth meeting

We are together for the fourteenth time to try to understand the lessons taught by Rudolf Steiner during the agriculture course held in Koberwitz in June 1924. We are commenting on the fourth lecture and concluded the previous meeting at paragraph twenty.

We followed Steiner in focusing upon compost in our efforts to understand what is happening during the process of transformation of organic matter into humus, a process similar to that of vegetalisation of the Earth but one that is not completed. This is an alchemical-type process that will be completed in the plant in the formation of the flower, fruit and seed.

This process of vegetalisation is continuously active in the soil - we can even call it *etherisation* - and it finds expression on the physical plane in the formation of plants and on the etheric plane as the separation out of an individual etheric body from the 'etheric ocean' We see this as the etheric or living component of the plant.

We have often said that the Sun, in the sea of the ethers, is the heavenly sphere that governs the process of individualisation. We know that the Sun 'blesses' the plant when it is shining in the sky, and also when it is on the opposite side of the globe during the night.

This dual solar action governs the plant phenomena that we call *heliotropism* and *geotropism*. In fact geotropism is only apparently an attraction of the plant to the centre of the Earth. In reality the plant communes through its roots with the Sun on the other side of the Earth.

This statement need not be too surprising because even orthodox science says that the Sun is the source of radiation and of sub-atomic particles. The smallest (gamma rays) can cross the Earth from one side to the other as if it were empty space. We are referring to more subtle forces that are not impeded by the Earth's matter and can arrive at our plants even when the celestial bodies that generate them are not visible in the sky. In light of this we can be even clearer why the plant is commonly called the *daughter of the Sun*.

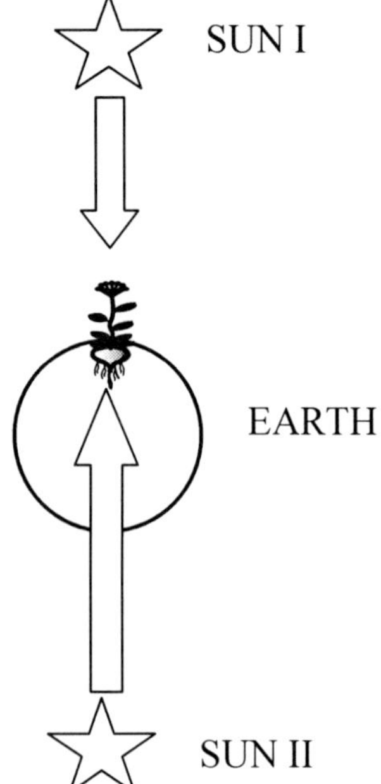

In the last meeting we understood the need to create the ideal conditions in the compost heap for the processes of transformation to proceed towards the formation of humus, and we will return to this in the fifth lecture.

Now Steiner will give us directions for making a product that could complete the action of compost, given that it is increasingly difficult to find sufficient quantities of good quality compost. We are convinced that through potentisation it is possible to create a product which is even able to replace the compost, in the sense that one can stimulate the process of forming humus in the soil as manuring does in a cruder fashion.

We can also consider it with the thought that the Earth has become tired of being exploited by humans. We have already talked about the fact that in ancient times farmers used a cycle of alternating cultivation and fallow periods that is extremely generous by modern standards. In these periods the land was allowed to regenerate from the effects of human greed, thus producing what it required and not what the farmer wanted.

Returning to the Koberwitz course, we are approaching the pages in which we will be introduced to the horn manure preparation now known as the '500'. These biodynamic preparations can be regarded as the bearers of healing moral forces for the Earth with which one can rectify the problems caused by human selfishness. Given that we humans have used the forces of the Sun and all the planets to our advantage we will see that Steiner has developed a preparation for each planetary force.

Some damage that people have inflicted on plants may seem, at first sight, to be forgivable trifles but if we ignore the surface view with which we normally relate to the world of nature, we can recognise even in seemingly insignificant actions an attitude of abuse against forces from the cosmos, and hence against the spiritual entities from which these forces arise. Without dwelling upon GMO where immorality is evident, we can even think of commonly accepted practices such as hybridisation. For example, one of the interventions upon corn has modified the angle of inclination of the leaves with the aim of more direct exposure to the rays of the Sun and therefore a greater stimulation of photosynthesis. It seems nothing but behind these actions is the desire to use the force of the Sun for purely selfish purposes.

If we begin to understand the influence of the preparations and then use them consciously we can begin to honour our debt to the Earth. This may be part of what is meant in the Lord's Prayer when we say: "*Forgive us our trespasses as we forgive those who trespass against us.*"

All this must find a wider acceptance in our hearts and the forgiveness of debts – pardoning - should become a rule of life because it is the beginning of overcoming the law of karma. By the logic of the physical plane all of us have the right to seek reparation for an injury, but Jesus Christ taught us to forgive, or to renounce our right to parity and thus begin to produce positive karma. Moving beyond the law of retaliation ("an eye for an eye..."), we reach liberation by forgiving.

Writing off a debt and thus waiving a right is a gesture of freedom and love, and therefore these two faculties are the attributes of God which man will bear for the entire cosmos.

Now we must understand well that the gesture of freedom inherent in the procedure required for the making and spraying of the biodynamic preparations is given in the dynamisation. The time actually devoted to the potentisation – from a strictly utilitarian point of view - is lost time that we might more profitably devote to other work, entrusting dynamisation to some mechanical tool or device. The time that we feel necessary for a thorough mixing of the preparation in the water is far exceeded and might better be used to prepare for spraying, which will take place in space. Once again, we come up against the two spiritual realities of space and time within our earthly existence. It is crucial to understand these for our evolution and to travel the road to spiritual perfection. From this perspective giving a preparation to our field becomes a cultic act.

Whoever uses the homeodynamic products does not fulfill the gesture of dynamisation and so must draw even more heavily upon their consciousness during spraying. From this point of view the simplified manual requirement is a definite advantage. On the other hand, whoever has the homeodynamic product must make the maximum effort to ensure that the preparation is supported by their highest concentration so that there will be no lapse of awareness. Precisely in order to allow individuals who use homeodynamic products to develop further awareness we recently wrote in a series of articles, published in the magazine ALBIOS, wherein is described the exact process of preparing a product, including dynamisation. In this way each of us - with a bit of goodwill – could prepare a potentised preparation.

For the dynamisation of large quantities of biodynamic preparations it is obviously necessary to get mechanical assistance, but it is equally necessary that the farmer never leaves the scene to let the machine get on with the work. The machine can eliminate the physical effort but should not reduce the time sacrificed for potentisation. For example it might seem useful to equip a machine with an automatic timer for the reversal of the direction of dynamisation, but let a person decide when to change over when the vortex is completely formed.

In other words the machinery must not lead to a lapse of consciousness. This is not only valid for dynamisation but for all the work that must be done in the countryside, in accordance with the principle that it is permissible in terms of time and effort to delegate the work to machines which it is impossible to do by hand, as long as you are master of how to perform the work manually. Man must maintain awareness of all production processes that occur in their land and the machines should only be a help to make it feasible and economically viable.

If we want to help the Earth we must practice agriculture that not only knows how to take but also give, so that the earth produces not because it is obliged under the influence of drugs (conventional agriculture), but produces as a sign of gratitude for our attention. We must become aware of our debt and repay it. It is within our abilities to treat the Earth through acts of love and perhaps the Earth will freely return this multiplied many times over.

Preparations are therefore an opportunity for a concrete act of love for the Earth: therefore they are a cultic act. Feel in the centre of our I, of our consciousness, that dynamisation is a process that makes a strong connection with the cosmos and thus with the Creator God. Remember that when we dynamise we create a vortex in which all the water molecules orientate themselves to certain stars according to a direction given by the preparation.

Do not forget that the three laws regulating the vortex are the same laws that regulate the movement of planets in our solar system – the laws of Kepler. Dynamisation creates a cosmic resonance system and all the water molecules are orientated towards a certain star particular to each preparation. The preparation is a means by which we can establish a link with the heavenly Father, represented by the Son, or the Way to return to the Father's house, using our consciousness, which is an aspect of the Holy Spirit.

We can now approach the twenty-first paragraph ("*This all points ...*") of the fourth lecture. To understand what is written there we can offer the drawing that follows.

We have highlighted the cow's horns, hooves and tail because, as Steiner said, these are the points of entry and exit of the forces. We have referred to a cow, but we could use any type of animal, obviously using different parts of the body. For example if we wanted to put a cat at the centre of our thinking we would not mention the horns but the whiskers.

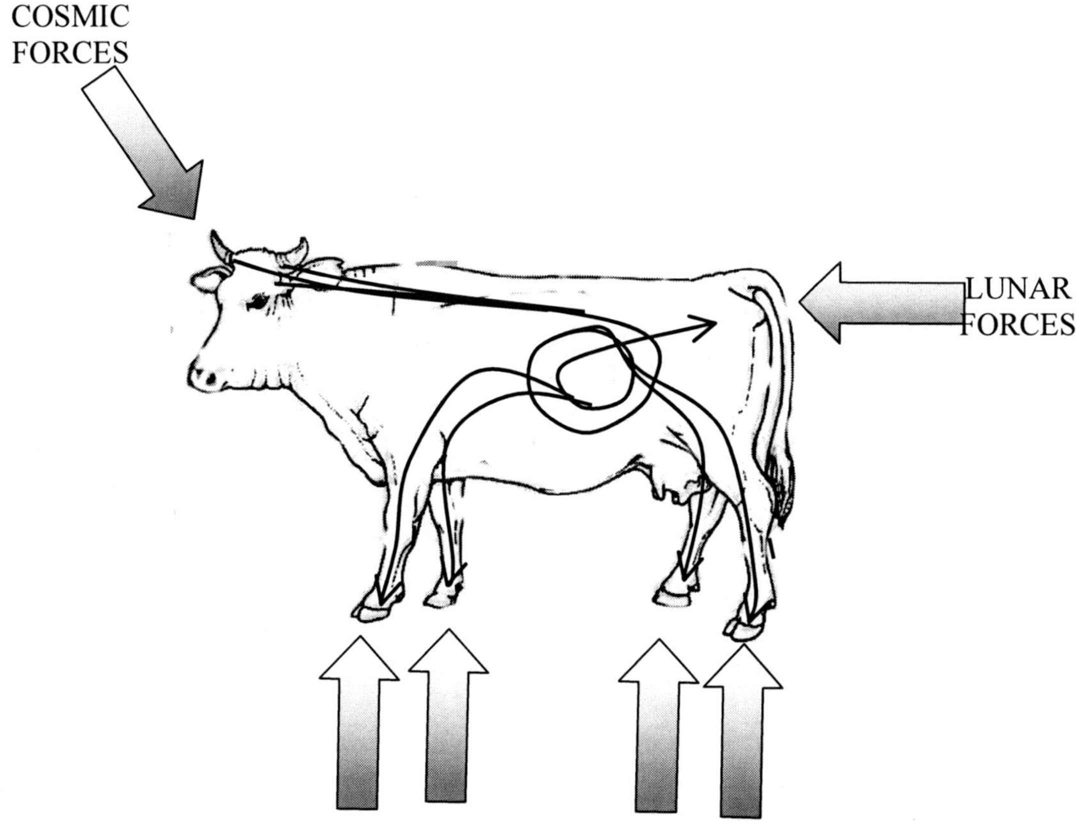

Steiner says that the horn collects cosmic forces. These cosmic forces come into animals, live within them, and then would exit. But they fail to depart from the tail because the forces of the Moon enter from that direction. Neither do they manage to get out via the horns which take in the forces of the Sun, because the horn's form makes it into a '*cul de sac*'. Even the route through the hoof is unsuitable because the earthly forces enter from there.

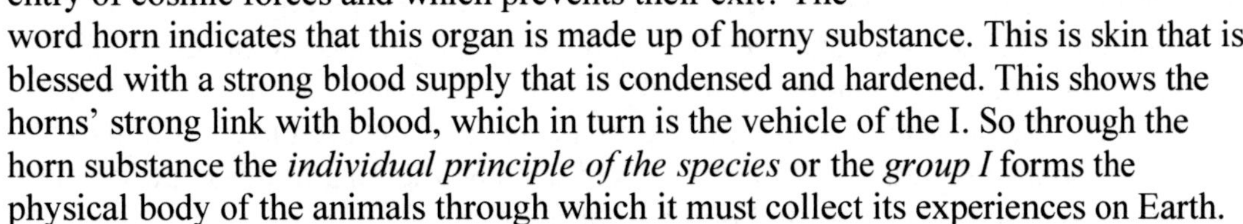

Therefore the animal can take forces from the cosmos with its front parts - horns, mane or whiskers depending on the animal - and these same forces remain securely within it. If this were not so, animals would be shapeless because their form is governed by the cosmic forces that are entering and departing.

But why is it precisely the cow horn that allows the entry of cosmic forces and which prevents their exit? The word horn indicates that this organ is made up of horny substance. This is skin that is blessed with a strong blood supply that is condensed and hardened. This shows the horns' strong link with blood, which in turn is the vehicle of the I. So through the horn substance the *individual principle of the species* or the *group I* forms the physical body of the animals through which it must collect its experiences on Earth.

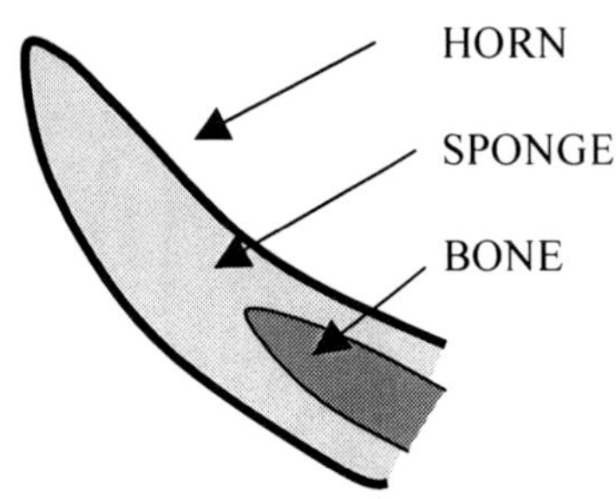

The cow horn is composed of three parts: the innermost is made of bone, the second part is formed of a spongy tissue, and the outer layer - the horn itself - can be forcibly separated from the two inner parts.

Therefore the bony part is attached to the skull and the horn is a prominence comparable to a root tip. The growth of horns is lunar and when the cow is in calf a ring is formed at the base of the horns. Thus the horns represent a metabolic process, pushed back, carried to the head, or more precisely to the back of the head. It is interesting to take note of this for all animals. The tusks of the wild boar undertake the function of horns and the tusks are placed in the front part of the muzzle, while the rhino horn is located in the central part of the head, and this gives a clear indication to the understanding of these animals.

It would be interesting to ask why these forces that reflect back into animals don't cause obstructions or block energy. In reality this does not happen because the incoming forces are used and therefore consumed for the processes of digestion and the forming of organs.

Now we come to twenty-second paragraph ("*Antler formation is something ...*") in which Steiner speaks of the animals that have antlers such as deer. Antler is profoundly different from the horn and the substance of which horn is formed. The most important difference is that the stag's antler is made of bone. We know that bones - even human bones - come from a condensation of warmth ether, while the light ether is the basis of the formation of animal muscle - meat.

The fact that here we have bones exposed to the light means the I has been brought to the outside while the cows' remain hidden inside. We know that the cow is an animal of gentle nature, slow or phlegmatic, while any animal that has antlers has a more extrovert and nervous character.

It is interesting to note that the stag grows antlers when the deer is pregnant and that the antlers drop after childbirth, as if the male accompanies the process of gestation. Pregnancy is a process concerned with the hindmost part of the animal, even if it is not a purely metabolic process. The cow is a predominantly metabolic animal mainly because it has enhanced digestive processes: just consider that it has four stomachs. The metabolic process, however, is a process of destruction: when we digest something we destroy what we have ingested and then assimilate it. The reproductive process is clearly a process of construction. Therefore the stag antlers must be viewed in the light of the 'construction' of the fawn, as if it were directly involved in the pregnancy of the doe. The formation of antlers corresponds to a more formative process than the cow. In the plant world it is yarrow that we use to create a link with the spiritual world to bring down formative forces. The cow horn is more linked to digestive processes and will be useful for the preparation of horn-manure, which is replacement for manure and which acts on the metabolism of the Earth. So horn-manure - the product of a horn related to metabolism - is used to manure the Earth. The yarrow preparation - which is linked to an animal that in this context could be defined as formative - serves to bring the form into plants.

As the horn is used to capture cosmic forces, the hoof is used to receive the earthly forces. The hoof could become interesting for making preparations in areas where there are no cattle. By now we have a reputation for our interest in the desert, and so we had to create a suitable preparation to replace the horn-manure for North Africa. We think it would be useful to consider a "hoof-manure" for which the animal sheath used is the camel's hoof.

Probably, in continents other than our own, it is not even correct to practice a type of agriculture based on preparation components from animals and plants typical of the European continent that bear etheric and astral forces typical of Europe.

So we come to the twenty-third ("*A cow has horns ...*") and twenty-fourth ("*This all points to ...*") paragraphs. For a better understanding of what is stated therein we must bear in mind that we find the maximum astrality in hollow organs. The most hollow organ we have is the stomach which is also the one with the most acidic environment. We have already pointed out that the cow has four stomachs. Then the cow has a long intestine well over 30 times the length of the animal and this is also hollow. The cow has within itself a volume of air of well over 200 litres. So we can deduce that the cow is a highly astralised animal. (Ruminating is an extreme of tasting which is a soul experience.) This is why what is expelled in the cow dung is impregnated with much more astral forces than those that entered the animal with the ingestion of grass. The ingestion of plants brings the cow a wealth of etheric forces that are mostly used by the animals for their vital functions and movements. The latter means both the internal circulation and the external movements are linked to the labours of the animals. So in between the journey from fodder to manure there is an impoverishment of vital etheric forces and an enrichment of astral forces, so much so that the uncomposted manure is a real poison for the Earth.

The second sentence of the twenty-fifth paragraph ("*Now suppose we take this ...*") contains a phrase that should awaken our interest in a particular way. It reads: "*What we would actually be doing is putting something etheric and astral into the Earth, something that by rights belongs in the belly of an animal and in the belly produces plant-like forces.*" The words "*plant-like forces*" means the process of vegetalisation that we have repeatedly mentioned. In other words Steiner is telling us that the manure can bring forces to promote vegetalisation of the Earth.

The sixth sentence deserves a word of clarification. It speaks of the watery and mineral components. The liquid element is an environment in which the alchemical ether is manifest and that encompasses the metabolic functions, cell division and cell proliferation, whereas the Earth element is the environment in which the Life ether is manifest.

All plant growth is outlined in the few words above: the impulse to life and the manifestation of life. To clarify further we can say that we are talking about the Earthly pole of the plant that we know is linked to the world of manure. Afterwards Steiner will speak of the other pole of the plant that is bound to silica.

In the last sentence of the paragraph Steiner tells us again of vegetalisation of the Earth when he said: "*Manure has sufficient forces to overcome even the inorganic earth element.*" To vegetalise the earth element one must be able to transform the inorganic to the organic, or the mother rock to humus.

Thus, we arrive at the twenty-sixth paragraph ("*Now, what we put into ...*") that begins with Steiner stressing that the manure must be put through a process of decomposition before being put into the soil. This is necessary because the manure must lose its previous form, which remains as if floating above its physical-material component. What should be brought into the fields is the set of formative-forces, for which manure is really only a vehicle, a physical support.

Towards the end of the paragraph Steiner says that the quality of animal manure is dependent on the balance of processes that take place there. The whole series of microorganisms and creatures that come and go in these composting processes are only indicators and witnesses that everything is proceeding in the best possible way. There is no point in bringing them in with the aim of improving the quality because if everything is properly set up they come along spontaneously. On the contrary, if the manure heap is not balanced it would be equally useless because these creatures would be inserted into an unsuitable environment for them; not only would they be unable to contribute anything positive but probably they would be sentenced to die. In the past there have been many fashions along these lines. One of the most famous involved the use of Californian worms that were promised to be the cure-all for every problem. The fact that there was so much talk shows that it was the usual commercial venture and not based on real knowledge. It is well known that someone, noting that fertile land has many earthworms, thought to assess the richness of the soil - and thus its market value - based on the number of earthworms present. It follows that the best way to increase the value of land would be to import truck-loads of earthworms. But much more useful are worm casts because it is extremely rich in living calcium and therefore is capable of conserving life. (This manure is particularly important for establishing dry-land agriculture.)

In the twenty-seventh paragraph ("*Next let us take the* ...") the specific exploration of the preparations begins, the first of which is the horn-manure or '500'. Steiner succinctly describes the stages of preparation without giving a precise depth for the burial of the horn. We think that Steiner is providing indications of a general nature that would be suitable for different geographical situations. We believe that, for our zones, the right depth is 120 cm. Also it is good to use calcareous soil that is rich in humus.

We should focus attention on the part of the paragraph in which it is said: "*Because the cow horn is now outwardly surrounded by the Earth, all the Earth's etherising and astralising rays stream into its inner cavity. The manure inside the horn attracts these forces and is inwardly enlivened by them.*" If we think about it for a moment, the etheric body of the Earth has withdrawn 120 centimetres underground, but there's not really much biological activity at that depth. Well, we think that Steiner was referring to the forces of Christ that became available due to the mystery of Golgotha. From this standpoint we can understand that what occurs in the burying of the biodynamic preparations is a process we might call 'Christianisation'. This is why the whole ritual of preparation and interment must be executed with an attitude of devotion and veneration, as a sacred act.

Steiner points out that to get a good horn-manure the burial should take place in a soil that is not too sandy. We actually use a calcareous soil but if you only have siliceous ground we suggest the incorporation of a considerable amount of compost around the pit in order to create an environment that is rich in vegetative forces.

In the twenty-eighth paragraph ("*Once the winter is over* ...") Steiner tells us how the first horn-manure made in Dornach had completely lost any unpleasant smell. With this sentence Steiner wants us to understand that the poor quality stinky astrality undergoes a transformation into high quality astrality. We know that this high quality astrality derives from the 'external' planets from which the four etheric forces flow, while the low quality astrality is linked to the internal planets, which in turn represent the world of desires and selfishness and are therefore linked to the emergence of pests. In a different worldview we also know that the interior planets is Purgatory, and the loss of any bad smells from the preparation can be understood as the transition into a condition that we could define as 'paradisical'. Thus the preparation has undergone a process of Christianisation: it has been purified, vivified and even resurrected.

We believe it is evident that the transformation within the preparation is not so much biological but spiritual, which then leads to an effect on the biological and physical. When Steiner, in the previous paragraph, speaks of winter as the period of the soils greatest vitality he refers to the spiritual vitality. Do not forget that Christmas falls right in the mid-winter and the birth of Jesus can be echoed in the birth of the child in us, the new man full of inner vitality. It is just at this time that we have the opportunity to create tools so that the Earth can be vegetalised and brought forward to its future incarnations.

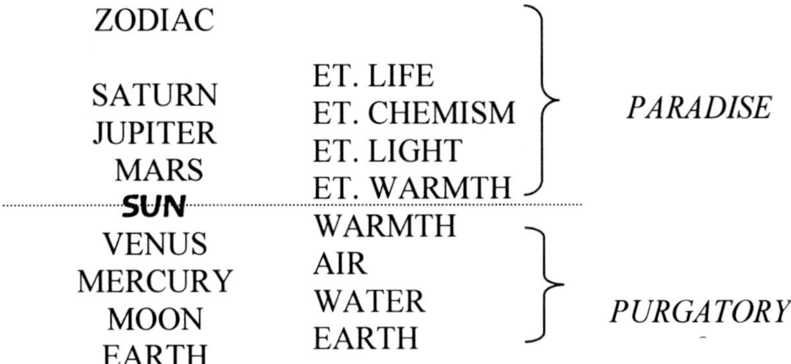

Steiner does not specify further how the horns should be buried. Our reflections have lead us to propose two alternatives, the first suited to a planetary consciousness, the second to zodiacal consciousness.

The first mode works with the horns arranged so that they form a circle. We have 100 arranged in the incarnating direction and four in the opposite orientation.

The idea with this arrangement is to avoid bringing in a onesided gesture. The 100 therefore represent unity at a higher level, whilst 4 is the number of manifestation of the way. In other words we can say that we want to connect to a higher level to get help at that level for the manifestation of life.

If we want to connect with the forces of the Zodiac - or we could say if we go to the 'house of the Father' - the horns are laid down oriented along the Pisces-Virgin axis with the horn tips facing the Virgin.

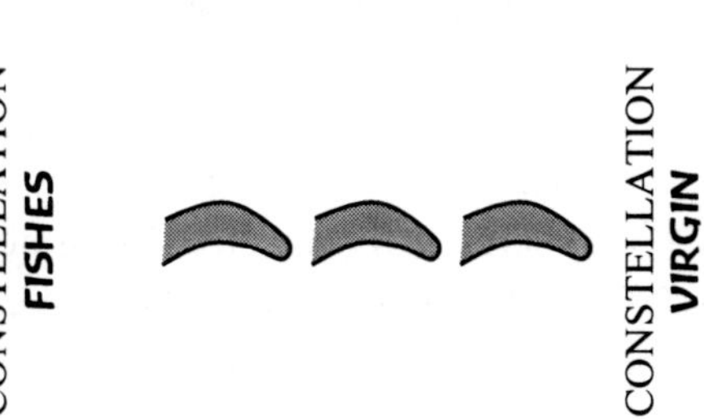

What we are saying is an extension of what Steiner gave us in Koberwitz in 1924. We should not think that what has been said at that time constitutes an insurmountable limit beyond which we can and must not go. After all, the same Steiner repeatedly exhorted us to develop what he was teaching.

For example talking about the horn-manure to

students outside the course, he indicated that a time would come when one horn would be sufficient to provide the necessary forces for an area that - with the 'traditional' doses – would require at least 150 horns. He spoke of a time when a certain process would have been activated. It seems clear that he was talking about a process within our consciousness that would permit the use of more advanced techniques.

With this in mind the use of horn-manure can be modified to increase its effectiveness for particular situations. For example in the case of land so poor or degraded that it fails to retain the messages carried in the homeodynamic preparations - even when distributed with the diffuser - it may be advisable to stir the horn manure in water which has been boosted with the addition of red grape juice. It is recommended to add three litres of grape juice to each forty liters of water used for dynamisation of horn-manure to spray on each hectare of land. The prospect is to increase the effectiveness of the preparation one hundred times.

For instance we have used grape juice successfully to fortify the nettle preparation to remedy lodging of a field of soybeans. The soybeans not only raised themselves up but grew a span higher than before. The grape juice is so effective because it draws upon the I organisation and therefore the individual principle of the species takes such a strong step into manifestation that it brings about the raising of the soybeans.

Returning to the twenty-eighth paragraph, the third sentence develops the concept of the change of the lower astral into the higher. Steiner, referring to the odour, uses the words "*concentrated*" and "*transformed*". In these two words is the principle of metamorphosis. In fact in the process of overcoming the form, a gesture is brought to its maximum expression – concentrated - and then it is transformed. For the metamorphosis between the leaf and the flower it is first necessary that the leaf develops its highest expression before the plant can express itself in the flower.

Metamorphosis on the astral plane signifies a passage from the internal planets to the external planets and this metamorphosis results in the transformation of the smell. In the next phase Steiner speaks of the '*immense astral and etheric energy*' that is the life forces of the external planets.

In the same sentence he speaks about dilution in ordinary water that should "*perhaps be warmed up a bit.*" Warmth is the energy that enables all manifestations of life. Using warm water opens this door so new life forces can flow. Perhaps we may insert and emphasise that if a farmer has enough enthusiasm - which is also due to warmth ether - one can do without warmed water. We note that the word enthusiasm means '*to be one with God'*.

This *being one with God* can help us take more further steps, and can even help us completely to bypass the physical media that we now use to bring the forces of the cosmos to the Earth and its plants. A first step in this direction is the use of homeopathy-like potentisation. This also should not be seen as an end in itself but as a transition towards a more direct relationship with the spiritual forces.

In the twenty-ninth paragraph ("*I have always found ...* ") Steiner provides indications of quantities – how much preparation is to be sprayed in how much water on such a soil area etc. It will not have escaped the careful reader's attention that these indications are extremely vague and we believe we can take proper notice of this if, once again, we consider this lack of precision as an indication that the quantitative aspects of biodynamics are of relatively little importance compared to other factors such as the awareness of farmers - which <u>is</u> fundamental.

Steiner finally gives indications on dynamisation. We have repeatedly spoken of this technique so we should not dwell on this once again. Just remember that during the creation of the watery vortex the liquid is arranged in internal surfaces one molecule thick. Well one litre of water can form a plane a molecule thick that can cover an area of no less than 400 hectares. Think then what surface is formed internally when we dynamise tens of liters of water. This large surface can be imagined as a parabola of a huge radar capable of collecting an impressive amount of cosmic forces. When you decide to change the direction of the vortex, the mass of water undergoes a collapse to return to the size of the vessel, and then it expands and collapses again and again. In this rhythmic alternation of expansion and contraction, comparable to systole and diastole, the connection is achieved with the solar dimension of life that characterises agriculture.

The need for alternate stirring directions is governed by the need to bring into the preparation both the solar influences we mentioned at the beginning. If one always dynamises in the same direction this will lead to the preparation holding only the forces of materialisation or only those of dematerialisation.

Remember also that the dynamisation requires one hour because an hour is the smallest temporal image of the twelve constellations and so it is the minimum period of time to establish a conversation with the whole Zodiac. Preparations are therefore not designed to link in with the planets, but with the Zodiac and perhaps even with the Father of the Zodiac which is the Milky Way. The Sun, which is a star, must be considered as an ambassador of the Zodiac in the planetary world.

The physical manure is connected to the Moon and thus is linked to the planetary sphere but horn-manure brings the archetypal force of humus and is therefore linked to the zodiacal sphere.

The next two paragraphs are easy to interpret. We can simply point out that in the thirtieth paragraph ("*Just imagine how little ..*"), when Steiner spoke of the pleasure of potentisation, he said that the farm household's sons and daughters could do this. In nominating the children Steiner refers to the current of heredity. But from another point of view the children are 'our inner children' that is our higher self, and this is invoked because the activity we are talking about is one of creativity. Even with the terms *'sons and daughters'* we can see the spiritual component (male element) and the soul components (female element), respectively recalling the zodiacal and planetary worlds.

Fifteenth Meeting

At our last meeting we talked about horn-manure and now, continuing with the fourth lecture from Koberwitz, we will prepare to tackle the second preparation proposed by Rudolf Steiner: horn-silica.

However, we believe it appropriate to develop some initial considerations to help us understand the scope of what Steiner is offering farmers and humanity in these pages.

We believe it is evident from the directions given for making the first biodynamic preparation that we are facing a new form of working to provide remedies to treat the diseases of the various kingdoms of nature.

Normally technicians identify the active ingredient in a plant that is capable of addressing a disease. They collect the plant specimens and make a decoction or tincture thereby freeing and strengthening the latent medicinal properties. But we have seen that Steiner does not only propose the creation of a manure heap in which natural processes take their course: the resulting material would be the bearer of forces of manure only in proportion to the degree of its development and transformation.

Steiner's guidelines for the preparation of horn-manure has shown for the first time a different methodology that is based on bringing together the various kingdoms of nature: we are guided to take the manure and a cow horn, put them together, bury them in the soil, and to leave them buried to concentrate the forces of the cosmos. We then remove them and put them in water and submit them to dynamisation for an hour. In this process the mineral kingdom intervenes during the interment, the plant is present in the water and in the origin of the manure, the animal kingdom is involved in providing the horn and the astral forces coming from the cow-dung and finally there is even the human input.

It should be obvious that we have described a process of synthesis, a process that is opposite to that normally used for the creation of medicines and natural remedies. The process by which we reach the identification and extraction of the 'active ingredient' is a process of analysis. It is based on analytical research of all the parts of a plant to recognise and discard all that is not thought to be of direct interest or that could be counterproductive, and to single out the sole principle desired. Not only does Steiner tell us not to remove any component but he tells us to place the base substance into an ever wider context – starting with all of the Earth and then bringing it into communication with the forces of the cosmos. This procedure is opposite to the creation because creation is a process of moving from an original Unity and coming to one of multiplicity. We take one fraction of this diverse multiplicity, join it with other aspects of diversity, put everything into Mother Earth and finally, in dynamisation, forge a link with the Cosmos. Remember that the laws governing the vortex through which we dynamise the preparations are the same laws that govern the movement of the planets of our solar system.

We can say that the suggested medicine is a medicine of the Father because it connects to all the forces of creation. Whoever has studied the Gospel knows that the way to get to the Father is through the Son, and in this context we can recognise the Son element in the Earth because for 2000 years the Earth has been the vessel in which the actions of Christ are manifest. The burial of a preparation means joining it to the forces of the Son or, in other words, Christianisation. But if this is such an important function we can not shy away from achieving it in the best possible way and to do this we have to study all the details, as befits any ritual in which every movement has a precise meaning. We recall in this regard what we said in the previous meeting about the identification of burial depth and the layout of the preparations in the pit. We have largely emphasised the need for these operations to be done with an inner attitude of devotion and veneration befitting a sacred act.

The enormous manuring capacity of horn-manure is precisely due to the fact that it has undergone the process that we described, which is not only the application of a technique to bring manure to a good level of maturity, but a process of spiritualisation that provides access to a new source of energy for life. The adjective '*new*' that we used is meant to express the concept of discovery and the possibility of appealing to the beings from whom Life originates, because we are well aware that these are the forces which gave rise to all creation and so are not new in an absolute sense. These restore an alliance that has been rejected by men so that we can resume talks without which the world may not have any future. Of course the conditions of the Earth are much changed from when life on Earth fell in the beginning and therefore the way in which man can be linked to the forces of Life must also change. The coming to Earth of the second person of the Holy Trinity should be interpreted precisely in the light of this need to re-establish a relationship on a different basis.

But we have to focus our awareness on the fact that our planet will not return to life with polluted fields of urea, potassium sulfate and every other kind of chemical fertiliser. Only by creating a new capability for representing all the cosmos and then allowing the Earth to resound again with the whole cosmos - the only true source of life - can we bring about the life we need. In what we have just said we can recognise the true principle of healing which is applicable to the mineral, plant, and animal worlds, and even to humans.

In this system of healing man is the link between the source of life and its manifestation and the healing forces of medicine will be governed by the awareness of the therapist. Therefore we can have a medicine which can resonate as far as the sphere of the Moon, another to the Sun, yet another with the Zodiac and so forth.

But we reiterate that a true medicine can never arise from a process of analysis, because analysis necessarily brings separation from the living context and it seems clear that life can only come from the reintegration.

The thoughts that we have formulated here are the key to understanding that all the preparations are no more than keys to restore a relationship with the cosmos, even if each preparation follows a different path and works with different forces, as we will see in what follows.

With Rudolf Steiner we will now look at the horn-silica preparation through deepening our familiarity with the process of dynamisation that, in our view, has not been properly understood even by the biodynamic world.

We will return to this point after commenting further on the "Agriculture Course " from paragraph thirty-two ("*Once gain take cow horns ...*"). We will consider both this and the next paragraph together in which, as we said, Steiner talks about the horn-silica.

We believe that these two paragraphs are not too tricky to understand. Recall however, that we made the acquaintance of silica and limestone in the first lecture and we had grasped their polar nature, while the second lecture showed how clay was inserted into this polarity in its mediating role. Then silica and limestone were presented in their relations with the planetary forces. In the fourth lecture the world of limestone is represented by dung and the world of silica by this fine quartz, which takes the form of a reversed bee cell because it is a hexagonal parallelepiped terminating in a pyramid that is also hexagonal.

Silica crystal

Silica, because of its transparency and its form defined by straight lines, represents the cosmic forces, while the terrestrial forces are represented by limestone. So in strengthening the growth of the plant, horn-manure strengthens the forces of Mother Earth and horn-silica strengthens the forces of Father Sky. These two preparations can be considered as the means to connect the plant with the two aspects of life, because the plant, like us, is the child of terrestrial and cosmic aspects.

Just as horn-manure was buried from Michaelmas to Easter so that it can be loaded with the forces active during the winter which are life's archetypal forces that descend during the Holy Nights (Life's archetypal forces), so the horn-silica is buried from Easter to Michaelmas to participate in the summer expiration of the etheric body of the Earth into the cosmos. Life on Earth during the summer yearns to be reunited with the origin of life and thus with the forces of the Father, and so expands into the cosmic distances. From a different point of view we could say that the horn-manure strengthens the forces of the Mother who is fertilised at Christmas, while the horn-silica strengthens the forces of the Father that all creation longs to join. This is why the plant first appears vital and vigourous, and only then completes the upper pole where the flowering and fruiting are the result of looking toward the cosmos.

We can understand the importance of horn-silica at a time when - between electromagnetic fields and various other types of pollution - layers have been created that tend to prevent the connection of the plant with the superior forces. The preparation is a true communion that allows communication to resume which has been made very difficult nowadays.

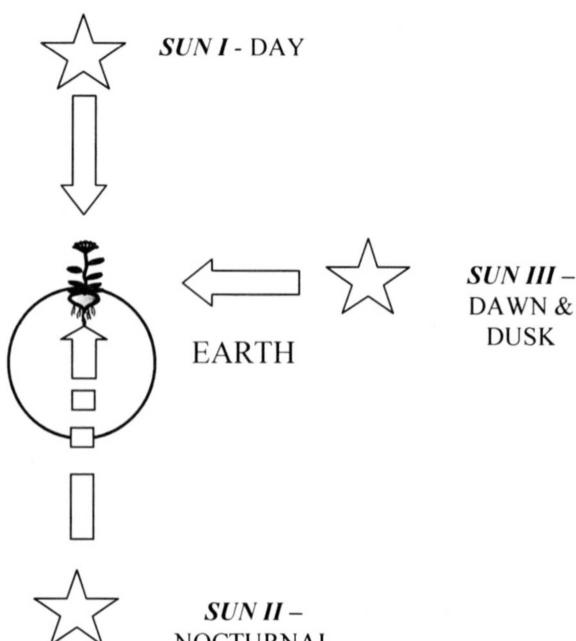

We recently argued that the plant looks toward the centre of the Earth (geotropism) and the Sun (heliotropism), but then we said that the geotropism is only apparent because in reality the plant intends to connect with the 'Night Sun' through the centre of the Earth. The plant lives therefore in this communication between the day-Sun and the night-Sun with these two preparations inserted to enhance this dialogue.

Of course, since everything in

nature is threefold we should try to find the 'third Sun' and then the third preparation that is able to reinforce the meeting between the two poles.

We can add another consideration to our reasoning: we learned that the horn-silica remains buried between Easter and Michaelmas, during the period in which the Earth is penetrated with light. In contrast the winter is the time of darkness. By now we also know that the ways to return to the Father's house were defined as the Mysteries of Space and the Mysteries of Time. The mysteries of space take place in light, while others, representing a development through inner work, take place in the dark and thus have a connotation of time. From this point of view the horn-manure preparation - linked with the winter, the dark and inwardness - assists the plant kingdom to find the road to the rediscovery of the mysteries of time and therefore has regard to evolution and the creation of new plants. Horn-silica, on the other hand, is linked to the forces of light and to following the path of knowledge and awareness. In yet other words, horn-manure represents the way of the shepherds and horn-silica the path of the wise ones.

Do not forget that the burial of the two preparations occurs during the periods around the equinoxes when the forces of Christ are active, linked to the constellation of the Fishes (spring) and that of the Virgin (autumn).

All these considerations lead us to conclude that in making the preparations the quality of the original substances is not as important as how they are buried and how we dynamise them, always being aware that the 'ritual' that is set up is like a sacred act.

We have already said that the farmer's act of going to the fields to spray the preparations is a devotional act that falls completely under his responsibility and has a meaning only if it is accompanied by adequate awareness and a proper inner attitude. At the end of the year or of his farming career, when the farmer looks back, the moments dedicated to these operations will probably be recognised as the most important because they are moments in which he is transformed from a farmer into an instrument for new life to come down to Earth. We are talking about forces that allow the Earth to become the new Sun.

We must bear in mind that the fate of the Earth is to become a Sun, but it is a destiny that will not happen if no one takes active steps that might somehow initiate that event. In this sense the farmer serves the evolution of the whole cosmos.

Returning to the creation of the horn-silica we can note that Steiner does not provide precise indications on burying the preparation. We suggest the choice of a siliceous soil, preferably in the mountains, and that the preparation is buried about 80 cm into the Earth. This should encourage penetration by the forces of the Earth's astral body, unlike horn-manure that needed to be penetrated with forces of the Earth's etheric body.

This is understandable when the etheric body of the Earth is compared to its astral body. During the winter the etheric body of the Earth is drawn inward and we can imagine that its limit is about 120 cm deep. During the summer the etheric body is breathed out, though not completely, and replaced by the astral body that remains in the more superficial soil around 80 cm deep.

This obviously applies to our latitudes [Northern Italy] because in different latitudes we must consider other things. This is why we suggest that everyone should make their own preparations in their farm, so the preparations may resonate with its particular etheric-astral organisation.

The export of preparations made in - say - Europe is a serious mistake that may render such biodynamics ineffective. Be that as it may nettles do not grow in the Canaries and it is difficult to find cows or deer in desert areas; to make a valid preparation in those locations it will be necessary to understand the various preparations and to adapt them to the etheric and astral situation of such places.

Repeating what we just said, in the following diagram we can represent the arrangement of the ethers around the Earth at noon which is when the Sun brings about their maximum chaoticisation.

In this depiction the inner circle is the physical Earth. Moving outwards we find the Life ether, the Alchemical ether, the Light ether and the Warmth ether outermost. We can imagine the action of the Sun as a chaoticisation of the subtle bodies of the Earth and therefore as the cause of the rotation of our planet: the rotation is a way to escape the chaoticisation caused by the Sun on the Warmth ether.

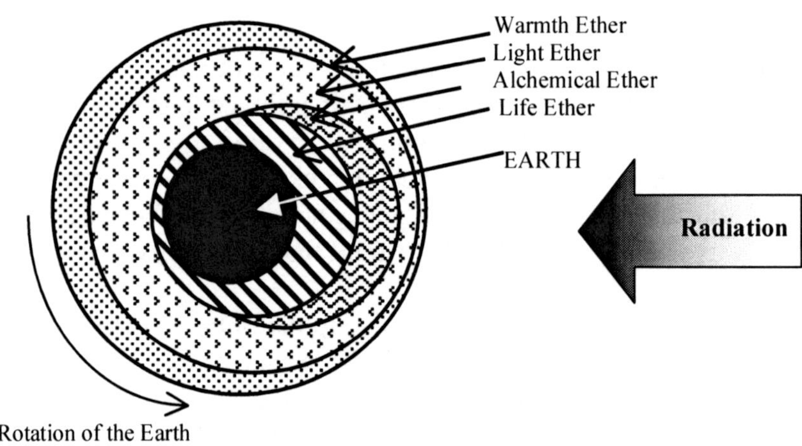

At this point we might consider something about dynamisation. We can begin by designing a container that must be flared, possibly with pads on the bottom, in order to encourage the upwelling of the water. The relationship between the width of the vessel and the depth of the fluid should be 1:1.618, which is the ratio of the Golden section corresponding to the proportion of the whole of sacred geometry and of the human body. From what we are saying it is clear that if different volumes of water are needed one should have tanks of different sizes. Use of the same vat with proportions above would not be possible for different volumes.

We all know that dynamisation requires the formation of the deepest possible vortex. The dynamisation that produces the vortex represented below is obedient to the famous laws of Kepler.

Let us try now to introduce some spiritual considerations to this type of dynamisation. First, we can see that when this kind of vortex is impressed upon the mass of water there is a rotating spiraling towards the centre.

The entire body of water receives an impulse to plunge into the depths of the vortex's centre and thus has a centripetal behavior. Light now enters the space vacated by the water. However, most of the body of water, especially in the lower part of the vessel, remains in darkness so this type of vortex makes sure that the preparation is mainly dynamised within the darkness. In other words, since Light corresponds to time and Darkness to space, operating as described above will achieve a dynamisation in time. It is a vortex that corresponds to the process of the Fall from Earthly Paradise and has a materialising effect. This type of vortex is particularly suited to the dynamisation of horn-manure that must resonate with the cosmos in a materialising way - bringing things to manifestation. By now we all know that the force of materialisation is typical of the so-called 'primary action' of the Sun. 500 incidentally has the function of bringing a quantitative impulse and from the spiritual point of view this may be called Ahrimanising in that it brings a separation from the spirit world. We can now better understand the importance that the preparation is buried and thus Christianised, but also the necessity that whoever works with these preparations has full consciousness of what is being done.

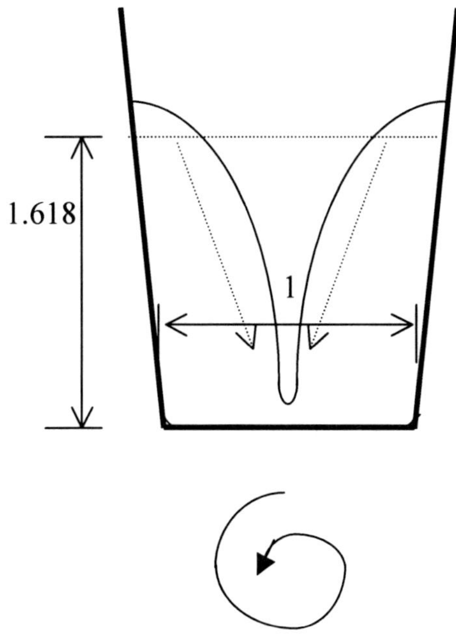

The proportions of the vessel and the 'dark' vortex

Do not forget that Rudolf Steiner said that the scientist of the future would come to the lab bench working with the same inner attitude with which the priest approaches the altar. We know very well that orthodox homeopathy uses vortices for production of the products, but the starting substances are not buried and therefore cannot be bearers of the processes of the evolution of consciousness. If we add to all this the fact that barcodes show the triple 6[25] - Ahriman's number – and are then affixed to the packaging of these products we have all the elements to understand what forces are conveyed through them.

We will take this opportunity to let you know that we have studied (and applied) a design that is superimposed over the barcode. This design is transparent so that it does not disturb optical scanning of the barcode, but which can neutralise the effect on the spiritual level. The picture below takes its cue from the famous words of Plato: "The soul of the Earth is nailed to the Cross".

---

[25] The first, last and central bars correspond to the figure 6 and assist the scanner to position the code and then read it in every orientation. For those who want to understand the influence of the bar code better we recommend reading 13.1 Apocalypse of John.

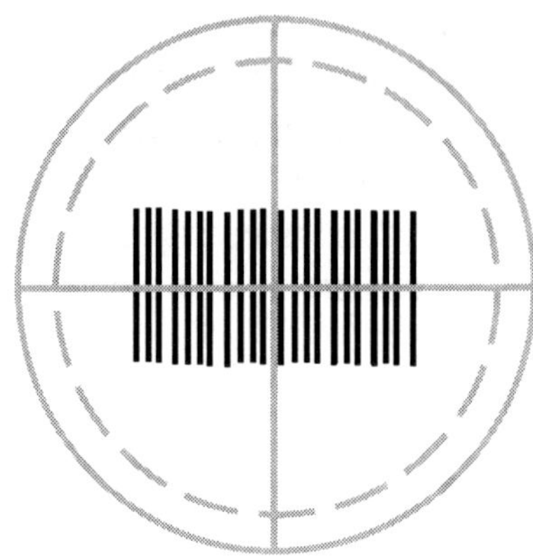

Neutralisation of the barcode

The second type of vortex occurs when the mass of water is not flowing toward the centre but towards the edges. The vortex formed has the shape of a cup. In this way water does not undergo a centripetal impetus to the depths but rises as in a centrifuge. This corresponds to a kind of dematerialising gesture, particularly suited to dynamise the horn-silica. In this case the light is dominant and so, in being encouraged to go to the periphery and towards the cosmos, it might be recognised as a luciferic action were there not adequate preparation.

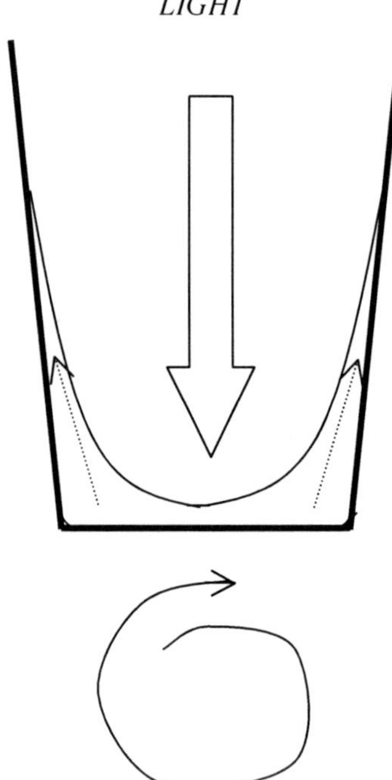

The 'spatial' vortex

What we are saying can take our minds back to the myth of Daedalus and Icarus in which Icarus, disregarding the warnings of Daedalus, adventured toward the Sun, ignoring that the wings were held together by wax that was not a suitable material for this adventure. This type of vortex is capable of bringing in a lot of light and so can be called a 'spatial' vortex.

To clarify further we can add that time, linked to Darkness, involves an inner evolutionary path and therefore has an affinity with Christmas and with the preparations which remain buried over the winter. Space on the other hand is the outward evolutionary path linked to Light and St. Johns, and preparations that remain underground during the summer. The preparation plants, as representatives of the plant kingdom, are inserted into these evolutionary paths and becomes bearers, through their own group I, of a message that enables evolution of the species in the Christian sense. This is the path that has been indicated by Dr Steiner for humanity to assist in the spiritual evolution of the Earth and its plants.

The fact that the plant is induced to improve flowering or rooting etc is totally secondary. This is merely the result of cosmic Life that is brought to Earth and made to flow more powerfully to the roots or the leaves or in the top part of the plant.

We can pursue the issue and consider the type of wood that is most suitable for constructing the stirring vessel for the two types of dynamising that we discussed.

The two impulses brought by the two types of vortices correspond to the two actions of the Sun, and knowing this we can be guided to two trees that bring solar forces: the lime and ash trees. The ash, for its posture and its physiological processes, will be most suitable to support the incarnative action of 500, while the lime will be fit for 501. If we want to be even more precise and effective one could think of using stirring sticks of the complementary timber in order to avoid one-sidedness.

We have seen two types of potentisation that resonate respectively in space and time, but we believe that we should always try to look forward, and it would therefore be interesting to find a method capable of potentising, in a single gesture, which combines space and time.

First you need to eliminate the bond formed within the vortex that brings one-sidedness. Then we can think of bringing both space and time - the laws of which were introduced into the mass of water by the vortex – into the same body of water. We propose that instead of moving the mass of water inside the container, the whole container is put in motion in order to connect its contents with space and time. In this way of working one must bear in mind that time is a kind of inner subjective experience, unlike space that can only be experienced externally. We think that time may be considered to be linked to purification of the soul, so much so that it is bound up with the number 7 which corresponds to the number of planets and which, in turn, represents the soul component of our solar system.

Space, and thus light, can be seen as related to the world of stars and to the longing to return to the origin of our system.

We can also say that time is a kind of 'Mary-like' experience, while space can be characterised as a Christic experience. In this way we found once again the frequently mentioned axis of the Virgin-Fishes that gives us protection and new life.

Now we can think of putting one of our preparations, such as 500, in a bottle and give this bottle a particular movement. This is a movement that traces the form of the water 'Vitaliser' that we have produced[26]. We can associate the idea of the darkness - and thus time - with the small inner circle, which we can see as the bearer of a purifying or action or as the activator of processes of purification of the land. We could also use the words *transformation* or *transmutation*. The bigger circle is

[26] This is available via l'Albero della Vita. Versions are available for single taps, 15mm supplies and for irrigation pipes.

associated with space and thus *vitalisation* or, better, *Christianisation.* Now we can better understand why the invigourated water creates a single centre - as is made evident by checking with the method of *sensitive crystallisation.* This shows the restored relationship with the spirit.

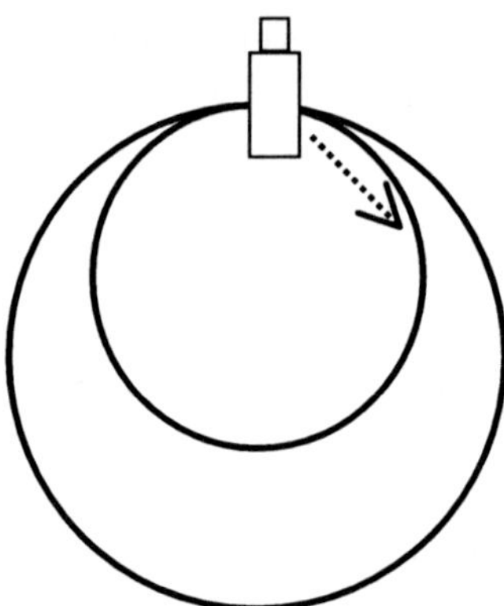

The gesture of the new temporal-spatial dynamisation

The figure that we have described is a third 'vortex' that no longer bears the Luciferic and Ahrimanic forces from which we must be protected, but which represents the new gesture which is capable of activating the forces of Mary and Christ.

In this figure, the inner circle is connected with the Virgin and the outer circle with the Fishes. The Virgin is a sign of the Earth while the Fishes connect with the idea of the 'Baptism of Fire'[27]. In the square of the elements with which we are now familiar, the route from Earth to Fire is the *via secca* or dry way through which the healing of Michael[28] moves. Currently all the preparations that we issue are dynamised with the gesture of the above figure. With this gesture we are both in the present and in eternity and therefore, through the door of the Fishes-Virgin, we can resound beyond the Zodiac to connect with the area where the 'germs' of Life originate[29].

To allow a better understanding of what we are saying, it's clear from the physical manifestation that we can recognise a single plant or what may be called the phenotype. On the other hand the genotype brings together all the plants of the same species. The archetype is the family to which a single plant specimen belongs and, further, there are the germs of the plant world. In other words, the phenotype is the single plant on the physical level, the genotype is the etheric plane, and the archetype is the idea. The existence of germs is generally ignored. Access to the world of germs is possible through the door of Mars, Jupiter and Saturn (the world animal, vegetable and mineral). But this is not enough because then it is necessary to trace these to their origin that is beyond the Zodiac, in the spaces between the Zodiac and the Milky Way. This is to say as far as the four quadrants of which we have already spoken elsewhere. Of course the opportunity to connect with the forces that are located in those areas is dependent upon a proper awareness, otherwise all efforts will be useless. We must realise that, like the ancient alchemists, we will be able to work in the outer world only to the extent that we were able to transform the world within us.

---

27 The Baptism of Fire is baptism with the Holy Spirit

28 See the Author's publication: "From the Healing of Raphael to the Healing of Michael."

29 Dr Steiner speaks of these 'germs' in his book Theosophy.

The remaining paragraphs of lecture four do not present significant difficulties. We will only emphasise that in the thirty-forth paragraph ("*You see, these things* ...") Steiner returns to indicate the method outlined previously - that the approach to any form of knowledge must follow a path from the general towards the detail. It is also an appropriate way for daily activity because Life always comes down from the Cosmos and we must always refer to the Cosmos. All of the preparations are bridges that are capable of linking the plant with the Cosmos, a bridge that must be open both ways so that the plant is in touch with the Cosmos and also so the Cosmos can learn from the plant.

At the end of the fourth lecture on June 12th, 1924 Steiner made himself available to answer questions from the delegates at Koberwitz. It is instructive to analyze at least some of these because they contain interesting ideas.

Steiner was asked the following question: "*Is it all right to use a machine to stir the manure for larger areas, or is that not permissible?*"

In reply, Steiner emphasises that it is profoundly different whether one mixes by hand or machine because: "*When a man works at a thing himself, he gives something to it which it retains.*" That is not a surprise for us, but it is difficult for the wider world to accept. Steiner speaks of subtle movements of the hand and we know that the hand is the archetype of all the animals' limbs. It is not coincidental that it is the only limb in nature that has an opposable thumb because it is actually an expression of human freedom. The use of the hand is equivalent to employing all of our soul forces. Above all it is the instrument to insert the specifically human: an element of freedom that we know is the other aspect of love. From this point of view when we spray a product that has been dynamised by hand onto a plant we bring the plant a seed of the far distant freedom and love. When we start to bring these dynamised preparations to nature we take a step towards becoming 'the tenth Hierarchy' and conquering freedom and love for all creation.

We no longer represent fallen man but the man who has picked himself up again, one that has transformed himself from a creature to a creator.

This must not, however, instill something like a sense of omnipotence but an attitude of extreme humility. We become mere instruments of the descending forces of the cosmos to redeem the Earth and infuse it with new life. Without humility none of this would be possible because with pride at the centre of our work our ego does not leave room for the forces of the Father and of the Son, who entrusted us with the conquest of Liberty, when in fact they have also entrusted us with the responsibility of enabling their activity.

It seems strange that the rhythmic repetition of a gesture can be considered free. However, the freedom does not lie in the repetition but in the decision to undertake an activity that is not obligatory. Repetition of an action for a certain period of time is just a way to imprint a message from the etheric: we should not forget that agriculture is the practice of cultivating plants, and they are only a physical and etheric body.

The last sentences of Steiner's response say that it should be fun to mix the preparation by hand so that ..: "*It could be something you do on Sundays after lunch*". It seems evident that this is not a reference to physical lunch because the timing does not coincide: horn-manure should be dynamised at around 6 pm. Evidently Steiner was referring to a different type of Sunday nourishment and to a new social reality based around these practices.

Further on in the question and answer session we can read: "*Where can you get the cow horns? Do they have to come from Eastern or Central Europe?*"

The advice of Steiner is to use horns of animals that live in the area over which one will spray the preparation. Then, making an example, he says that having made the horn-manure with an American cow horn perhaps you should compress it more. This is because in America the *Sal* forces are dominant, so if we were to make a preparation to be used in Asia we should avoid compression and leave it fluffy, as airy as possible. Anticipating questions extended from this issue we can say that the above principle applies when potentising as with homeopathy, with a different degree of filling the dynamisation flask. The development of a preparation for European areas requires filling the container approximately two-thirds full, while the development of a preparation for America would be fuller, and in Asia the liquid should not fill the vat so far.

Another interesting thought is explained on page 114. "*Is it important who does the work? Can it be anyone at all, or should it be an anthroposophist?*" The answer is obvious in the sense that we have repeatedly emphasised consciousness when we talk. In the initial phase of his response, Steiner says that with meditation one enters into a special relationship with nitrogen - where the cosmic imaginations live. Well we want the right idea or archetype to enter the plant, which enters through the nitrogen, and therefore, using the words of Steiner, we want the plant to enter into its image. The image of the plant can be helped to enter with preparations and especially with those

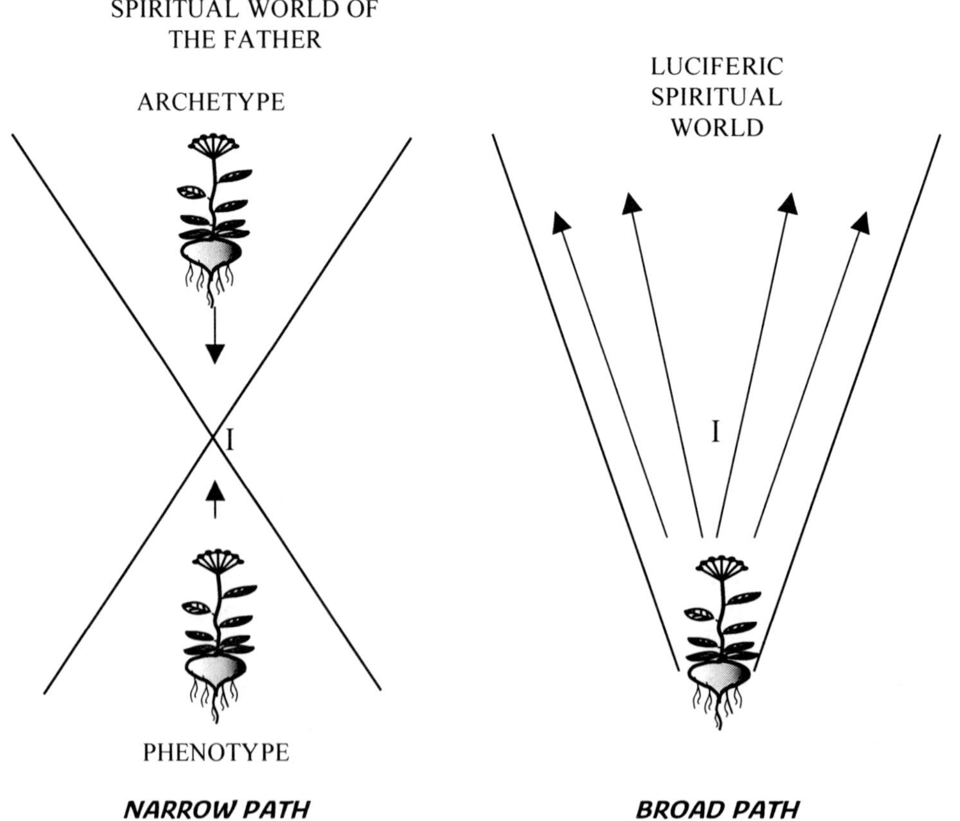

of the thirteen Holy Nights, but also simply by meditating. This seems like a confirmation of all our reasoning on the preparations, which we have always called 'crutches' to support a consciousness that is not strong enough alone. Sooner or later these preparations will have to be abandoned so we can begin to walk alone. To meditate means to focus on life incarnate, in order to broaden our awareness and ensure that the higher planes bring down the real Life, that of the Father, to penetrate the manifestation of life that is the plant.

This meditation is called the 'Narrow Path' in the Gospels. Traveling this narrow road requires a metamorphosis which first involves the intensification of a principle followed by its overthrow. We become the centre because we reached our most intense concentration upon some manifestation and then turned our consciousness to the higher worlds where we find the true Life.

In contrast the broad road exploits our given natural abilities, dependent on our karma or other reasons, enabling direct connection to the upper planes of Life. It can be atavistic, opening up old chakras, or - if cultivated indiscriminately - can lead to a strengthening of the link without the development of consciousness. Following the broad path one is clearly not obliged to pass through the 'eye of a needle' which makes the fruit of suffering available because of troubles which have been conquered. The pilgrim on the narrow road must, as advised in the Gospel … "*be ye therefore wise as serpents, and harmless as doves.*" The comparison with the purity of the dove is easily understandable; to understand the first comparison one must think about the way in which the snake becomes free of its old skin like removing a wetsuit. The snake will wedge itself between two stones so that the old skin is removed by friction. Even our rebirth requires a change of skin in addition to purification.

The 'paradise' that is easily reached via the broad road is not the paradise of the Father but that of Lucifer. It is also accessible with the use of drugs and the opening of chakras in a manner linked to the past.

There is nothing that can be obtained down the broad road that cannot be achieved on the narrow road, only this requires much more effort. However, it is the only one that avoids being lost once we have crossed the threshold of the spiritual world ('the eye of the needle') because we are aware of every step we have taken.

Fifth Koberwitz Lecture

Sixteenth Meeting

As we said, the fourth and the fifth lectures form the heart of the whole Koberwitz course. In the fourth lecture Rudolf Steiner presented the first two biodynamic preparations and now he will present us with the others. Steiner, in our view, is putting forward two levels of alchemy: the first is of a personal nature where man is called upon to initiate processes of profound transformation in harmony with spiritual laws. The second involves the forces of the Earth to complete the work begun by the human spiritual transformation.

Steiner thus allows us to approach the unknown world of the preparations and leaves the door ajar. At the same time he gives us indications of a whole method that, if properly and thoroughly understood, enables us to consider the creation of many other preparations. We do not believe that Steiner described the biodynamic preparations intending to establish the limits beyond which further intervention is prohibited. Rather he has given archetypal indications of an alchemy by which man can stimulate the transformation of the Earth. Man is not the protagonist in all this because the transformation of the Earth will be accomplished by all the Beings of the universe. However, we cannot escape from our role of activating a series of processes that cannot proceed without the participation of our consciously directed will.

When we create a preparation we must be aware of the fact that we are producing something unique that can resonate with all spheres from the mineral to the most spiritual including the organic-biological and that of the soul.

Nobody in the history of mankind has achieved the like although Paracelsus and Lullo before him had tried to take the road then traveled by Rudolf Steiner. But neither of these great people were of comparable stature. Then we were still guided and inspired in our thinking by the spiritual world, but now humanity requires knowledge and freedom.

Today we are in an age where almost everything is easily achieved thanks to the enormous progress that technology has made, but never has it been as difficult as it is today, even on a practical level, to create a biodynamic preparation because it has become almost impossible - without great efforts in which the will is put through severe tests - to procure the necessary materials including cow horns, stag's bladders, and so forth.

As we dip our toes into the fifth lecture, we must bear in mind that in the fourth we were oriented within a great polarity between the forces of the Earth sustained by horn-manure, and the forces of the universe supported by horn-silica. Starting from this polarity we can build up a picture that will enable greater understanding of the role of the preparations that we are going to meet. To this end we will represent the horn-manure with blue-violet colour (a colour that characterises the Earth element) and the horn-silica with yellow (Air element). We can imagine that the intermediate area is occupied by all the other preparations and so is coloured green – a colour that characterises the plant world.

Or, if we see the polarity between light and darkness in the polarity of the horn-manure and horn-silica, in the intervening space we may envisage the colours, one for each preparation. According to Goethe's theory, the colours are indeed the product of the marriage of Light and Darkness (if we include any relatively opaque in the latter category).

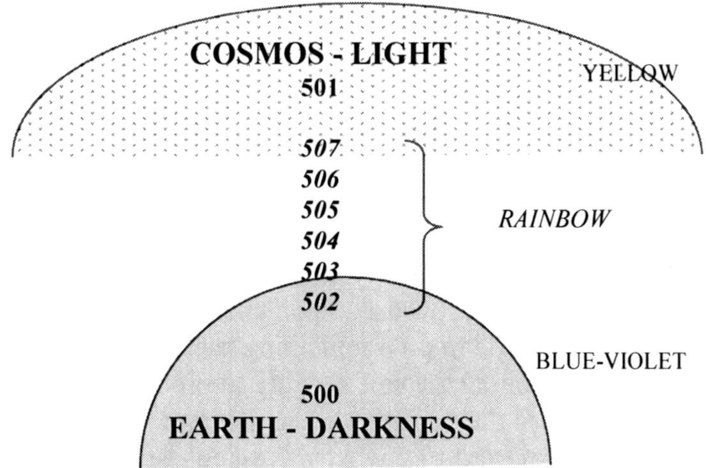

The preparations we are approaching are known collectively as the 'compost preparations', but we will see that their insertion into a compost heap is only one of their many uses.

We are entering a world which is the synthesis of all worlds and so we must be prepared to take many points of view and to seek many fields of application of the principles that we will meet. The knowledge of alchemy that is entrusted to us finds in our individual will the prerequisite for reaching the world, but we can also find our inner limits.

The preparations that Steiner is proposing are the flagstones for the route that leads us to work with nature, but these stones can lead to much more with a deep understanding of the method that was proposed at Koberwitz.

The transcript of the fifth lecture was entitled: "*Observation of the macrocosm, the task of spiritual science: building the heart of the soil and the growth of plants.*"

It is an extremely explicit title in which the word order establishes a sequence of activity for those who want to practice a certain type of agriculture. First of all you must learn to observe the macrocosm. Attention must then be paid to improve the heart of the soil and only at the end to the growth of plants. Normally today those involved in agriculture are primarily concerned with plants, so much so that the soil is regarded only as an inert support medium and of course there is not the slightest chance of farmers raising their eyes to the heavens.

Once again there is a suggested approach to agriculture which must take its point of departure from general aspects and only at the end come to the analysis of particulars without shredding up our fields into infinitesimally small bits.

Do not be surprised that in the first few paragraphs of the fifth lecture Steiner dwells upon reaffirming these methodological aspects. Our own experience is that these principles cannot be stressed enough, because our daily mundane experience impels us in a totally opposite direction. It is extremely easy to fall into the trap and seek solutions in the infinitely small.

We will not defer any more and begin to comment upon the first paragraphs of the lecture given on the thirteenth of June.

Particularly worthy of note in the first paragraph is a statement: ".... *the etheric vitality must be retained within the realm of the living.*" It seems superfluous to point out that this statement excludes any type of mineral-chemical intervention where, once again, the soil is considered as an inert support and nutrients are brought in the form of absolutely lifeless synthetic minerals.

In the second paragraph ("*Now, as I said ...*") Steiner reaffirms that the plant is the completion of a process of vegetalisation that the Earth is not capable of completing alone. However, this is a concept that we have already largely covered.

The third paragraph ("*Let me start by saying ...* ") returns to the concept of farming as an exploitation of the Earth. In this regard we have already mentioned the fact that farmers leave the soil fallow periodically so the soil can regenerate itself. The modern approach to agriculture is normally based upon taking all that you need from the fields, and when you *do* decide to give something back to the Earth this is always done with a view to maintaining the soil's ability to produce more and thus to reap yet more again. Steiner stressed the necessity of beginning to reverse the relationship with the Earth and that we begin to restore what was taken away, "*in order for [manure] to acquire the capacity to properly vitalise the depleted soil.*" What Rudolf Steiner presents us with is a moral relationship with nature. The preparations must be seen as a way to bring those forces needed for the Earth's own evolution and which continual cultivation depletes. To this end it is very important to make the preparations in the correct way and their influence must be rightly understood, because a moral act must take place with an awareness of what is done. Even with all the best farming techniques if our sole purpose is to make money does this align us with the moral vision that we talked about? Only if we act with the interest of the Earth uppermost do we make it an act of love.

In the fourth paragraph ("*For example, scientists of today...*"), through the example of flies, the errors are highlighted in an approach that seeks a solution to a problem in a purely analytical way. Unfortunately this is the dominant approach in science today.

We are now in the early days of January 2003, a period of the year in which flu usually gets a grip on the population. According to the common view, the flu is regarded as an enemy to health and fought with vaccines, because it considers the virus to have brought the problem, more or less like the flies in Steiner's example are considered to be the cause of dirt. The word '*influenza*' should help us understand that this disease is governed by the influence the planets have upon our body. The flu is a sign of an incorrect relation with cosmic forces that calls for a purifying reaction whose sign is the fever. Raising of our temperature indicates a strong action of the I within the lower bodies. It may seem to be just a linguistic coincidence that the second month of the year is called *Febbr-aio* (the bearer of fever). Remember also that the second of February is the celebration of Madonna of the candles, Mary who purifies - or Candlemas - which falls 40 days after Christmas. The flu is a sign that man has become a little more cosmic and so a little more open to the forces of heaven. In this light, vaccination is shown to be totally detrimental especially because it acts through a person's blood and takes us away from a connection with the cosmos. Besides concerns about vaccinations are shared by a growing number of people who also belong to the scientific establishment, but unfortunately the economic interests that gravitate around the business of vaccines affect the whole system of health. Vaccines are still promulgated as essential for human health, and it is also the subject of similar propaganda campaigns for animal health. Veterinary surgeon are continuously offered these products although most veterinarians will recognise their futility. We can only repeat again and again that the solution to these problems must be sought in the infinitely large and not the infinitely small.

The fifth paragraph ("*Thus, when animal manures ...*") makes clear that micro-organisms, considered the agents of the transformations that occur during the process of aging of manures, are really nothing more than a symptom of what is happening. These symptoms can be very useful for diagnosis but cannot in themselves be considered particularly essential because once again the forces that determine the process should be sought in the great and not by assigning causes through atomistic observations of these small beings of nature.

It is difficult to accept this point of view because it runs counter to a conception that has been drummed into us since birth and is the point of view that is normally used in a context that affects our fears, which is that of health. Normally, a doctor is concerned with treating symptoms at the level of the physical organism and not removing the real causes of the diseases themselves that, as we have seen, are acting on a completely different level.

Maybe our scientists, who engage in the study of small particles with increasingly expensive methods and disproportionately poor results, could profit enormously from recalling what is written upon the 'emerald tablet' of Hermes Trismegistos who says: "*That which is above is as that which is below, and that which is below is as that which is above, to perform the miracles of God.*"

We must also, in using the preparations, emerge from the limited view and begin to sense that we are interacting with all the forces of the universe for the good of the Earth and our plants.

The sixth paragraph ("*Of course, making a statement...*") states a principle that we have repeated many times during our meetings: that it is useless to know what is right if nothing is done to achieve it. Similarly it is pointless to criticise those who use wrong systems in the various fields of life if we are not able to offer an alternative that would not only avoid those mistakes but also be technically and economically viable. Indeed insisting on criticising the negative without offering viable alternatives only irritates people and is therefore counterproductive.

The seventh paragraph ("*A second result...*") contains an apparent contradiction if it is interpreted as meaning that the minerals are able to vivify the water. In reality Steiner does not refer to this or that mineral but to the chemical compounds that, as children of the Alchemical ether, are linked to the world of water and are able to stimulate the processes of the world of water represented by leaves. These types of mineral elements are not able to raise themselves to the Life ether. In fact through the water, in which they produce chemical reactions, they are not of the level which brings life to the Earth. Our effort should instead be directed precisely to bring life in the soil through the formation of humus, the presence of which, amongst other things, will reduce the demand for water.

With the eighth paragraph ("*Now in current agricultural...*") he concludes the part of the fifth lecture that we can consider as a preface to the real central issue, in which essentially he has repeatedly suggested that one looks to the large scale, because preparations are the gateway to reach the planetary forces and the zodiac. On this basis we can now proceed with the subsequent paragraphs.

In paragraphs nine ("*What, then, is the real...*") and ten ("*We must ensure ...*") Steiner lists two series of substances. The first list includes nitrogen, phosphorus, potassium, calcium, chlorine and iron. The second has silica, lead, arsenic, mercury and sodium. The first group of elements is commonly recognised as being of fundamental importance to agriculture, so that their absence on the ground can undermine prosperity, while the other group is recognised as having a stimulating function at most.

Steiner points out that this view is incorrect. It is due to the widespread experience that if the substances belonging to the second series are depleted it is rare that a farmer would have the negative consequences which are evident should the farmer fail to reintroduce the elements belonging to the first series as he works upon the land.

Actually those that are defined as stimulants are of primary importance and are more important than the others which we do not notice only because they are directly and freely given from the cosmos.

Incidentally, we note that Steiner uses the term '*stimulant*' and '*fine dilution*', which are often in the vocabulary of those who practice homeopathy. Also note that with regard to the two series of substances, Steiner speaks of the first as a "*free gift from heaven*" through the rain, and then of the other radiated from the cosmos to the ground followed by absorption by the plants. We can then place in our memory these two possibilities of communication with the Earth that can be clarified – simplifying - as one path of communication through water and another through air.

The other group of substances should be brought in physically, and so through the Earth element. What is missing is Fire that is represented by humanity with our direct intervention as the fourth possibility of communication. Indeed, given the progressive deterioration of other means of communication for the many reasons we now know, the fourth firey path becomes more important. In this latter path communication is no longer passive but becomes active.

What we have just said allows us to draw the diagram below, upon which we will elaborate as our discussion develops:

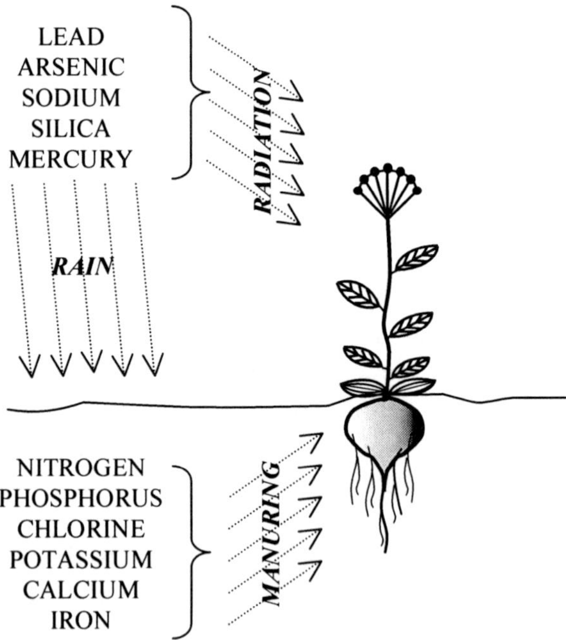

We have depicted that the substances drawn above the Earth are acquired through the rain and the air, whilst the substances listed below must be brought in physically.

A person can, through haphazard (chemical) fertilising, make the land incapable of absorbing substances from the cosmos. It seems particularly relevant that Steiner says that plants build up their bodies thanks to the absorption of those substances and thus we are talking about the cosmic nutrition of plants. So we can enrich our diagram showing the two nutritional paths: the cosmic one from the sky, and the earthly, obviously from Earth. It seems obvious that if human activity inhibits the first, we must intervene deeply with the second, confirming what we have often repeated. Obviously, if Steiner spoke today he would not merely talk about "*haphazard*" fertilising, but also of electromagnetic fields, hybridisation, genetic manipulation, and so on. We believe that even badly made organic fertilisers can act in the *haphazard* way defined above.

The twelfth paragraph ("*For this reason...*") contains the first steps towards healing. Immediately on the first line are the words: "*the manure must have proper care.*" Then in the second sentence Steiner says it is "*a question of infusing the manure with living forces, which are much more important to the plants than the material forces, the mere substances.*" This last clause should be indelibly engraved in our memory. The predominant mentality maintains that the important aspect is the matter and never speaks of forces. Here we are told that what counts are the forces that surround the world of life. This highlights the importance that those forces can flow, because the plants are daughters of the universe and not of the ploughed earth that is only an intermediary.

The homeodynamic method provides a means to eliminate the haphazard element mentioned by Steiner and this is the homeodynamic preparation 'Purifier and Harmoniser'. Now we can understand a bit more that the extraordinary results we sometimes achieve with this preparation depend on the fact that it is capable of a purifying role upon the previous haphazard fertilisation and harmonisation with the macrocosmic complex.

The twelfth paragraph mentions the homeopathic method and these references have formed the basis for our method. Of course it is necessary to practice a form of potentisation which is suitable for the plant kingdom, which is obviously different from the human kingdom; let us not forget that both plants and people are formed along a vertical or Solar axis but with the opposite orientation and therefore to use just the one way of thinking for healing them both may not be the appropriate.

Thus, we arrive at the thirteenth paragraph ("*When it comes to manuring ...*") that would be especially important to quote in part: "*But we must also experiment with other ways of giving the manure the right degree of vitality and the right consistency. We must enable it to retain of its own accord the proper amount of nitrogen and other substances that it needs in order to bring vitality to the soil.*" The key for understanding this phrase, in our opinion, is: "... *retain of its own accord*[30] ..." This means that the manure or compost must undergo a transmutation. So we can summarise what we have just read as follows: the upper world must bestow its forces with the rain or by direct radiation and the world below must be able to implement a transmutation. Remember that nature, or if you prefer the world of life, is the stage for a continuous alchemical-type transmutation just as the chemical reaction dominates the inorganic world. The organic world lives completely separately from these inorganic laws. It is therefore *not* so important to ensure that the land contains the various substances in the right percentages, but it *is* important to bring that soil transmutative information. Our preparation 'Pro humus' is the 'distillate' of this

---

[30] The Italian says ".. arrivi da sé ...". This gives more of a flavour of arrival of something not previously there, rather than retention of what was already there.

transmutative information for the soil that can thenceforward do with less compost.

The fourteenth paragraph ("*I am going to ...*") should not present significant difficulties so we will limit ourselves to anticipate Dr Steiner and reveal that the preparation ingredient that cannot be replaced by anything else is the nettle.

In the fifteenth paragraph ("*To begin with ...*") Steiner speaks of potassium and its importance. Potassium is already known to promote the growth of plants especially in their trunk or stem. Saying this, however, he also cites carbon, hydrogen, nitrogen and sulphur that, with oxygen, we know to be the components of protein. At the end of the paragraph Steiner clarifies that what he is proposing is indeed the way to prepare potassium to relate properly within organic processes that lead to the formation of protein.

Potassium, on the level of planets, is linked to Venus and the Moon, while at the zodiac level it is linked to the forces of the Virgin. We know that the Virgin constellation brings forces of generation and purification. Therefore potassium should have the ability to connect to life and purify it so that new life can be generated.

Please note that elsewhere we have said that the Virgin is the constellation behind which you can see a cluster of more than 2,500 galaxies. This concentration of galaxies can be considered like a "placenta" from which the solar zodiacal system has drawn its life. This placenta was connected with the Zodiac by an ideal umbilical cord that ended at the constellation of the Virgin, which can be considered the door of life for our system. Similarly potassium can be considered the gateway through which the life of the universe can arise in the world of matter.

It is, however, important that we understand the importance of purification so that new life can arise. The new life in the soil that we are considering is humus!

To reinforce the idea of the link between potassium and the constellation of the Virgin, we can say that the Virgin is represented in eurhythmy by a figure that recalls the gesture of an embrace and expresses the letter B. B has a gesture of protection and support, and we just saw that potassium is involved in the plant mainly to the formation of structural tissue. Moreover, the word 'tree[31]' starts with the syllable *Al*, which expresses a connection with the spirit world or, if you like, the divine protection. This is immediately followed by the B, which characterises the perennial or trunk aspect of such plants. We have said that the B shows a gesture of protection and this letter occurs three times in the word 'father[32]'. So the bark, which covers the trunk and the tree branches, has such a function of protection for the plant.

Incidentally note that the E is linked to the forces of Mars which expresses itself in the conquering of the inorganic and in the pattern in which the branches leave the trunk (which is clearly visible in the E whose essence is given by an x with a small cupola above - ⚲). The R represents the movement of the sap and the O the final crown.

In other words, potassium is the substance bound to the forces of the Virgin Mary. What we are saying is wonderfully and artistically reinforced by what Raphael has created in the Sistine Madonna.

We will now go, with Steiner, into alchemy through the creation of a preparation that is the bearer of the message of purification and vivification so everything of which we have just spoken can materialise. In other words, what comes passively

[31] 'Albero' in Italian.

[32] 'Babbo' in Italian

down from the cosmos through the rain or through the Light can become active, through people, thanks to the preparation Steiner is about to give.

In the sixteenth paragraph ("*Take some yarrow ...*") Steiner introduces us to knowledge of yarrow by making a wonderful description. Yarrow is a plant rich in potassium and sulphur. If potassium is connected to the forces of the Virgin, sulphur is connected with those of the Twins (Gemini)[33]. We have already met the constellation of the Twins when we talked about the cross of the Earth elements and its two axes Virgo-Pisces and Twins-Archer.

It is practically unique that the yarrow succeeds in using the sulphur which is not a central element in the life of the plant world, but instead in that of the animal and human, in the same way that sulphur is the substance that enables relations and balance between the components of protein. In the animal and human worlds, yarrow is able to improve everything that stems from a weakness of astral body. This ability derives from its connection with Venus and with the Virgin. Let us not forget that our astral body has a point of connection with the physical body in the kidneys which are the bearers of the forces of Venus within us. From this point of view the yarrow may be useful in treating weakness of the visual apparatuses, disturbances of blood pressure (when they are of renal origin), dizziness and more.

In the plant kingdom there is not, as in humans, an organ which incorporates the forces of Venus, because the plants have no incarnated astral body, so yarrow focuses its operations in liaison to the group-soul of plants. We emphasise that the yarrow is presented as an exceptional plant that is extraordinarily useful in fields just by its mere presence.

In the seventeenth paragraph ("*Now here is what ...*") Rudolf Steiner explains in detail the procedure for making the preparation. We note that after insertion in the deer bladder, the preparation must be exposed to the air during the summer and then buried at a modest depth during the winter. Steiner is still talking about cosmic nutrition although this is not immediately obvious. In fact, the exposure to the air during summer exposes the preparation to the radiant 'nutrients' of the cosmos while the burial at a modest depth exposes it to the elements carried as the rain percolates into the soil.

Paragraph eighteen ("*Then take this ...*") speaks to us of the ability of the yarrow preparation to radiate its influence, even if distributed in small quantities within a manure pile as large as a house. The indication of quantity is vague but serves to give the idea of its great power. We also find confirmation of what we said a moment ago about the purifying and vivifying influence of yarrow because purification is analogous to '*refreshing*', and '*enlivening*' clearly relates to vivifying.

It also confirms that the incorrect way of farming only apparently impoverishes the soil in regard to earthly nutrition. In reality the theft that is perpetrated relates to nutrition from the cosmos. So with the yarrow preparation we can restore the dialogue because the deer bladder filled with yarrow flowers remains for a few months exposed to forces of the universe and then is interred for six months. Of the two forces within the preparation the dominant one is the second acting through the rain in our drawing. The radiating fraction acts mainly during the Holy Nights when the yarrow is used in the form of the 'Opening to the Group I' (Homeodynamic product G01).

Our drawing can now include references to the yarrow preparation (or 502) and to sulphur.

[33] See "Agriculture as Spiritualization of the Earth: the Christianization of Nature", by the author. Not yet in English (2008).

Paragraphs nineteen ("*Now – since it is always ...*") and twenty ("*Here we need to have ...*") deserve a combined response. In them we are presented with the animal sheath that is the stag-bladder in the case of the yarrow preparation. Steiner reveals why such an apparently strange organ is used. We believe it is necessary to grasp the fact that the stag is a very astralised animal as can be observed from his very nervous behavior. The deer is an animal that lives in very close connection with the territory where the astral forces dominate and for which the stag antlers are excellent receivers. As astrality has its point of contact with the physical organ of the kidney, it is the organ most suitable to strengthen the yarrow preparation's influence. However, since the kidneys are not hollow the organ that is most closely tied to the kidney, from both spatial and functional points of view, is indeed the bladder. The kidneys have the typical form of astrality - like a bean - but because they are not hollow, the forces pass around the outside and are conveyed through the ureters to the bladder. Then the bladder, being hollow, can gather and concentrate these forces. It is no coincidence that urine is rich in nitrogen and that the urinary function is highly influenced by changes in emotions, even in humans.

Let us now consider another point to help us understand the scope of what we have just read. In the first paragraphs of this fifth lecture Steiner has apparently only introduced us to an understanding of yarrow. In fact, in his customary and not completely explicit way, he has already introduced all the other preparations. We have just talked about the yarrow preparation (502) and we have located its position on our sketch when we talked about the descent of forces from the cosmos by way of the rain. The second method of descent through direct radiation is favoured by the dandelion preparation (506).

Iron is twinned with nettle (504), while calcium will enter into preparations made from chamomile (503) and oak bark (505). The preparation for potassium is yarrow again (502), while phosphorus is the essence of valerian (507). Nitrogen is not tied to any one preparation but this is not an oversight. Recall that in the thirteenth paragraph ("*When it comes to manuring…*"), Steiner said that the manure can '*… retain of its own accord ...*' thanks to stimulating the living processes of transmutation that introduce as much as is necessary. Further study reveals that living calcium and potassium can transmute into nitrogen. Obviously we mean a nitrogen which is the child of nature's own alchemy, not the inorganic nitrogen which is widely used in conventional agriculture. Now we can also understand better why it can become superfluous to fertilise soil with manure. Manure is only a means to bring the forces that can more effectively be brought in with preparations.

At this point we can place the result of these considerations onto the drawing and, indeed, of further considerations which we will complete later.

Let us now expand our vision of what has been said. Lead and arsenic are two elements that recall old Saturn, while silica and mercury recall old Sun. If we remember what we have said on many occasions, old Saturn is the bearer of the soul-quality of sacrifice and that giving is a quality of ancient Sun; well the elements in our drawing are linked to their sacrifice and continuous self-giving to Earth.

Evil was born on the old Moon and we find these forces in nitrogen and phosphorus under the surface of the Earth, where the phosphorus represents Lucifer and nitrogen should be the destructive Ahrimanic karma of phosphorus.

Potassium and calcium represent the Earth. Let us not forget that we have always said that excess potassium collapses the structure of humus, and calcium bears the forces of greed and the gesture of contraction.

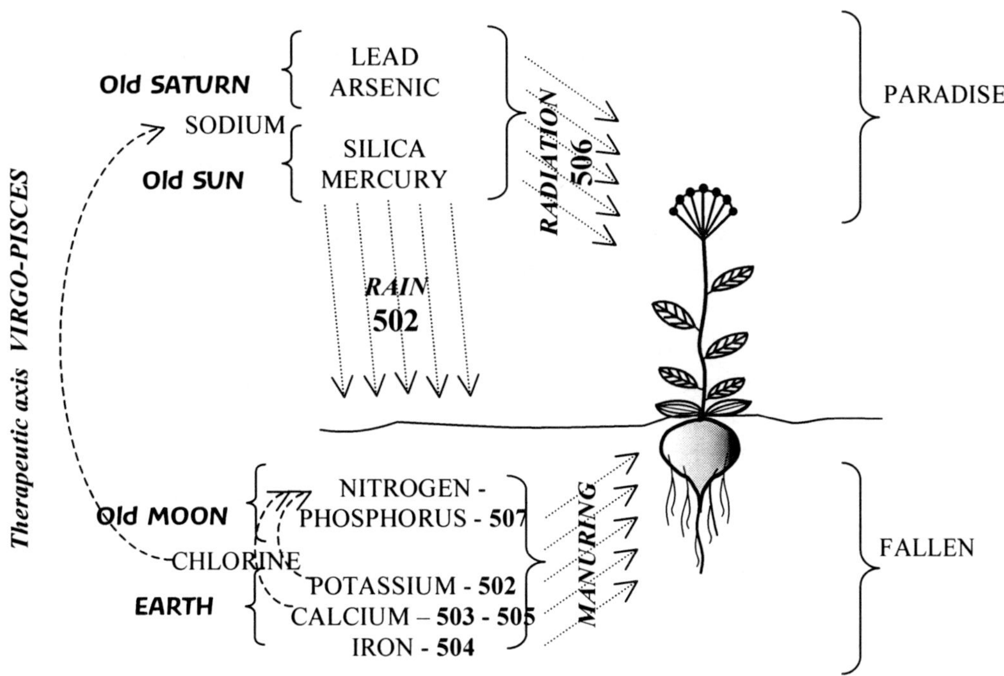

Therefore, in the list of substances which Rudolf Steiner produced the entire evolution of the Earth is present, and it is wonderful to see how all the forces that led to the evolution of the universe participate in building the smallest blade of grass, and how the transformation of that blade, which we know will become a new plant, will require the intervention of the cosmos.

In the drawing that we have built, the world represented above the ground is the world before the Fall from the earthly paradise, and obviously the lower part is the world after the Fall.

From another point of view the upper part of the drawing represents Abel, and Cain the bottom. Cain is the one who must work the earth with the sweat of his brow, but we believe that the time has come to see Cain in a different way. If we draw inspiration from the Bible, we see that Solomon (from the stream of Abel) realised his famous Temple but its actual construction was entrusted to a man whose name was Tubal-Cain. In a similar way today Cain must accomplish the transformation of nature. Incidentally Cain means 'He who thinks for himself' or 'new man'.

In this drawing is the whole history of humanity and in the middle is the 502, the preparation that opens the door between Heaven and Earth, between macrocosm and microcosm.

The purification, of which we spoke a moment ago, is the joint task of 502 - that we defined as Mary-like - and of Michaelic 504. If we remember that Michael is called the 'countenance of Christ' we can claim to have restored the now famous Virgin-Fishes axis. Bearing in mind that Sodium[34] is the ambassador of the Virgin

[34] See Rudolf Hauschka's 'Nature of Substance' for the substance-constellation associations

and that Chlorine is the ambassador of the Fishes we have also finished placing the substances mentioned by Rudolf Steiner in the first paragraphs of this conference.

What we have just said opens new horizons because between the Virgin and the Fishes can be inserted - according to a different point of view - the whole subject of healing and it is therefore evident that there is a substance suitable for therapy in the upper part of the drawing - Sodium - and one from the bottom - Potassium. Just bring to mind the famous Sodium-Potassium 'pump' in the cell membrane.

If we deeply appreciate all this wonderful architecture, which emerges from looking at the large scale, we can also have less interest in the single plant because it is only the consequence of the balanced flow of forces that we have just seen. Today it is humans who must worry about the flow of these forces, and that is why Steiner has taught the guidelines for making preparations.

We would like a fact to be firmly grasped - that in the lecture that we are commenting upon, the teaching of Steiner passes through a preparatory phase to the practical phase, in which we are given precise indications for concrete work. Clearly what we are saying is just the beginning of the application of preparations … just consider the use of 502 to fight pests. We wish to remind you that on another occasion we emphasised that Mary's mantle or cloak covers nine layers above and nine layers within the Earth. It is curious to note that the term 'mantle' to indicate the various layers within the Earth is also used by orthodox science.

If we look carefully at the picture of the Sistine Madonna we find a representation of what we are saying. The right side (as one looks at it) inside the mantle of Mary is coloured indigo and represents the interior of the Earth. Polarically (on the left) Raphael has represented the head of the child Jesus, in perfect harmony with what we have represented in the plan that we have built together. The upper part of our drawing corresponds with the blue cloth of Mary.

The two drawings mentioned are reversed compared to nature and this is because we studied the plant that we know to be reversed compared to what is appropriate for humans. Steiner, in the second lecture, said that the farm is comparable to "*an individuality standing on its head: we only look at it correctly when we imagine that with regard to a human being, it is standing on its head.*"

We repeat that the yarrow preparation allows the union of what is above in our drawing with what is below. It is an extremely *mercur* and therapeutic preparation because a disturbed relation between above and below is the origin of all disease.

When we put a seed in the soil the yarrow preparation allows the ancestral archetypal force (spiritual principle) and the idea (soul force) to approach fully and to manifest completely in the plant that will emerge. The yarrow, in other words, is the door of life from the point of view both of the substance and the form.

A full understanding of this preparation requires, however, a grasp of at least four points of view. The preparation of yarrow, seen in its connection with potassium, acts on the physical level on the roots and supporting tissues bringing a vertical impulse in the woody part of the plant. From this point of view all the plants that are unable to carry themselves such as the vine and cucurbits betray a lack of these forces. We have said that yarrow is linked to Venus that acts on the physical level in cellular nutrition and maximising the impact of food.

On the etheric level 502 favours the whole cosmic nutrition of the soil. In this sense it is more associated to Mercury II, the formative forces of vortices.

On the astral plane it promotes flowering. 502 is the basis of the homeodynamic product that we called 'Pro Flowering'. On this level the descent of the idea into the seed is also favored.

On the spiritual level 502 connects the plant with its Ego Group.

We conclude with an expression of wonder at the greatness of this preparation that is a door to all of the planes of existence. It is a path through which the unmanifest may become manifest.

A key feature of yarrow, which is the base of 502, is to maintain its form in time (even when dead and dried out it keeps its form) and this fact says that the inner being of yarrow is beyond time and space.

Seventeenth Meeting

At our previous meeting we explored the yarrow preparation and we stressed its key role in assisting the dialogue between the plant and the subtle alchemy that comes from the sky, through various substances, to be accepted and reworked by the Earth. One can certainly argue that the Yarrow preparation is the most alchemical.

Also in the meeting we were able to grasp the profound link between the preparation and the forces of Mary. We can add that this preparation in particular corresponds with the forces that are associated with Mary Magdalene in the Gospels.

We wish to reiterate that the work we are doing is certainly bringing other layers to the cycle of lectures held in Koberwitz in 1924, but it has no intention to be exhaustive. Indeed, to all those who have interest in the thoughts of Rudolf Steiner, we extend an invitation not just to look at the world of the preparations in connection with the field of agriculture, but to see in the preparations an archetype of the relationship between the world of matter and the world of the spirit.

To stimulate reflection we can notice that there are eight preparations, just as there are eight moments in the yearly etheric cycle of the Earth when the etheric forces swap[35]. These moments have corresponding preparations and we can associate these with the seven Sacraments in view of the fact that Easter and Michaelmas (the two equinoxes) can be assimilated into the same sacrament. The preparation of yarrow, in this context, is the sacrament of baptism that is both the first and a necessary preparation for all the others. Moreover, to use a language closer to the Anthroposophical world, we can say that the baptism involves the transformation of part of the soul more tied to the physical body and to instinct, namely the *sentient soul*[36]. In this regard, returning to the parallel between the yarrow and Mary Magdalene, we note that she had a sister, Martha, and a brother Lazarus; Martha is the *rational soul* and the *consciousness soul* is Lazarus. In this family one can grasp the archetypal representation of the three aspects of the soul.

We talk about archetypal representations because they form a model from which we cannot deviate in any evolutionary path. Actually without conversion and an openness to accept the new forces that descended from heaven no transformation is possible. The preparations are the bearers of these archetypal forces in the world of agriculture and, I repeat, yarrow is just the first step towards conversion and openness. So we approach Mary Magdalene who, moreover, was a great alchemist.

We insist on this aspect of the soul because the sentient soul is characteristic of the Italian people. We note that it is the lowest facet of the soul, but it can be transformed into our highest aspect – Spirit Man. The best means that the Italian people have available for inducing this transformation is the pursuit of art. It is no coincidence that it was the Italian people that initiated a historical period known as the Renaissance and that Italy has kept 90% of the world's artistic patrimony.

Now we will move on from yarrow and meet the world of chamomile (preparation 503). We will start with the twenty-first paragraph ("*Let's take another example.*") of the lecture given on the 13th of June 1924. Chamomile is a plant that is related to potassium, as is yarrow, but above all chamomile contains calcium which has the

---

[35] See the annual 'Agricultural Astronomic Calendar' from l'Albero della Vita

[36] See 'Spiritual Scientific Glossary', by the author and Fabio Montelatici, in translation 2008

power to attract into the compost those substances that can stimulate the growth of plants. Within the manure these substances will be able to attract the unbound and widespread Life into the soil. We can say that chamomile is an enhancer of fertility because, thanks to the action of calcium, it is able to stabilise the nitrogen in the soil. We know that nitrogen is the bearer of cosmic images and the fact that it is stable means that a plant can have, at every stage of life, a model that inspires it to grow strong and balanced.

If you wished to establish a parallel with chamomile in a phrase of the Gospel you could say that its activity can be reduced to: "*Let not your heart be troubled*" directed by Jesus to the apostles (John 14:1).

The fact that the plant can always count on a stable model means that it is immune from external influences that would take away its natural way of being. It would not be attractive to the world of alkaloids, nor is it necessary to defend itself against the risk of hybridisation by impure pollens. In this sense the chamomile has a bond with Venus, the planet through which one can access the forces of the Virgin.

Chamomile can also be useful to humans. Any natural-remedy expert knows that chamomile can be used to quell stomach ache (sphere of Venus) and has a mild therapeutic action in respect of sleeplessness if it is caused by an excessive excitation of the astral body in the head region such as could result from a prolonged activity of thought. Looking at it from a chemical point of view, we could say that excessive activity of thought leads to the formation in the brain of a type of precipitation called 'oxalates' made of mineralised calcium. In these conditions chamomile, which contains living calcium, has the ability to dissolve these formations.

We can rephrase what we have just said: chamomile can act at the level of the head by disintegration of the results of too much thought. Then we can deduce that it can also restore order in the root zone of a plant in view of the correspondence between the root of the plant and head of a person.

Chamomile, dissolving any hardened soil that can be considered as blocks for free Life, allows the flow of life in the plant unburdened from all kinds of suffering. We reiterate, however, that the positive effect on the growth of plants is not a direct effect of Chamomile, but the result of chamomile bringing order to the soil. This begins with the activation of living calcium that is able to stabilise nitrogen. This stabilised nitrogen allows a constant and orderly descent of cosmic images that, in turn, harmonise the ground. The harmonious soil is highly acceptable to the plant that is then able to grow under these optimum conditions.

We now have all the elements necessary for a better understanding of our homeodynamic preparation 'Pre-Sowing', the basis of which is the chamomile preparation. Chamomile is also a key component of the preparation 'Pro-Bunching', because when a plant is cut it needs a strong connection with its own image in order to rebuild, and there is nothing better than chamomile to strengthen this link. We can add that bunching requires action by the forces of Mercury, and in the world of preparations chamomile represents the connection with that planet.

Once again the words of Rudolf Steiner allow us to understand that dealing with agriculture is not just a question of dealing with plants, but with all the vital processes that take place in nature and that stem from the life of the cosmos: Agriculture in Steiner's vision, is not just for the farmer, but it is an activity to promote the evolution of the Earth, plants and animals. Only as a byproduct does it provide farmers with the means to sustain their existence.

In the first sentence of the twenty-second paragraph ("*We may not simply say ...*") it is emphasised that the simultaneous presence of potassium and calcium within chamomile is of great importance: "*Yarrow develops its sulphur forces in exactly the amount needed for working on potash*", and so may be the bridge - thanks to the coordinating role of sulphur - between the free astrality and the Earth. Thus Chamomile, because of the fact that it also contains calcium, is in a position to absorb any excess astrality brought in by yarrow and enable the plant to fulfill its vegetative role. In normal agricultural activities excess astrality can result from several different causes, which often result in a disturbance of the nitrogen processes. Therefore the monitoring of astrality undertaken by chamomile can also be useful as a regulator of flowering and fruiting, where there is the danger that these might be initiated too early.

Of course, the amount of sulphur within chamomile is different from that in yarrow, as a result of which the preparations are called to establish harmonious relations between different substances or to activate various processes.

To take advantage of chamomile's ability to inhibit catabolic processes and thus to halt early flowering, we have made and tested preparations that are designed to prolong the life of cut flowers and green leaves. Remember that plants containing alkaloids all show early flowering processes so Chamomile can be the basis of a remedy against the formation of alkaloids. If we think that the formation of alkaloids is a sign of Lucifer within the plant world we can sustain the validity of what we have said about the protection of the soul.

Paragraph twenty-three ("*Now we must go further ...*") considers the effects of human and animals ingesting chamomile. These considerations, from which one could also draw therapeutic indications, are brought forward with the primary intention of identifying the animal sheath for the creation of the preparation. Each preparation arises from a synthesis of the four kingdoms of nature. Since chamomile barely affects the bladder this organ is not considered, in contrast to the development of the yarrow preparation that was created using the stag bladder.

However, chamomile does act strongly upon the intestinal walls and from this fact Rudolf Steiner develops his indication to stuff the yellow flower heads of chamomile in a length of bovine intestine. The choice of this animal is guided by the fact that the cow's fundamental characteristic is centred in the stomach and intestines. The cow has four stomachs, and its intestines are twenty-two times the length of its body.

The long intestine of a cow reminds us of the earthworm which itself can be considered as a kind of itinerant intestine. The internal glands of the earthworm secrete limestone and in the animal kingdom the earthworm is what the chamomile plant is in the plant kingdom – suggesting again that it is not just coincidental that it is so important for the vitality of the land. In the approach which seeks a synthesis of all the kingdoms of nature and which inspires the whole field of the biodynamic preparations, there can be no better pairing of the chamomile - with its principle of nitrogen stabilisation as a vehicle for Zoe – with the earthworm. Once more in this case the preparation is buried to make it a bearer of the influence of Christ.

In the remains of this paragraph and in the twenty-fourth ("*After that, all that ...*") Steiner gives more practical details for making the '*precious sausages*' of the chamomile preparation. Once again he emphasises - as if it were necessary - the need for any agricultural intervention, but particularly when creating the preparations, that everything be kept within the context of life because the forces capable of supporting life can never arise from something dead.

In the next sentence Steiner tells us that the forces of humus must act deeply on these sausages, so they should be buried not too deep in a humus-rich soil. What Steiner suggests, in his long-sightedness, is to develop a real 'spiritual earthworm' because he understood that the earthworm would become increasingly rare and there is little choice but to replace it with something that was able to bring the same forces.

He also mentions the snow, which in this context represents a door of Life, which would best be solarised and therefore vivified with the cosmic images. Note that for the alchemists humus represented the Sun within the Earth and, therefore, even when Steiner is talking about snow exposed to the rays of the Sun he is still talking about humus. The preparation is to remain underground throughout winter, because it is dominated by an etheric component that must also accumulate high quality astrality.

The end of the paragraph summarises what we have said so far. It clearly says that the well-made preparation is able to stabilise nitrogen in the manure, so that it acquires a strong ability to vivify the Earth that, in turn, can exercise an extraordinarily stimulating effect upon the growth of the plants. This is a fertilisation with cosmic images that, in addition to growing healthy and strong plants, will also allow the eaters of those plants to resound with their own cosmic images, in the same way that eating foods that resonate with the low quality astrality leaves us little choice but to resonate with low quality astrality.

The twenty-fifth paragraph ("*To our modern way ...*") concludes the part of the lecture dedicated to chamomile.

For our part we should summarise the effects of the chamomile preparation on the various planes. On the physical plane it activates the calcium process. This process works by reinforcing the etheric and making stable nitrogen. The consequence on the physical level of these actions is a greater growth of plants (Venus) and an enhanced cell division (Mercury).

On the etheric plane the chamomile preparation eases communication between the diffuse Life we know as *Zoe* and incarnated life (*Bios*). In other words this preparation refers to the etheric sphere of Jupiter and connects it to the soil. We can imagine that, under undisturbed conditions, the cosmic and planetary activities we are describing perform effectively alone. However, because everything is now made much more difficult by the consequences of unwise human activity against the other kingdoms of nature, we need something that can restore connections or reopen the doors that were closed. This something can be the biodynamic preparations.

We believe it is appropriate to pause for a moment to consider why the shape of the chamomile head resembles the form taken by bee colonies in winter (It: '*glomere*'). The bee is strongly linked to the forces of Jupiter. Well, as chamomile regulates the astrality in the soil, the bee regulates the astrality in the air, establishing a link, nexus or - if we prefer - an affinity between bees and chamomile. It is a similarity that can be confirmed only with imaginative observation, but which can be supported by the fact that the inside of the flower of this chamomile species is empty. Thus the chamomile has a flower that, through its axis of symmetry, shows a strong link with the etheric. However, it also shows within it that it has the gesture formed by the high quality astrality. In the bee, on the contrary, we recognise a peak of astrality, but also the ability to act upon the etheric. The bee sting poison is a remedy against arthritis and osteoarthritis, both of which are physical symptoms of a collapse caused by a lack of vital forces. Thus in chamomile there is a lot of ethericity with some high-quality astrality, and in bees we find a lot of astrality that can stimulate the etheric plane.

Returning to Jupiter, within humans the forces of this planet act in the liver, whose actions affect the intestinal functions, and remember that the intestine (the other function of Venus in us) represents our inner earthworm. From other investigations we know that Jupiter, on the ethereal level, has a special relationship with Venus, so we could say that the accounts balance.

At this point we could push our considerations to embrace the sphere of the zodiac. We could ask ourselves which constellation puts everything in motion, an activity which in the planetary spectrum belongs to Mercury. The first of these constellations is the Crab, but this is more directed towards activity within the Zodiac. The second is the Bull that strongly directs its action towards the Earth. This allows us to say that preparation 503 allows resonance with the constellation of the Bull to bring the forces on Earth. If we think that the Bull's associated substance is nitrogen we can better understand one of our homeodynamic products that is called '*Pro Nitrogen*'. This amplifies the forces of chamomile. We note that understanding these few paragraphs which Steiner dedicated to chamomile has enabled us to find solutions which make it possible to promote bunching, holding back weeds, pest control and stabilisation of nitrogen in the soil. We could say with Steiner that "*this all sounds quite insane*", but we have preferred to press on to verify that when these thoughts are brought into agricultural practice they really work!

On the astral plane the chamomile preparation acts to block the momentum of the plant to bolt to seed, that is to jump into the sphere tinged with astrality, and this is due to the powerful etheric which it develops. This function manifests its link with forces from the sphere of Venus.

All that still remains is to clarify the action of chamomile preparation on the spiritual level, namely with the sphere of the group I. In this context, the preparation develops an extremely subtle and important role (as, incidentally, do all the others). It facilitates the connection between the group I of the plant with its (ideal) centre in the Earth, namely with the seat of consciousness of all group I's.

With the twenty-sixth paragraph ("*As I said before ...*") Steiner begins to discuss the nettle preparation known as '504'. Steiner says that the Nettle is the '*greatest benefactor of plant growth*', and that it is irreplaceable: it is not possible to use a different plant to make a preparation with similar characteristics. At most one can use dried nettle if we were to find it necessary to make the nettle preparation in places where nettle did not grow.

Nettle also contains potassium, calcium and sulphur, but also "*it has a kind of iron radiation that is nearly as beneficial for the whole course of nature as the iron radiations in blood are for us*". Iron is one of the main components of haemoglobin that carries oxygen throughout the body. Oxygen - which is the representative of the etheric - also brings the hydrogen, which is linked to the world of the spirit. We have often said that the action of the spirit upon the etheric determines the state of the immune system. The radiation of iron can lead the impulse of the spiritual life to the different organs. The nettle, a plant, has the capacity to carry the spiritual life in the plant kingdom and we know now that the spiritual element in the plant kingdom is represented by the group I.

Within the twenty-seventh paragraph ("*Stinging nettle is a real ...*") Steiner proposes an affinity between Nettle and the human heart. To understand this combination we should consider that the heart is the central organ of the rhythmic system and the arrangement of leaves along the nettle is strongly rhythmic. Also in the chest formic acid is formed which also gives the nettle its familiar sting. It is not a coincidence that to bring blood to a certain area of the body it was traditional to strike that part with nettles. This therapeutic practice is not important because it stimulates the blood to transport certain forces, but because its rhythmic aspect is linked to heart function. We recall in this regard that, in the anthroposophic vision, the heart does not perform the role of pump, but on the contrary, acts to regulate the otherwise free flow of blood. It is through this restriction that an individual and rhythmical component is introduced into the flow. The blood is actually put into circulation by the movement of muscles. The movement is led by the astrality and muscles are the physical bearers of astrality. The heart is a muscle, but it is also linked to the I and therefore has the ability to regulate the excessive astrality. Many diseases of the heart depend on the fact that the I was no longer able to dampen down the excessive and low-quality astrality. This excessive low-quality astrality is also responsible for the deformation that can sometimes occur in the heart, because the principle of form is linked to high-quality astrality, which is clearly lacking in such heart disease.

Shortly afterwards Steiner makes a rather curious assertion. He says that, when necessary, nettles help to free the land of the excessive presence of iron. It seems strange that a plant which itself contains so much iron might liberate soil from its excessive iron, but it's not so odd if we think that the soil iron is mineral, while nettle is a plant. Therefore nettle brings an impetus of vegetalisation to the mineral iron thus bringing it 'into movement' and the nettle then transmutes it. Moreover transmutation is a normal phenomenon in the world of the living and it is only marginal in the mineral realm whose transformations are predominantly chemical.

The influence of nettle is positive even in the case of an iron deficient soil. Even when present only at the edges of the field ("*in out of the way spots*"), it is able to bring its forces.

We wish to point out that the successive preparations show an extraordinary logic. The first, 502, is the preparation that creates a link with the life of the most distant cosmos. 503 enables the humus to bear the forces of life. Finally if the humus has excesses the nettle within 504 is capable of putting it into order. This may be understood as a series of interventions in which the next preparation completes the task of the previous one and sets the scene for the succeeding one, through the lifetime of the plant and supporting all phases of development until the maturity of the fruit and seed. At each stage of development the plant undergoes a metamorphosis that includes a transmutation of substances and all these transformations can follow each other with harmony and balance thanks to the support of the preparations.

The nettle follows the work of the chamomile. If the chamomile has brought life into the soil and has supported the part of the plant linked to water, the nettle brings the plant the impetus to overcome the vegetative stage and to rise vertically. So if the plant shows signs of the problems caused by excessive earthly vigour (excess of humus, of water, or of manure) nettle is potentially able to bring remediation. Not surprisingly the nettle preparation is the main ingredient of our preparation '*Anti-lodging*'. This problem, especially in cereals, is caused by a weakness of the base of the plant and the loss of elasticity brought about by excessive vigour, often caused by over-abundant nitrogen fertilising.

An interesting application of nettle for humans would be its use in the treatment of sclerosis of the blood vessels. Of course, the development of a preparation for humans requires an appropriate potentising technique, but will not fail to give satisfactory results. Do not forget that preparations are archetypes and so they can be applied in all realms of nature, obviously through adequate preparation.

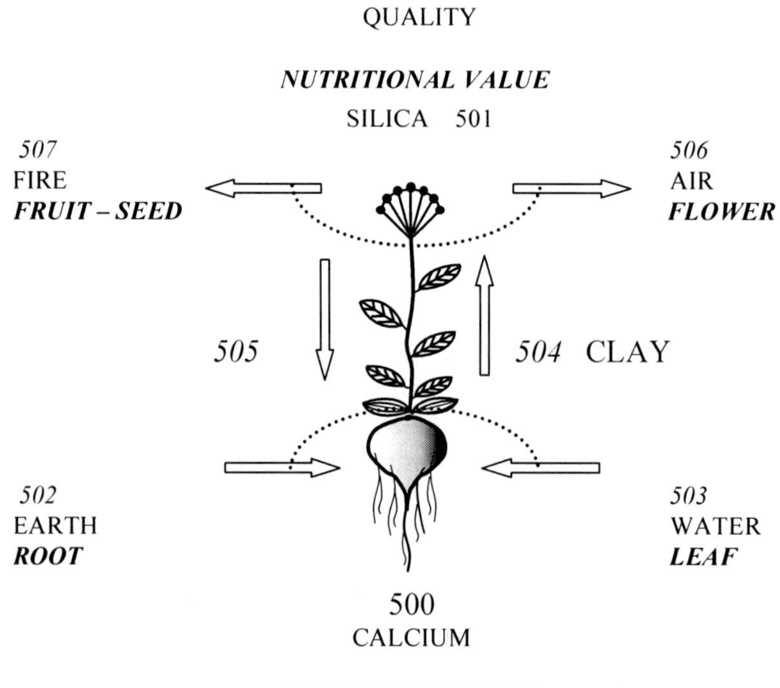

Scheme of the influence of forces and preparations in the plant

There is very a interesting sentence in the twenty-eighth paragraph ("*If it should ever become necessary ...*"): "*Nettles like iron so much that they draw it out of the soil and into themselves, and although this does not get rid of the iron as such, it at least undermines its effect on the growth of other plants.*" This obviously does not refer to activity on the physical level and it is difficult to explain in physicochemical terms. If the excessive iron cannot act it means that in some way it is isolated or 'encysted'. This effect of isolation is the one that the etheric produces on mineral substances if it fails to bring them into the flow of living processes. The laws and rules that govern the world of life are opposed to those that govern the merely mineral world. Therefore, something that remains purely physical remains alien and must necessarily be isolated. Another way to prevent something mineral from undermining the integrity of a living organism is expulsion. When we drive a splinter into our skin our etheric body reacts by concentrating water (the medium of the etheric) in that area, which later brings about the expulsion of the foreign matter. But if the splinter had gone in too deep it would be encysted and its ability to cause damage would be reduced.

At this point we can tackle the paragraph twenty-nine which begins with some interesting terms: "*Now, to improve your manure still more, take whatever stinging nettles ...*". This time he says "*improving*" not "*revitalise*" as he did with chamomile. In this 'improvement' is contained the observation that there is something that does not have an adequate quality. The nettle in our drawing has been placed in a central location, as *Mercury*. Together with the oak bark preparation in the drawing above, this preparation is in the position corresponding with clay. But only the Nettle is put in the middle of the heap just to underline its particular function to restore order in all processes of the compost or manure.

Immediately after we are told that we do not use any animal sheath to make this preparation. We believe that this is due to the fact that the nettle is linked to the heart so there is nothing suitable for containing it. We would like a synthesis of all the organs, so it is right that its container is the entire Earth. To identify an 'organ of the heart' in the human organism we must think about the whole body because blood reaches the most distant cell.

It also seems strange to suggest that the nettle should be buried with a bit of peat. The peat in this case protects the nettle from the influence of electromagnetic fields. At the same time it transmits a message to the nettle to become a defense against the same electro-magnetic fields. Remember that magnetism is a corruption of the chemical ether that is linked to Mars. Mars is the planet that has iron as its metallic ambassador on Earth and nettle is notoriously rich in iron. Therefore, nettle preparation can weaken the effects of an electromagnetic field. From this point of view, therefore, the message carried by peat would be most desirable. We can identify a different dynamic since nettle is linked to the ascending sap, corresponding to the arterial circulation which - if it does not play its role properly - will not allow the I to fully express itself. Electro-magnetic fields act as a force opposed to the I principle so the combination nettle + peat becomes ideal to strengthen the ascending current (ascending sap, mineral salts, I) and defend it against adverse forces.

Although peat has this very positive application, it should not be used indiscriminately because it also has the property of absorbing vital forces. Hence its use in nurseries is strongly discouraged. It could, however, be used advantageously within architecture on the outside of a construction. Peat could be used as thermal-acoustic insulation.

The thirtieth paragraph ("*When you add this ...*") presents the nettle as the alchemist directing all processes of transmutation within the compost. Thanks to the nettle preparation, Steiner says, the compost literally becomes "*inwardly sensitive and receptive, so that it acts as if it were intelligent and does not allow decomposition to take place in the wrong way or let nitrogen escape or anything like that.*"

The nettle can be compared to a conductor who directs all the instruments so that they unite their voices to produce a sublime harmony. The earlier comparison with the heart now seems even more fitting; it is the heart that knows everything that happens in every organ of the body. We could, however, take a further step and think that a blood vessel and a nerve ending serve every cell. Thus there is information returning within both the blood flow and the nervous system. Well the nettle acts upon both currents: the blood with fertilisation and the nerve with the latter aspect of bearer of rationality. In this sense the nettle can be assimilated into the human I. The I is actually the organisation that governs the behavior of organs and cells and adapts them to work together to fit the requirements of the various situations in which the whole organism finds itself. The I organisation is that reasonable being that keeps the internal balances and has the heart as its associated internal organ. Nettle can be regarded as nature's reasonable being. Within an agricultural organism we can recognise a body with similar function to those that the nettle undertakes and this is the woods, or in a different context, the aromatic plants.

The regulating ability of the nettle is evident from its form. In its lower part nettle is strongly linked to the Earth in that it has an extremely disorganised fasciculate root and a square-sectioned stem with a purple base. Moving up the plant we find the green colour in the leaves that are very regularly arranged (water). Its bloom, which we could call 'orderly and proportionate', suggests a close affinity with the Air element. Finally, formic acid, along with the stinging hairs, demonstrates the presence of Fire. We have thus found another way to see that the nettle is a summary of the forces of nature.

In the order in which Steiner presented these preparations the Nettle (504) comes after Chamomile (503) and we can see that the nettle plant grows upon a strongly vegetative root that brings the gesture of Mercury (503) from which a very orderly plant emerges.

From what we have said we can also draw therapeutic indications: for example, the observation of the manner in which the root of the nettle grows allows us to understand that this can be of help in water retention. Furthermore, we stated that Mercury brings movement and that the root of nettle is also linked to Mercury.

The end of the paragraph says that the manure that has been made "*intelligent*" by nettle may in turn make the soil "*intelligent*" so that it can individualise itself for different plants. This '*individualise itself*' means that it can find genuine ways so that the plant is connected with its I, whose consciousness is in the soil beneath the roots. (It's *being* is with the planets and stars.)

In summary, we can say that the Nettle performs the following tasks:

- On the physical level it brings verticality, gives structure and puts order in the soil's organic matter (action related to Saturn)
- On the etheric level it brings order to the vegetative force and pushes this to rise towards the top of the plant where they will support the nutritional value, transforming its impulse into quantity and quality. The rise of lettuce as it moves to flower and the lifting of wheat are examples of the forces of nettle, which can then be used as a remedy when these developmental stages are deficient.
- On the astral plane it frees the soil from the excessive presence of iron. This influence is on the astral plane because the excessive iron is a too strong link with Mars, causing a gesture of closing in upon itself and consequently hinders the rise of sap, which cannot reach and feed the more astral top of the plant.
- On the spiritual level, nettle individualises the ground for each plant and thus promotes the connection with his Group I.

The thirty-first paragraph is the last one dedicated to the nettle. If we think that the Koberwitz lectures were given in 1924 this paragraph should make us think, because Steiner, in a situation which in our view might seem like paradise, says that the other methods of fertilisation adopted will "*tend to turn all the first rate agricultural products to mere stomach-fillers. They will no longer have real nutritive power for human beings.* " More than eighty years have passed since then and it seems impossible that we have not yet understood that there is no more time to waste and that we must regain possession of a new and different relationship with plants and with nature.

Meeting Eighteen

We will continue reading the fifth lecture at Koberwitz and grappling with the world of the preparations that, I remind you, is a door to an unlimited view into the cosmos.

The preparations we will discuss now may be considered even more 'noble', not so much because the other preparations are less important or less elevated, but because with the next preparations we will enter the sphere of Light and cosmic Life. With these one can bring cosmic nutrition into the living realm.

Remember that the biodynamic preparations are not specific to the plant kingdom but affect the whole of life, even if we study them in a farm. Perhaps at this point it would be useful to retrace the steps we have taken so far in this world.

We already know horn-manure or 500 that embraces the terrestrial sphere, and its complement, horn-silica or 501 that embraces the cosmic sphere. 502 or yarrow (linked to Venus) is the bearer of forces related to the Earth element and 503 or chamomile (tied to Mercury) is the bearer of forces of vigour and is tied to the Water element. 504 or nettle (tied to Mars) brings the impulse to the ascending sap after having put the lower sphere in order. Now we are about to meet the oak bark preparation 505 (linked to the Moon).

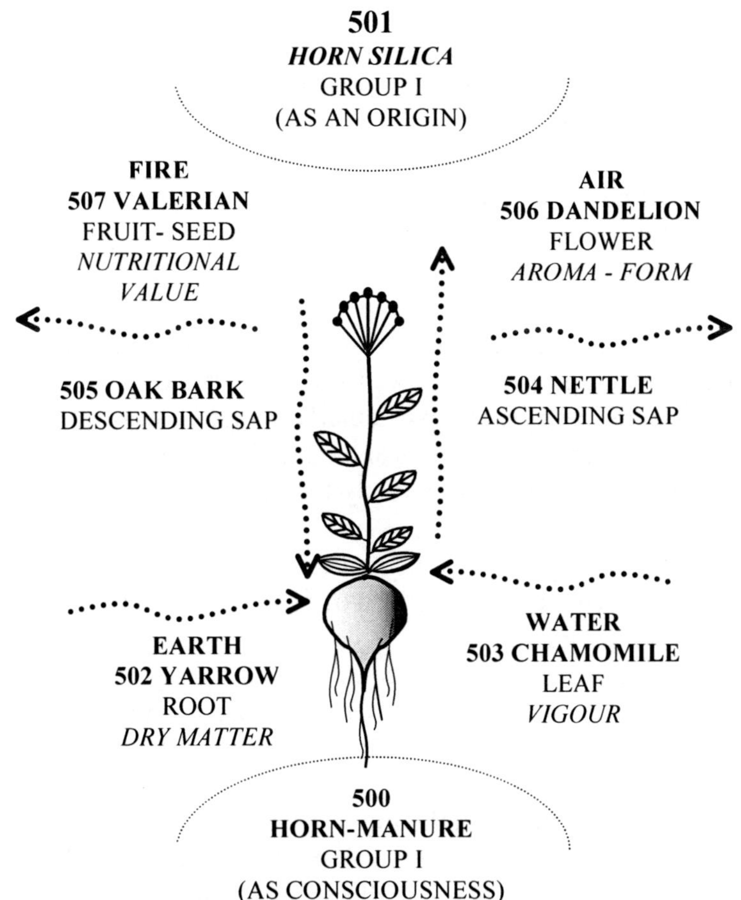

We will see that the nettle, as well as bringing order in the metabolic sphere, carries the impulse to verticality and with that the transition to flowering. However the forces of vigour should not rise to the flower as they are in the metabolic sphere, because the forces of vigour only stimulate the constructive processes that are called 'anabolic' in biological sciences. At the top of the plant catabolic processes must dominate and these bring in 'destructive' processes. What rises from the lower pole of the plant must be transformed. We could picture to ourselves that the forces must pass the scrutiny of a 'guardian' who only lets the vegetative forces pass that are ready for transformation into quality. This is accomplished by the oak bark preparation. At the human level such a guardian corresponds to consciousness.

Let's begin reading paragraph thirty-two ("*Now it can happen ...*"). Here Steiner speaks of plant diseases and emphasises that knowledge of a disease, extreme precision in description, complete mapping of physiological pathways and bio-chemistry - all this does not automatically mean that one is able to heal that disease. In the plant-world successful treatment can often be carried out with a '*universal remedy*' or a generic medicine if it resonates with the cosmic periphery. Most diseases, says Steiner, can be alleviated with "*a rational method of manuring*."

If we think that this statement was made in 1924, we can understand that the problem has grown out of all proportion between then and now.

We said that humus is the result of a process of vegetalisation of the soil that has not come to a conclusion, and we have also said that this process finds its completion in the plant. However if the humus is not to become an Ahrimanic system of forces derived from Zoe, but is instead just the result of altered processes, the plant will do nothing but bring these altered processes in which it is inserted to a conclusion, and express them as a range of diseases.

What we have just said can be extended to the whole range of work that goes on in the fields because we know that even with green manuring, the habit of cultivating when the ground is not resilient, but is too wet or too dry, develops unbalanced conditions in which plants cannot grow healthily.

The "*rational*" which Steiner refers to is not the technical knowledge gleaned from books, but a more general concept of balance based on knowledge of the laws governing creation. Returning to humus, we have often stated that it is preferable not to manure and give up on the idea of a particularly abundant year rather than take on the problems that have troubled our plants and our soils for so many years, requiring us to spend lots of money and effort to remedy the mistakes we could have avoided if only we had been a little less greedy.

In a thoroughly messed up situation the nettle is not able to restore complete order and therefore, whilst carrying out its function of pushing life upwards, can only bring up the mess that we caused ourselves.

It is here that we discover the importance of the oak-bark preparation (which we have just identified with the qualitative element) because it brings a healing impulse that acts as a brake upon excessive vegetative forces by regulating their 'ascent' to the upper parts of the plant.

If one wished to make a parallel in the human being we should more properly speak of metabolic forces, which in excess will cause the 'thumping' type of headache. Such headaches, although clinically not particularly serious, are able to greatly reduce the ability to function of even the most robust people, greatly interfering with their ability to exercise their will.

The plant does not get headaches as indeed it cannot get the flu, however under the same conditions mentioned it manifests an astralisation that permits the appearance of fungi and especially the formation of alkaloids. Now we know that alkaloids are toxic and that their presence betrays poor quality food. These are an expression of a type of inflammatory disease that gives the plant a kind of hallucination.

The plant can also have diseases that bring a kind of sclerosis, in which the forces of death or hardening are dominant, and which tend to make the plant too woody. We refer to '*rogna'* (Scabies), canker, and more generally to diseases that occur in excrescences. Such diseases result in viruses.

The oak-bark preparation is able to control the forces of the lower part of the plant and to prevent the ascent of tumultuous processes, caused by over-fertilisation or immature lifeless manure and foetid (astralised) manure, all of which are often recommended by those who deal with organic farming.

We can also say that the inflammatory processes resulting from the descent of excessive poor quality astral forces from the cosmic pole of the plant call up what is formed in the soil as a result of astralised manures. These also come back up the plant attracting the external astrality. If we can bring some order to the soil with proper manuring and if we have high quality seeds we would have resolved most of the problems that are found in agriculture today. They would actually only occur in exceptional cases.

Thus we arrive at the thirty-third paragraph ("*That is how ...*") which asserts that a "*rational method*" is necessary to put a certain amount of calcium into the soil. However, this alone is not enough because calcium is useless unless it is first inserted into a living process and so becomes living itself. Living calcium can become a bearer of healing. We have often emphasised the fact that nothing should be put in the ground before it has been 'pre-digested' and this often leads us to take a critical stance against those techniques used and considered very positive and eco-compatible. We refer in particular to green-manuring because, as we have so often said, the inclusion of fresh organic matter into the soil requires the soil to take on a difficult digestive process, requiring forces that are then unavailable to support plants.

In paragraph thirty-four ("*Now one plant ...*") we meet with oak-bark for the first time and having just read of the need to enrich the fertiliser with calcium, we are told that the ashes of this part of the plant are as much as 77% calcium. Steiner then brings to our attention what he has already mentioned in the ninth to twelfth paragraphs of the fourth lecture about the parallel between the living part of the soil and tree bark. He makes us appreciate how the bark of the tree, despite being a waste product, is made of substance which is still in a living state and therefore has the ability to bring order "*when the etheric body is working too strongly*".

This is no longer a living calcium such as is found in chamomile and which acts as a powerful stimulator of vegetative processes. However, it is a calcium which - although still living - has been expelled from these processes and therefore is able to act as a regulator and moderator.

In a nutshell we can say that in cases where 503 is too powerful in its efforts to strengthen the etheric and 504 is not able to regulate this sufficiently and thus pushes this excess up the plant, we would find a favorable condition for the mentioned diseases. In this context the 505 - which dampens down excessive vigour caused by 503 and raised by 504 - can play a regulating role.

The damping of the forces of vigour involves an increase of 'awareness' and the plant begins to discern which forces of vigour are positive in their environment but negative outside it; consequently the inappropriate aspect of these forces is rejected back towards the bottom of the plant and are not left to rise to the top indiscriminately.

The fifth sentence of the thirty-fourth paragraph ("*Now, one plant that ...*") reads: "*Calcium in any form will kill off or dampen the etheric body and thereby free up the influence of the astral body*". Calcium represents the physical world and therefore death, while the etheric is life. Thus through the opposition of the physical and the ethereal, calcium always has the tendency to kill life. But the calcium we are talking about is still integrated within a life process so it does not kill but only moderates. We have many examples in nature of similar regulatory functions. For example, the earthworm with its internal calcium-secreting glands not only has the function of promoting the vitality of the soil, but if there is too much ethericity they will dampen it down. Similarly, but on a higher level, the bee distributes astral forces but where there are too many it disperses and thus stabilises them.

We will take this opportunity to point out that in nature there is not only one type of limestone, but one can recognise seven. So first there is mineral limestone, and then chamomile which is living calcium of plant origin. Calcium of plant origin but which is dying is oak-bark and there is living limestone in the animal which is from the earthworm. An animal calcium which is being transformed is the shell of the crab, a dying calcium from the animal world is found in the bones and, finally, there is the calcium which is capable of bringing new life which is preparation 500. Each of these types of calcium is the bearer of different forces.

Returning to the oak-bark preparation, this is the first in the sequence of preparations that draws in cosmic forces and we could say that it is this that allows organising beings of the cosmos to work upon plants.

This is a very beneficial preparation in sick times like ours. Today the causes of illness are so numerous, and not all are of agricultural origin so that they often escape our attention and are not taken into account in conducting effective healing.

We might take as an example the situation at the smallholding, 'La Nuova Terra' in Codroipo, Italy. To get to the property one drives for about twenty kilometers along the perfectly straight 'Napoleonica' trunk road and the land is where this road ends. The road can be seen as a channel that conducts all the forces generated on the road and dumps them on the smallholding, causing all sorts of diseases. For this reason we have decided to plant oak trees as a blocking screen between the end of the road and the farm.

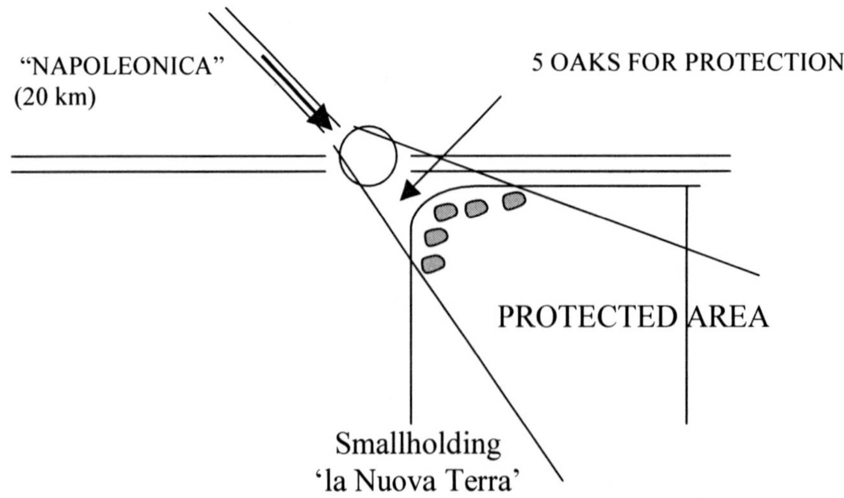

Protective influence of the oak trees

At the end of this paragraph is an expression worth working upon. Steiner asks what makes it possible to channel a hypertrophied etheric without generating shock in a plant. "*We need to use calcium in the particular form in which it is found in oak bark.*" We note that he does not say to use calcium which is the oak-bark, but uses the term "*calcium in the particular form…*" to indicate that we are not interested in calcium as a chemical element, but in the system of forces in which calcium is embedded. This structure is the form, and is linked to light, and Steiner invites us to appeal to light that can be found in the oak-bark, or to the cosmic that works through it. Once again I stress that we are commenting upon a genuine agricultural alchemical text.

In the thirty-fifth paragraph ("*For this purpose* ...") the procedure for the creation of the preparation is set out. First we see that Steiner suggests using a piece of bark that can be gathered with ease by breaking it off by hand. So don't take the bark off the trunk but from a low hanging branch of approximately 5-7 cm diameter whose bark can be broken off by hand. The 505 is linked to the forces of the Moon II (excarnating Moon), and this allows us to understand that bark must be seen as a excretion of the skin.

Then we find the indication that the crumb-like bark must be pressed into the cranium of a domesticated animal. He specifies that it is of little importance which animal's cranium is used, but he specifies that it must be 'domestic' and this, of course, can not be unimportant. The fact that an animal is domestic actually means that it has renounced its natural way of being, and its proximity to the human means it is acquiring a universal character. The old-time wild dogs preyed upon sheep and now, renouncing this way of being, have turned into shepherds' dogs and protect the sheep. They have been able to adapt to the needs of man and become an indispensable support in a situation of objective difficulties, as in the case of dogs for the Blind. Similarly this preparation supports the plant to take within itself - into the area of flowering and fruiting - only those forces (quality) that will be able to be transformed in order to nourish the higher realms of nature, and not only enough forces (quantity) to produce the seed suitable for a new life cycle. We could say that 505 marks the transition between selfishness and altruism for a plant. It corresponds to the 'Guardian of the threshold' in human spiritual development.

Before putting the cranium into water it is good to wrap some wire around it to prevent the cranial bones from coming apart and dispersing their valuable contents. The burial is best about one meter deep in a clay soil, covering it over with a layer of peat. Peat isolates the preparation from the forces of the other planets, including the Sun, and allows the cranium to remain in the lunar sphere. Besides being the 'mummified' bearer of life, peat brings a further gesture of resisting overabundant life that comes from the world of 503.

The oak-bark collected in the autumn is energetically different from that collected in the spring. For the preparation the autumn bark must be preferred. The spring calcium collects etheric forces, while autumn is more influenced by the forces of death. Let us not forget that in the spring, calcium - with its forces of attraction - draws in cosmic nitrogen.

Finally, for the completion of the preparation, it is necessary that the area in which they are placed is submerged by muddy and preferably putrid water that can further reinforce the excarnative Moon influence. Paragraph thirty-six ("*We have now added* ...") should present no significant difficulties in comprehension so we will refrain from comment.

Before moving to the next preparation it is useful to summarise the action of 505 in the four levels of being:

- on the physical plane, it reorganises the land from excessive ethericity;
- on the etheric plane, it orders the vegetative forces and removes the excess. One could also say that it acts against hypertrophication and therefore constitutes a prophylactic measure against fungal diseases;
- on the astral plane, it promotes the transformation of vegetative forces into nutritional value:
- on the spiritual level, the plant prepares its excarnative or catabolic processes.

At the level of the sacraments for the Earth, 505 represents Confirmation.

The thirty-seventh paragraph begins with the statement: "*Now, we still need something else, something that will draw in the silicic acid from the whole cosmic surroundings.*" We can deduce that it is not enough just to rely upon the oak bark preparation to draw silica from the cosmos - the description we have just finished reading indicates that 505 helps to convey the forces of silica down the plant. However, 505 is not sufficient to attract them from the cosmos. In this way Steiner introduces the dandelion preparation.

Steiner tells us that the Earth is losing the ability to draw in silica. We don't notice this in part because this phenomenon occurs very gradually, but mainly this is because we have lost the sense of the importance of silica. As we increasingly turn our attention to microcosmic phenomena it becomes devoid of meaning that silicon is 42% of our planet. On the contrary we are concerned about the lack of substances that are found in the soil at an insignificant percentage. The fact is that we are no longer able to appreciate what it means to lose the ability to draw down the silica of the cosmos because we have lost the ability to receive the Life that comes from the cosmos - the life as a gift of heaven. Bring to mind that over the last century cereals have lost about 30% of their silica content.

This problem is not only true of the soil. Plants also have lost the ability to feed themselves from the forces of the cosmos because they have been forced by chemical fertilisers into communicating only with the soil that is located in the immediate vicinity of their roots. And what about animals that are increasingly deprived of horns that are their organs for perception of *Zoe*?

The role of people must be to speed up the natural process caused by the aging of our planet that is intended, however, to sharpen our ability to relate to the forces of Life. Whilst in part this makes urgent use of a remedy - the dandelion preparation - it certainly also requires a revival of our awareness.

The next paragraph ("*In this regard, ...*") opens with the observation, which really seems too optimistic in retrospect, that in 1924 it was becoming possible to talk with scientists about transmutation of the elements without embarrassment. We take this statement as an affirmation for the coming years given that even today there is enormous distrust towards these issues in the scientific establishment. The paragraph gives detailed indications of the possibility that calcium and potassium, if they are able to work properly within an organic process, can be transmuted in nitrogen. This is extremely important because it allows us to understand that probably the process of legume-mediated nitrogen fixation is not limited by the presence of bacteria, but it is important to be within a proper organic process. This remains the basis of a research project: to induce a weed seed to stimulate the formation of humus and nitrogen-fixation in the soil, which at this point we can call the 'transmutation of nitrogen', thanks to the presence of humus. We intend to ensure that this does not happen in two years as at present with the remediation, but in a matter of two or three months. The need to work through a plant is dictated by the fact that sometimes the soil is so dead that it cannot respond to the stimulus of the homeodynamic products, but it is much easier to react positively to a message carried by something with an innate affinity such as a root.

Shortly after Steiner links plant growth to the substances associated with the four elements, and to sulphur.

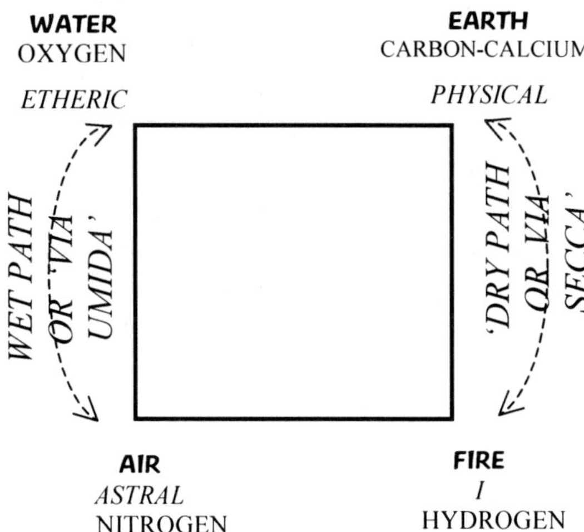

Relations between the four elements

Steiner highlights the fact that there is a relationship between oxygen and nitrogen, and between calcium and hydrogen. As we have said in the past[37] the relationship between them has been established and identified, according to the alchemists, as the '*via humida*' and '*via secca'* or the wet and dry paths. This ratio is approximately 1:4.

Note that Steiner ascribes calcium and not carbon to the Earth elements and this is because carbon, the hub of organic chemistry, is the process (Luciferic) down which we are 'fallen', while calcium is the second fall, governed by Ahrimanic forces. We can say that our fall, the work of Lucifer, coincided with the densifying of our 'diamond being' into opacity, whilst the second fall, which entailed a penetration with ahrimanic forces, resulted in the formation of limestone (bones), and we must now confront this reality.

The above can be represented in a semi-mathematical form that must be read in a qualitative way:

$$O : N = C : H$$

But it can also be: $N = \frac{HxO}{Ca}$

Or: $N = \frac{HxO}{Ca\ (K)}$

[37] See "Agriculture as Spiritualization of the Earth", by the author – not yet in English (2008).

What Steiner is saying may be presented simply in the assertion that nature is a real and masterful alchemist, and so there is no sense in thinking of intervening in imbalances of soil and of plants with chemicals as they are found in mineral nature, but these must be inserted in processes of transformation and transmutation which is possible only in the world of the living. What is increasingly missing is people's renewed knowledge of alchemy and how to establish a new dialogue with the forces of the zodiac.

We have already clarified in the past that the essence of physical substances on Earth resides in the Zodiac. So to talk of nitrogen, potassium, calcium, sulphur and so on actually means to refer to specific constellations of the Zodiac. So when we read: "*Under the influence of hydrogen, lime and potash are constantly being transmuted, first into something resembling nitrogen, and then into nitrogen itself*", in order to truly understand this we have to move away from the field of chemicals because in such a laboratory these changes will never be verifiable. This is why it is so essential that we make ourselves a living soil where the forces of the cosmos will be welcomed. We know that potassium is the substance that belongs to the constellation of the Virgin, as calcium belongs to the constellation of the Scales.

Potassium actually belongs to the world of lime, although chemically it is not calcium, but it expresses the same gesture of contraction. We can note with interest that the constellation of the Virgin is traditionally represented with the Scales in her hand. It is also curious that it is necessary that hydrogen is like a catalyst that mediates for the transmutation we are talking about to occur. Hydrogen is the most labile substance and is mainly found in the upper layer of the atmosphere. So once again we can understand that the soil must develop the ability to establish a much wider connection than one confined to its immediate physical surroundings. We can also note that the catalyst for subtle transformation, hydrogen, has the atomic number 1 which can resonate with the atomic number of any other substance. Moreover Hydrogen is the first element that was created.

It is also interesting to note that in the formula is the relationship of hydrogen and oxygen representing the sphere of water, the basis of life, and we have already said that all these transmutations only occur under the rule of life. (Note well that the formula that we place as the basis of our reasoning is not a correct representation of what we are saying, because it is a chemical formula. Alchemical formulae and symbols rest upon a crucially different basis and understanding.)

Paragraph thirty-nine ("*Silicic acid, as you know ...*") introduces the seventh preparation based upon dandelion. The paragraph opens with the statement that silica is transmuted within plants. We are also told that silica is needed in order for cosmic forces to be attracted and used by the plant. Perhaps the fact that the Earth is made up of 42% silicon - and is therefore living - is revealed as clear and important. But in order for silica to do this it must be placed in a "*thorough interaction*" with potassium, not with calcium. Meanwhile, we note that the relationship between silica and potassium expresses a relationship between the constellations of Aries and the Virgin. We have often stressed the importance of the Fishes-Virgin that we have presented as the axis of healing, as the bearer of the forces of Christ and Mary. In this case the forces referred to are those of Mary and the Lamb, the axis that is the manifestation of Life. We can understand that the key to everything we are saying is the constellation of the Virgin, that is Mary, without whose intervention neither the manifestation of life nor healing is possible. To attempt a further element of clarification we can say that between Aries and the Virgin the *light zodiac* is arrayed, so named because it corresponds to our consciousness.

The centrality of the Virgin has already become clear in our meetings, when we had the opportunity to say that the Zodiac originated from the Virgin and more specifically from the star called Spica which the Virgin is holding in her hand, from which everything is derived - the Zodiac and so also the solar system. Imaginatively we might think that the star Spica, which is represented by an ear, is *sown* in the cosmos and has given birth to the whole Zodiac.

Returning for a moment to what we said in relation to a previous paragraph we noted that silica can only enter into a proper relationship with potassium and Zoe with the assistance of hydrogen. Hydrogen has the constellation of the Lion as its reference. This is the constellation under whose action old Saturn was formed which is the beginning of the whole system. The Lion in us is the area of the heart, which is the centre of our biological, soul and spiritual life. In other words we can say that hydrogen is the being that allows *Zoe*, originating from the Virgin, to become life on Earth.

Note that Steiner states, in this specific case, that the connection must be established with potassium and not with calcium. Potassium develops its connection with the unbound and widespread Life (*Zoe*), while calcium is linked to embodied life (*Bios*). In 502 the significant influence of potassium was to establish a link with the cosmos - with the assistance of sulphur. For 503, which is used to stimulate vigour and therefore the organic facet of life, it was important to establish links with both potassium and calcium. From this point of view the absurdity of using potassium fertilisers must surely be clear. Remember that Steiner asserts that fertilising should bring life to the soil and thereby stimulate the vegetative aspect of the plant.

The fortieth paragraph ("*We can readily find this [in] a plant* ...") reveals that the plant for the development of this preparation is dandelion and provides the first indications for its use. The forty-first paragraph ("*Collect the yellow heads* ..") continues with signs for the creation of the preparation. The paragraph ends by explaining the role of the dandelion preparation in the process of cosmic nutrition. Plants are nourished by the cosmos through the flower. But the flower belongs to the part of the plant where (biological) life occurs with less power. This allows us to see in dandelion a valuable tool to bring the cosmic Life where it manifests itself with extreme effort, as in arid areas. In such situations 506 can become part of the instrument to bring life even where it is not possible to do this with classical biological measures. If you look at all the spiky light-filled silica-rich plants that grow in the desert this confirms the correctness of what we have just said. (Obviously we can transform the deserts of the world only when we have transformed the desert that is within us.)

The paragraph provides an indication of the animal sheath for the development of the preparation - the bovine mesentery. The mesentery is the membrane that surrounds the intestines of animals below the diaphragm and, when cleaned of its fatty component and held up to the light, shows a structure formed by ribs that together form a series of irregular triangles. We can see it as a periphery of the metabolic area of the animal.

We note that dandelion is a plant that, in the form of its successive parts, continuously alternates between cosmic and earthly gestures and therefore can be seen as a bridge between these two poles. For the specific purpose of this preparation, collect flowers at the stage at which they are just about to expand so that the opening continues after collection.

It is worth pausing a moment to understand why we use the mesentery. The stem of the dandelion flower is hollow. The Air element is the most similar to Fire and leads us imaginatively towards the seed (Fire) from the flower (Air). The forces of Fire are brought to the flower by insects. The mesentery is the boundary of the body that mainly contains the expression of Air (diaphragmatic area) and Fire (reproductive area) and can be assimilated by the insect world.

In other words: the flower that goes to seed thanks to the forces brought by insects, corresponds to the realm of the internal organs, to the expression of the Air and Fire embodied in the mesentery (intestines), and so the mesentery has the function corresponding to insects[38].

At this point we can make a consideration of a general character: the preparation based upon the dandelion develops its main influence in the region of the flower that we have always linked to the Air element. We also know that the chemical element associated with Air is nitrogen. Just now we stated that 506 is the preparation that promotes nutrition from the far distances of the universe. However, in the past, we have also said that soil-nitrogen plays the leading role in bringing the cosmic images[39] and enables the plant to expand the spatial environment from which it draws nourishment, greatly exceeding the limits of the soil physically occupied by its roots. In both cases, of course, it is not nutrition on the physical level, but one we might call a 'subtle' nutrition of Light. Everything will - we hope - be better explained in the following paragraphs of the Koberwitz course.

In the forty-second paragraph ("*Even plants, in order ...*") Steiner begins talking about a kind of sensitivity of plants. We know that plants cannot be sensitive because they do not have an incarnated astral body. Actually Steiner speaks of "*a certain sensitivity*" thereby referring to elemental beings that build the plant and remain connected to it. In subsequent sentences it is emphasised that the plant "*delicately permeated and enlivened by silicic acid*" is sensitive to all that surrounds it. Moreover it is easy to induce a plant to look only in its immediate vicinity for what it needs and there is a clear reference to conventional agriculture in relation to which Steiner expresses an opinion: "*which is not good, of course.*" We understand that when Steiner says that the properly aware plant can find what is suitable and benefit from what is in a meadow or in a forest nearby, he refers to the elemental beings. In another way this would happen just for the physical part of the plant

For 506 we believe we can provide a useful summary of the functions performed upon the various planes:

- on the physical level, it enhances the action of silica in the plant and also improves the quality of pollen;
- on the etheric level it increases the plant's nutritional value;
- on the astral plane it connects the plant with the edge of the universe;
- on a spiritual level, allowing the plant to connect with its associated planet and thus with its group I.

At the level of an Earth sacrament the 506 corresponds to marriage.

In paragraph forty-three ("*And so it seems ...*") Rudolf Steiner lists the five 'compost' preparations examined so far. Highlighting them in this way he emphasises the fact that their combined action is able to bring forces into the compost and manure

38 See 'Man as Symphony of the Creative Word', by Dr Steiner.

39 See commentary on Lecture 3 in this volume.

to support all the functions of the plant.

The rest of the fifth lecture is devoted to the valerian preparation and, strangely, it is only given these 10 lines. This does not mean that the preparation in question is unimportant but only that the Steiner concluded that, after exhaustive explanations about the other preparations, the delegates at Koberwitz should have been able to understand the need for this preparation without going into details.

Valerian has the characteristic of emerging above all other annual plants growing around it. The fact that it is always a span higher than the plants that are around makes it appear as a border of herbaceous vegetation, similar to how Saturn marks the boundary of the solar system. In fact Valerian is tied to Saturn and is the bearer of warmth forces and of hydrogen. The root of valerian - although fine - expresses a gesture of closure as if it were a head. This is comparable to the head of a person in which the nervous system must be closed in and not expand elsewhere in the body. This part of the nervous system must be closed so that it can connect with the highest spiritual beings through thought. The connection with the nervous system governs the use of Valerian as a remedy against insomnia.

The juice, which is simply pressed from the valerian flowers, is preparation 507. After being diluted in hot water and dynamised for at least 20 minutes, it is sprayed over the top of the heap to form an *etheric skin and* at the same time encloses the compost and makes it able to open to communication with the cosmos. Note that the valerian preparation with the dandelion and horn-silica preparations are the ingredients of our product called 'fruit plus'. It is a polarity to our 'Pre Sowing' which is composed of horn-manure, and the yarrow and chamomile preparations.

'Fruit plus' connects with the cosmic nutrition whilst the 'Pre Sowing' connects with the centripetal forces of the Earth. In the middle we can put the clay preparation. (Refer to paragraph 16 of lecture 2 of the Koberwitz Course).

At this point it is worth an aside to take inspiration from what we have always said about the three vital processes represented by a lemniscus. We have often talked about *Sal, Mercur* and *Sulfur* as the three processes that interact with and permeate life.

The influence of valerian is outside this scheme because it is linked to phosphorus. Phosphorus is the substance associated with the constellation of Cancer and Cancer is the constellation that puts in motion the forces of the Zodiac in a series of vortices in which the end of one cycle immediately precedes the start of the next.

We know that *sulfur* is the process of expansion that returns to the domain of life through *Mercur*. *Sulfur*, at the level of the chemical elements corresponds to sulphur, which we can imagine as *dirty* light because it is yellow and contains a tinge of the darkness toward which it turns along the lemniscus. The phosphorus, however, corresponds to white light through which it 'departs' from the lemniscus with its centrifugal momentum and without any possibility of return: thus it is death.

But also at the bottom of the lemniscus there is a chance to get back upwards to continue in the processes of life, or fall out and to meet death. This 'downward death' is linked to ash and then potassium.

The fact that valerian enables escape to the world of imponderables becomes interesting if we connect it with the aging of the plant that links the plant back with its origin through the seed. It is quite obvious to everyone that the fact of obtaining a seed comes with the death of the old plant and the opportunity to create a new plant, compared to the previous one, is a new manifestation of life. The phosphoric process is essential to achieve high quality seeds. It is no coincidence that the valerian is itself a good seed bath, even if we dare to believe that those we have made are better.

The aspect of strong ties with the world of quality is confirmed by the fact that the juice of valerian, obtained by the squeezing the petals, should not putrefy and can remain unchanged for 2 years.

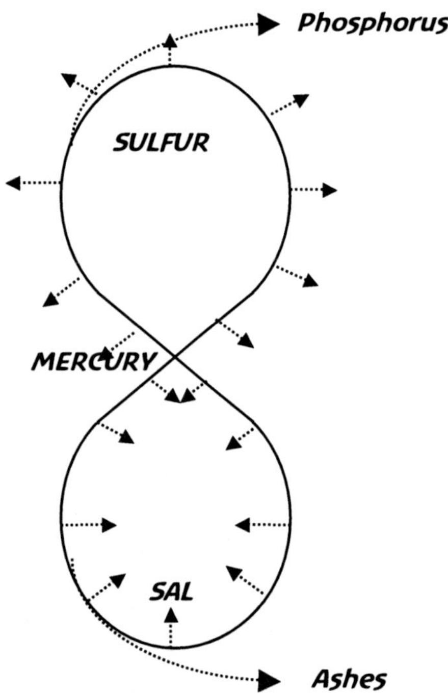

We can recollect the action of 507 on the various planes:

• on the physical level, it brings warmth and increases the resistance of plants to the cold. Our 'Pro heat' is made with valerian and the drone (the bee is linked to the world of the dandelion). The Valerian is a great activator for many processes;

• in etheric terms it moves the warmth ether. On this level Valerian acts in a special way on making the root able to connect with the cosmos, or in different terms, it allows the cosmic forms to enter the plant. Our 'root plus' contains Valerian and oak-bark;

• on the astral plane, it brings resistance to cold providing the impetus to continue to live despite the contrast represented by cold;

• on a spiritual level, it connects the plant with the I consciousness in the centre of the Earth.

At the level of an earthly sacrament valerian is the extreme unction.

Sixth Koberwitz Lecture

Nineteenth Meeting

We have finished reading and remarking upon the fifth Koberwitz lecture in which we were presented with the biodynamic preparations. These are the tools to bring new life to the land. Now we will tackle the sixth lecture that should help us to understand disease.

The fourth and fifth lectures are dedicated to healing because in them Steiner presented the remedies to restore health when nature is sick. The sixth lecture is the one in which we deepen our knowledge of the causes of the diseases to be treated. The title that was given to this lecture is: "*The essence of weeds, and animal pests, and the so-called plant diseases in nature*". We learn the appropriate methods of intervention to restore the environmental balance in a way that might prevent the manifestation of pests.

In some ways this lecture echoes the second. We hear once again about planets, cosmic and earthly forces and, more generally, of the forces of the cosmos in the plant world.

The first paragraph is an introduction to the new lecture and contains nothing particularly difficult. In the second paragraph ("*I would like to begin ...* ") by contrast, we jump into the heart of the discussion, because we are immediately presented with the so-called 'weeds'. A weed is normally defined as any plant that grows where we do not want it but we must admit that this is not very helpful for a correct classification and definition. Arguing in this way we come to consider a plant as a weed in one environment when in a different situation we cultivate the same plant. As an example we might consider the Jerusalem Artichoke either as a persistent weed or a prized crop.

Steiner also makes us see that labeling a plant as a weed is the logical result of our utilitarian point of view. From the point of view of nature what we disparagingly call weeds have the same right to grow as other plants that we consider useful. Indeed for the soil and its natural balance they are more useful than cultivated plants because they are actually the spontaneous response to a specific need of the soil which we rarely take into consideration when we decide to sow a certain crop.

Considering the fact that the next two pages will be difficult to understand, we will try to bring order using a sketch in which we will represent what we read to help us grasp the connections and relations. We begin with our usual plant sketch and then highlight the fact that there is a stream of forces that comes from the sky, passes through the Earth and enters the plant from the roots. This stream is clearly composed of forces that emanate from celestial bodies when they are on the opposite side of the Earth to us. These forces are divided into two currents that we will represent at the bottom of the drawing.

The upper part of the picture, on the other hand, represents two different streams of forces that descend directly upon the plant. The plant therefore is crossed by four currents of forces: the two we have drawn at the base of the drawing come from the inner planets (Moon, Mercury and Venus), the other two, of course, from the external planets (Mars, Jupiter and Saturn). Naturally we will place the Sun in a position between the inner and outer planets.

Please note that this drawing shows all the forces affecting the world of plants. But bear in mind that each species is dominated by just one or a few of these planetary forces. However, for the moment we will disregard this because we are interested in understanding the imbalances in these forces that bring about the weeds and pests.

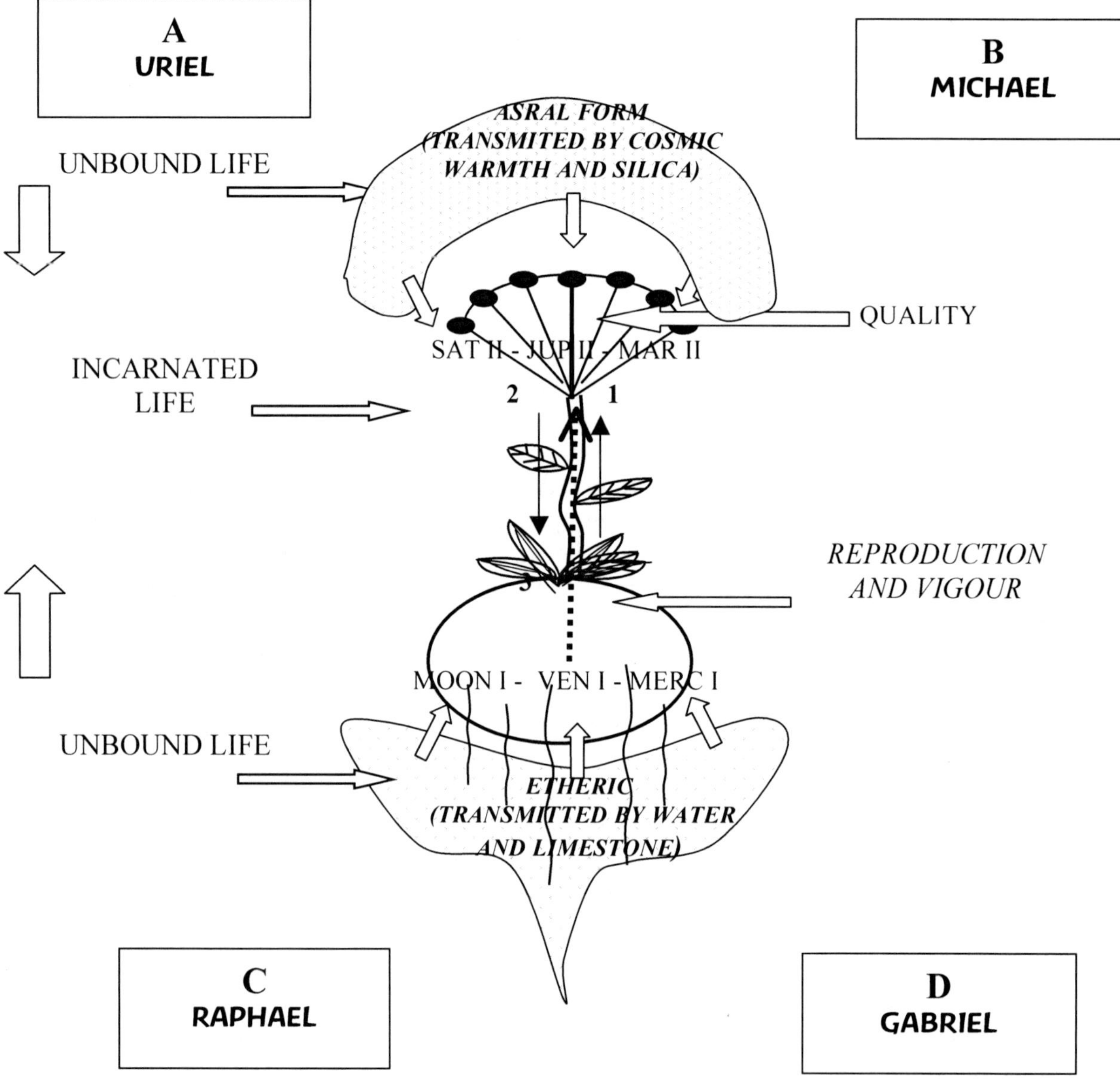

The action of the planetary forces and archangels on the plant

We also have tried to highlight in our sketch that the forces that come from below rise through the ascending sap, while those that act directly descend from above, obviously using the descending sap.

Do not overlook the fact that, for the umpteenth time, we have represented the plant as being threefold, composed of lower and upper poles and a mediating central part (Sun). The lower part of the plant is more influenced by the etheric forces - those linked to the growth and vigour - while the upper part is more closely linked to the astral forces that initiate flowering and bring quality. In other words we can say that life-processes dominate at the bottom of the plant and death at the top. We know that flowering, the prelude to fruiting, marks the approach of the plant's death, especially for annuals.

The four currents of forces that we are talking about (A, B, C, D) correspond to the etheric forces which we have repeatedly mentioned. But if we want to be complete we must have the courage to recognise that these currents are the actions of the four Archangels that weave in the cosmos. Those who wish to go deeper into this approach are referred to Steiner's cycle of lectures entitled "The Course of the Year in Four Cosmic Imaginations", in which the work of the four archangels acting in the living (plant) world is described.

We invite everyone to reflect on the fact that practicing agriculture, in the light of what I have just said, entails an interaction with the archangelic beings and we would imagine that the dignity of this activity, often seen as humble and sometimes even humiliating, might be somewhat raised.

Building upon the square (ABCD) that we are sketching we can say that the forces that rise from the bottom find their manifestation in all processes of reproduction and vigour. In the opposite pole is the world of quality, manifesting as perfumes, flavours, therapeutically active ingredients, medicines and in a broad sense everything that has the ability to nourish. A plant is worthy of feeding humans only when it can come into contact with these superior forces. If this link is particularly intense the plant acquires the characteristics of a medicine (specific link with Mars), perfume (connection with Jupiter) or essential oil (connection with Saturn).

The system of forces that envelops the whole being of the plant may be regarded as 'diffuse Life', or free Life, the unbound ethers or *Zoe*, from which derives *Bios* the incarnated life (the etheric body) of the plant.

We can complete our sketch by showing the archangelic forces that we described. The activity indicated by the arrow at the bottom right (see p 217|) is due to the archangel Gabriel, the lower left to the archangel Raphael, that shown in the upper left is due to the archangel Uriel and the last one to the Archangel-Archai Michael.

The activity of Gabriel, the angel of the Annunciation, takes place primarily during the winter (Christmas). Raphael acts primarily in the spring (Easter), Uriel acts in summer (at St. John), and Michael completes the cycle in the autumn (Michaelmas).

As far as planetary connections are concerned we can say that Gabriel is mainly linked to the Moon (reproduction), Raphael to Mercury (linked to the biological aspect of life), Uriel to Mars and Michael to Saturn. Michael also brings therapeutic forces but through links with the forces of the spiritual world.

If we wanted to sketch more comprehensive archangelic influences we must bear in mind that *Zoe* and *Bios*, despite being polar from one point of view, are closely linked because *Bios* would not exist without *Zoe*. Similarly archangelic influences work in the unbound etheric forces, but also in the incarnated etheric - the etheric bodies. Bearing this in mind we can say that the 'free' action of Gabriel (working from the outside in) manifests within reproduction in the descent of archetypal spiritual forces. Acting within a body Gabriel presides over metabolic processes related to nutrition. Raphael's external activity leads to the healing of nature, whilst internally he presides over the respiratory process, which we know is also a therapeutic process.

Similarly, Uriel's external role is to carry cosmic thoughts, the wisdom of nature. Internally he reinforces human thoughts giving a valuation of the sense of the times that allows us to understand the meaning of what is happening in the world. For example in this present time (May 2003) we must be able to grasp the fact that Mars, once known as the god of war Aries, is presently located nearer to the Earth than at any time in the last 73,000 years. Suddenly we might realise why we are witnessing an increase of all the conditions favourable for war. On the other hand, the transformed impulse of Mars brings the power of initiation so we all have a special opportunity to place ourselves in a state to receive initiation. Of course it is necessary to overcome the instincts of our ego that in this sense may be considered the relevant initiation trials. The times in which we live are characterised by the full manifestation of Christ in the plane of life and it seems quite understandable that the forces of evil are trying to distract man from this great event - and is there a better way to achieve this than by making us fear for our own lives? We can also interpret, using this key, the advent of the terrible diseases that afflict humanity lately from cancer, AIDS, and the recent atypical pneumonia. From this point of view we must take positively the eclipse of the Sun that there will be on May 31, 2003. The eclipse is a wonderful opportunity for those who commit to an initiatory journey because the eclipse signifies the nullification of all our points of reference. Enveloped in total darkness we can have the experience of initiation. Of course in the past an exterior eclipse was the right occasion, but now the eclipse should be brought inside.

Returning to the archangels we should not forget that Michael brings the forces of self-knowledge externally, and internally brings the forces of free and conscious activity - liberation (It: '*libera-azione' or 'free activity')* which should allow us to find the right answers to the questions: who are you, what are you doing, where are you going?

As we have said these four forces act within the plant kingdom, where this "who you are" inspired by Michael finds its highest expression in plant essences, in which we can witness the 'miracle' of Water that becomes Fire – as oil must be considered. We will only briefly mention the fact that oil is widely used when anointing through which many sacraments are administered, and that the word Christ means 'the anointed one'. To complete this circle remember also that Michael is called 'the countenance of Christ.'

In the third paragraph ("*We have seen ...*") there is an apparent contradiction. Speaking of forces from the interior planets acting through the Earth Rudolf Steiner adds that these are the "*.. ... forces we need to take into account when we want to trace how one generation of plants leads to the next.*" The apparent contradiction lies in the fact that we commonly think of reproduction as the function of the flower and fruit, not the lower pole of the plant. Actually the force of reproduction does not lie in the seed, but in the root. It is not by chance that a plant develops the root first. The seed brings only an echo of the reproductive strength of the root to such an extent that even visually a seed seems like a stone suggesting its relationship with the earth element. For 'reproduction' we must not only understand the process by which one body creates another body, but also what happens to individual cells and which shows its broader aspect in the vigour and growth of the plant.

Immediately afterwards Steiner refers to the forces at the top of our sketch followed by a sentence that is a reminder of the second lecture. We are reminded that the activity of the internal planets is highly influenced by the effects of "*the lime in the Earth*", while the action of the outer planets is "*influenced by silica*". This indication is also shown on our drawing to emphasise that limestone and silica are the two ambassadors or, if we prefer, the two guards on the mineral plane of the forces we are discussing.

We should note here that Steiner is directly describing only the activity of Gabriel and Uriel. At this point we do not find any mention of healing. We know that Raphael brings healing to those parts of Gabriel's forces that have been altered and therefore become pathogenic. We can consider the therapeutic action of Michael upon Uriel's activity in a similar way. We can say that Raphael is involved in healing biological processes (the breath controls metabolism), while Michael's intervention is necessary for a therapeutic input to the quality processes governed by Uriel.

The fourth paragraph ("*Nowadays we are not in the habit* ...") notes that the loss of knowledge of these four forces had serious consequences for international agriculture culminating in the spread of *phylloxera* in vineyards. This resulted in the destruction of all European vines. Only by grafting European vines onto American stock was it possible to reintroduce the vine into Europe. The vines in America had been able to survive *phylloxera* because the land on which they grew was strongly calcareous and limestone, with its contractive strength, brought the cosmic forces down to the roots. Here they opposed the responsible aphid which was bound to earthly forces. Obviously the methods used in the late 19th century were still organic so resistance to *phylloxera* has been maintained for a long time. The totally unnatural contemporary methods have brought the situation to a critical level. From what we have said it should follow that a remedy against *phylloxera* should be to bring light into the root of the vine. Therefore, the forces of Uriel as well as those of Raphael were not enough.

So we come to the fifth paragraph ("*Let me show you this* ...") where Steiner maintains that everything that comes from the inner planets, which works from the bottom up and that is manifested in the formation and growth of the seed, mainly affects annual plants. We have identified those forces and their movement along the (diagonal) axis connecting Uriel to Gabriel. From this we can deduce that what concerns the perennial plants moves along the other axis, the one that connects the influence of Michael and Raphael. (We must bring to mind once again that the examples that we bring, especially using drawings, are dictated by the need to isolate the various activities to understand them in detail. However, this division and schematic clarity does not exist in the real world where all activity interacts simultaneously and in a harmony conducted by high spiritual beings – the Spirits of Harmony.)

In paragraph six ("*On the other hand* ...") it is stressed that the stream of forces descending directly onto the ground comes from the external planets. We are referring to the Urielic current which enables the formation of full, round apples. Michael then brings this apple to the highest quality. So when big and beautiful fruit are obtained through forcing the apple (as is typical within conventional agriculture) it is a forcing based upon the effects of the current of Uriel. Now we can easily understand why these apples do not have a flavour in proportion to their beauty and why they certainly do not have a high nutritional value. To establish a planetary reference we can say that the flesh is the result of the forces of Jupiter, but the quality depends on Saturn.

The last sentence of the paragraph should not be forgotten because it is of paramount importance. It reads: "... *taking these different forces into account is the only way to gain insight into how to affect the growth of plants.*" You cannot do otherwise than begin to study the relationship of the plant with the cosmos and then try to put in place the conditions so that some forces or other are brought down depending on the needs we have. The phrase "*is the only way*" leaves no room for breadth of interpretation.

The seventh paragraph ("*Now a large number of plants ...*") is devoted to a better understanding of the forces of the Moon. The rest of the so-called weeds are the result of an excess of the forces of Gabriel through the Moon. Obviously in the weeds there is a strong component of Mercury, which is expressed in regrowth, and Venus that is expressed in metabolism. Precisely because of their highly developed metabolic abilities weeds are plants that can live upon little.

Steiner says that weeds are often the strongest medicinal plants. Not long ago we said that the medicinal plants have a strong link with Mars, and we know that Mars is the memory of the evolutionary phase called old Moon. The connection of weeds with Mars is highlighted by their exceptional vigour and the influence that they exert on other plants. The active ingredients of medicines are an expression of the strength of Mars that is able to unlock disease states and bring them into balance.

The paragraph presents the Moon as the celestial body that reflects sunlight back to the Earth, in the process giving the sunlight lunar characteristics that act to reinvigourate everything on Earth. When the Moon and Earth were still united there were impressive demonstrations of life on earth. It was the epoch of dinosaurs.

Steiner then tells us that now the Moon acts upon the Earth "*to intensify the normal condition of the Earth, so that growth can be enhanced to reproduction.*" In other words the forces of the Earth can work to achieve asexual reproduction (such as cell division) which we could consider as a sort of cloning. The Moon, however, is necessary for sexual reproduction. In Earthly reproduction the element of fertilisation is absent. However, lunar reproduction requires a male kind of input, while Earthly reproduction involves only the female element.

We might add that if the Moon reflects the rays of the Sun and given that the Sun is the home of Lucifer, the Moon will reflect a luciferic influence onto Earth, thus altering nature. Besides, the sunlight reflected from the Moon is no longer yellow but greyish. To better understand the relationship between the Sun, Moon and Earth, we must bear in mind that, although the Sun is many times larger than the Moon, the differing distances from Earth of these two celestial bodies is such that the lunar disk can exactly overlap the Sun as happens in a total eclipse of the Sun. We may think that the Moon is able to shield the Earth from the excessive luciferic influence. This reasoning completely changes our view of the Sun and the Moon. We normally consider the Sun in a positive way and the Moon as detrimental, but now we are faced with a view that gives the Sun a negative value and a positive Lunar estimation. Clearly we must learn to grasp two different points of view because it is also true that the Moon has received the worst part of humanity but, on the other hand, the Moon is also the 'home' of angels. The Moon is considered by Steiner as the corpse of the solar system, but it is able to stop the luciferic influence upon nature which is manifested in the growth of plants with heliotropism. The plant grows upwards because Lucifer, who is of an astral nature, seeks complete assimilation of the etheric of plants and in so doing draws them to himself.

We can still add, in an attempt to bring a bit of clarity, that Lucifer has three residences: Venus (which in ancient times was called Lucifer), the Sun, and Sirius. Venus accommodates the etheric-planetary dimension of Lucifer, the Sun holds the astral dimension, and finally Sirius - which belongs to the constellation *Canis Major* and thus the Milky Way - holds the highest spiritual dimension. In the threefold vision the spiritual Lucifer is that of Sirius; that which acts from the Sun is Lucifer's soul - his astral nature that wants to appropriate the etheric body of men and plants. The astral-Lucifer has already gotten hold of our thoughts and our breath. Although we said that our breathing is linked to the activity of Raphael and therapy, the fact remains that Lucifer has seized our breath which is the main cause of our being subject to death. In this sense Lucifer is the one that takes our last breath. The least-part of Lucifer, which acts as Venus, influences our metabolic sphere giving the pleasure of having a 'full belly.'

But even the Moon is threefold consisting of the hardening forces from retarded beings - those who had no capacity to evolve further upon or have burdened the evolution of the Earth; at the same time angelic entities weave their existence on the Moon as we have said; and finally archangelic entities that are able to adjust the luciferic influence coming from the Sun. These latter forces are archangelic in nature because they have to be able to oppose the luciferic forces which are of an angelic nature.

This last observation has interesting implications for us when we compare the subjective size of the Moon, that is the seat of the angels. We must develop an archangelic conscience which is formed above all when it comes to developing a community. This is also why it is useful to study in groups.

Even Sirius is three-fold! Actually Sirius is a system of two stars, a bright one that represents the cosmic force of Lucifer, and a dark one that represents Ahriman. The union of the two parties represents the Asuras.

If we return to the seventh paragraph and the Moon we can consider that the Moon has a 'synodic' cycle during which, for the perception of those who are on Earth, the illuminated portion of the surface grows (waxing Moon until it is full) and diminishes (waning Moon until the new Moon). Sometimes the relative position of the Sun, Moon and Earth results in us seeing only a small section of Moon illuminated by the rays of the Sun, and this small crescent is at the bottom of the lunar disk.

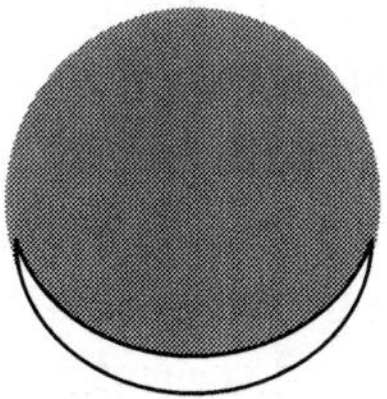

The sickle Moon

The situation described and shown in the drawing recalls the situation during a partial eclipse of the Sun when the new Moon covers part of the luciferic light of the Sun. This allows a particular spiritual experience. In this case it is not the sunlight that is blocked but the reflected light of the Sun, as if a spiritual force (black force) removed the possibility for the darkened part of the Moon from reflecting the luciferic sunlight. This image shows what alchemists called '*the black opera*'. In a different cultural interpretation this is the Black Madonna who has her feet on a sickle Moon and crushes the luciferic serpent's head.

Now we understand that it is mistaken to assign a luciferic attribute to the Moon. Actually when the Moon reflects the rays of the Sun it does not itself become luciferic - to the extent that the image above can even become the symbol of the victory over Lucifer.

Another meaning of the sickle at the bottom of the Moon takes us back to the Holy Grail, the cup that collected the blood of Jesus. From this it is clear that the therapeutic strength of the sickle Moon, when it doesn't appear in the lower part of the lunar disc, is greatly reduced. All this can be experienced within us. The alchemists called this '*the marriage of the King and Queen*'.

The eighth paragraph begins with such an obvious sentence that it seems pointless: "*When a living being grows, it gets big.*" Perhaps we are used to the fact that Steiner does not often say unnecessary things: even in this sentence there is a hidden meaning, as long we consider it spatially (Light) and not in terms of time.

Here it is made clear that growth uses the same forces that act in reproduction only with less intensity. In this regard we have already clarified that size is an earthly attribute, while the Moon is necessary for reproduction.

In completion of what was said in the previous paragraph we are reminded in the ninth paragraph ("*As I said before ...*") that the Moon is a mirror not only for the Sun's rays, but for the forces from the whole cosmos. Precisely because these are all reflected from the Moon, affecting the plant world not only for phases of growth but also reproduction, we can appreciate the fundamental importance of sowing in relation to the Moon.

The tenth paragraph ("*Now for a given location ...* ") does not present particular difficulties in interpretation.

The eleventh paragraph ("*You see, under certain circumstances ...* ") Dr Steiner describes the ideal conditions for weeds to grow, which in the drawing we have made correspond to Gabriel. The twelfth paragraph ("*However, when we take ...*") explains how efforts against the proliferation of weeds requires that we work with the current of Uriel. It is therefore necessary to work the soil so that it becomes unsuitable for receiving the lunar forces. In other words Steiner is telling us that to counteract the lunar forces one must use the forces that induce the formation of seeds, by preparing them in a way he will explain in the next paragraph.

In the thirteenth paragraph ("*So if we see ...*") Steiner suggests that we use the seeds of the weeds to combat their own growth. Seeds conserve the force that rises up from the root and which allows growth and reproduction. He tells us to perform certain operations in a certain order; first light a wood-fire and then burn the seeds on it, retaining the ashes. We think it is useful to suggest that the wood should be full of Saturn forces such as that of conifers. It is also good to be using a container that allows the passage of air, because a lack of air makes it difficult to incinerate the seeds. For this reason the container must have holes about one centimeter above the base to prevent the wind blowing the ashes away. It is not good to use paper, even for lighting the fire because it will not burn well and will leave unwanted residues. If you want to use a tin you should burn it really well before burning the seeds. Steiner does not say that the ashes should be made during the Full Moon or even better if Mercury forces are abundant to remove weeds' strong ability to multiply, but we will give these as our suggestions.

Steiner then adds that the positive effect against weeds requires a commitment of four years given the relationship of weeds with Mars, the planet that induces aggression.

In the fourteenth paragraph ("*You see, here you have an example ...*") Steiner even suggested the possibility of having the wild plants only in the positions that we want them. The use of weed ashes, which now seems more like witchcraft, once was in common use. The well-known St. John bonfires were set up with the weeds and after the bonfire the peasants took home the ashes that were scattered in the fields as necessary.

The beginning of the paragraph refers once more to the "*effects of smallest entities*", in which we note yet another wink towards the world of homeopathy. Moreover, when Steiner indicates the amount of horn-silica for use on a hectare of land (note that this is four grams) it seems obvious that the effect is not due to the weight of the preparation but to the subtle effects or the dissemination of forces through the substances. Of course we are referring again to the four archangelic forces mentioned above. Naturally the weaker the awareness of the farmer the greater will be the amount of substance needed, and we must admit that the forces we are dealing with are virtually unknown to most. We must begin to grasp these forces in the various realms of nature and we could start by asking where they are localised in humans. We feel we must formulate some thoughts for consideration.

We can indicate the fields of archangelic activity placing Gabriel to our left and Raphael to our right.

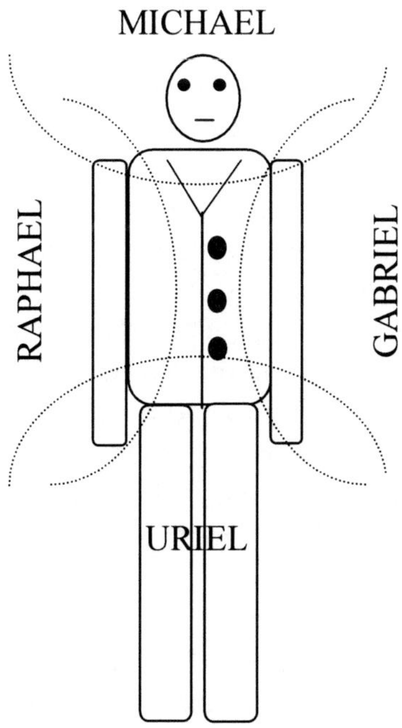

The spatial orientation of the activity of the four archangels in a human

Our left side is connected with our more creative feminine side in which women are more developed. The most rational is to the right, the most influenced by Raphael. These two forces can combine at various levels, for if they unite at a low level we have physical reproduction. If they join at a middle level, after purification of the interior planets, we can have the experience of enlightenment that we could also define as the birth of the Son of Man or Jesus in us. If the meeting takes place at the top level we can have the experience of initiation or the birth within us of the Son of God - Christ. This highest union is within the sovereignty of Michael. Uriel oversees the lower union.

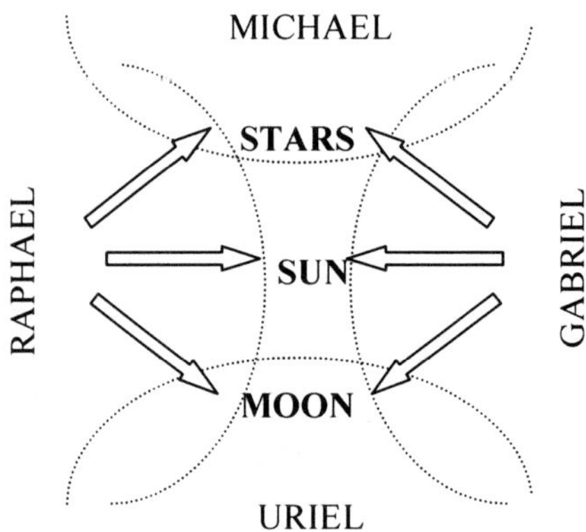

The three levels of the combined forces of Gabriel and Raphael

These thoughts can lead us to a level along the evolutionary path that is located between illumination and initiation. If we had the strength to follow them all our inner growth could reap its benefits. It does not matter if you forget at the level of the physical brain because every thought we conquer leaves an indelible path at subtle levels. These are thoughts that were cultivated in the mystery schools that anticipated by a long time the lessons that today are attributed to Christianity. For example, in Chartres one loved the Trinitarian Mary. Indeed, under the crypt was the Black Maria bearing the Child in her arms and he was even honoured by the Druids about 3,000 years ago, long before the birth of Jesus. Then there was the Madonna of the altar and finally Our Lady of Glass, representing the Virgin Sophia. Moreover, even then they tried to bring the forces of the Spirit into nature and this can only happen through Mary whom the Greeks knew by the name of Demeter.

All this, which belongs to the sphere of 'Platonic' initiation, must be brought into agricultural practice, of course, taking care to adjust the instruments to the needs of the world today.

So we come to the fifteenth paragraph ("*In all these instances ...*"). Here Steiner tackles the problem of demonstrating the accuracy, in terms of outcome, of the biodynamic method. Today we live in a materialistic world and any statement must be demonstrated experimentally. Steiner also suggests that experiments are made, but we must go about these in the right way because otherwise one runs the risk that the experiment proves precisely the opposite of what we want to demonstrate.

Our experience teaches us that it is very difficult to prove, with physical experiments, effects that do not belong to the level of manifestation. When the demonstration of an immediate increase in quantity of production is simple, it is less easy to demonstrate improvements in the quality of products. 'Subtle' laws are different from those found on the physical level and it is therefore easy to commit mistakes during the test. We have made mistakes in this sense when we tried to demonstrate the ability of our product to eliminate pollution from salt in the soil and atrazine in water. Our mistake was to experiment upon normal soil into which salt had been mixed. We only understood later that we had to use soil that was already contaminated with salt so that it already constituted a whole. We did something similar with atrazine in water. Sometimes it is more difficult to find the right way to set up a test that proves the validity of a product than to devise and make the remediating product itself.

However, beyond the needs of the world, Steiner makes us see that; "... *spiritual scientific truths are true in and of themselves, and do not need to be confirmed by other circumstances or external methods*." The limit is the inability to raise our properly educated thoughts up to the level of the spiritual worlds. Otherwise we would not have doubts; only the weakness of thought and its fall on the intellectual level requires physical demonstrations of spiritual truths.

Twentieth Meeting

Let's continue to comment upon the sixth Koberwitz lecture from the fifteenth paragraph ("*In all these instances ...*"). It starts with a discussion about physical evidence to prove the validity of spiritual scientific assertions. If we have sufficiently well-educated thoughts we would easily recognise that such truths "*are true in and of themselves*", whatever the results of experiments on the physical level. Obviously, the problem remains the adequacy of the manner in which the *evidence* is brought forth because there is a real danger of experiments being conducted that draw completely wrong conclusions. This is not so much because the truth is not true (!) but only because it was sought in the wrong manner. Clearly if we test the mechanical function of an organ like the heart we get only what the machine is capable of measuring, namely electric and magnetic fields etc. We will not gain an assessment of the life or functions of the heart but only an indication of the strength of its electrical parameters. In other words machines - which are really 'holes' for the spiritual world - can only measure forces that are similar to itself, namely those that hinder life. Hence the nonsense of what medical science (and agriculture) practices.

In paragraph 16 ("*We have spoken in general terms ...*") Steiner introduces the subject of animal pests that he develops in the seventeenth paragraph ("*Let us ...*") with the vole or mouse as an example. First, it is worth noting that the vole is defined as a "*the farmer's best friend.*" But the characterisation of *friend* contrasts sharply with all the methods which are normally used to eliminate the vole from the fields. These systems have the sole purpose of killing and are sometimes very cruel, such as poison or the deliberate introduction of a serious disease. Steiner does not hesitate to call these methods "*inhumane.*" In the next sentence this little creature is termed "*innocent*" just to highlight the error when we consider mice to be an enemy that we desire to destroy: the mice, like all pests, merely serve to highlight the fact that the soul of man harbours unworthy sentiments and give us the opportunity to improve. In this sense these pests are really our friends.

Perhaps to many it seems wicked that mice are injected with typhoid, but in everyday agronomic practice similar techniques are frequently used. For example, to combat many pests *Bacillus thuringensis* is often used without realising that this brings low quality Moon and Mercury forces to our fields. The accumulation of these forces in the soil and plants inevitably leads to the emergence of life-forms which are particularly related to these planets - fungi, weeds and viruses. So in addition to having lost an opportunity to understand a problem rooted in our soul, we have also created the conditions for further infestations in our land. Perhaps by trying to understand the message that the pests bring us and by being grateful for them, we also put ourselves in a position to formulate thoughts that allow us to free the 'pest' from the influence of its spiritual negative partner that would otherwise prevent it from experiences suited to its development. If we can do that, then maybe we can also 'ask' the pest-cum-friend not to damage the crops and it will be happy to please us to reciprocate for our input.

It is worth noting that interventions permitted for organic certification can have equally unpalatable biological effects. For example, pyrethrum is often used to combat insect infestations. This is a product derived from the 'Dalmatian chrysanthemum' - *C. cinerariifolium*. Thanks to the pyrethrins in it this product paralyzes the nervous system of cold-blooded animals. It makes no distinction between useful and harmful animals so it even eliminates the earthworms, the beetles and other useful insects, and bacteria that live in the soil. This chrysanthemum is extremely expensive today, so the product that is sold now is composed almost exclusively of synthetic pyrethrins. When this is put on the field it causes death and a strong compaction of the soil, even if it is an 'organic' product which in theory should not damage the vital processes. Farmers who are convinced that they are acting for the good of the Earth and of life cause serious damage and act contrary to their intentions.

Even those who apply the biodynamic method (or, worse, the homeodynamic method), do not do all that should be done to support the evolution of the Earth and animals. In all of our courses we continually repeat that the use of preparations is a great help to restore the fundamental balance and to deal with special emergencies, but our interventions cannot be limited to preparations. We must look to achieve a stable balance on our properties, and this is really possible only through the implementation of an agricultural organism. Unfortunately we must say that this is rarely done. We fear that this is due to an insufficient increase in the level of consciousness, which means that the preparations are treated in the same way as chemical or biological sprays. The most practical and convenient aspects of the method have been adopted but rarely do we take the pains to create the conditions necessary for a balanced expression of life. Let us strive to keep in mind that the task of a farmer includes the production of good food and respect for the earth, but above all to assist the evolution of the Earth to become a Sun through the process of vegetalisation and the formation of humus.

The eighteenth paragraph begins with the sentence: "*All of this experimentation and regulation is very superficial.*" This highlights the fact that if we continue to grasp the reality only in its outward physical appearance, we will continue to attempt to assist without really knowing the root of the problem and we will only meet with fatal failures. Proof of what we are considering is the very strange fact that many of the most important scientific discoveries are obtained by accident, even when they do not stem directly from 'mistakes'.

This paragraph states that effective action for the removal of voles requires the cooperation of farmers working in nearby farms. These must be in a position to understand the profound reasons that determine the method chosen for fighting pests and even the unconventional practice of agriculture.

In the nineteenth paragraph ("*Now what you have to do ...*") Steiner begins to describe the method that will result in repelling the field mice. There is an error in the second sentence of the paragraph, probably due to the translation [into Italian]. It refers to "*the sign of the Scorpio*". We believe it is correct to refer to the constellation Scorpion and not to the sign. This is because the signs can be seen in two ways: on one hand there is a theoretical division of the zodiac into 12 equal zones. Therefore these are no longer a reality although they coincided around the time of Golgotha with the position of the visible constellations. Actually in the 2000 or so years since then, the constellations have advanced more than 20° in the sky due to the so-called *precession of the equinoxes*. Therefore, today they represent the perceptible reality of the zodiac as the 'inertia' of creation. For the operation that we wish to undertake with mice we will then look at the visible constellations knowing that in our evolutionary future we will use our internal Scorpio and act therapeutically through it without the need for external markers.

This paragraph confirms that a different method should be used to repel animal pests than was described for weeds. For weeds we saw that it was sufficient to incinerate the seeds during the full Moon, and ideally also it would be better to consider the influence of Mercury. For animals it is necessary to refer to the Zodiac.

We have to admit, despite repeated attempts, that when we followed the indications in the next paragraphs we were not able to obtain an effective product. We are convinced that Steiner, in this instance, has given indications suited to the level of the listeners he was addressing, or that he has given indications with the aim of protecting the true and correct procedure because of the danger of indiscriminate dissemination of this knowledge. Actually it is true that this could also be used to wreak very serious damage. We are convinced that when Steiner prepared a product against rabbits (as he did during the Koberwitz Course) he used a different approach that we have also followed and which has proved very effective. Now we will follow the directions given by Steiner and then try to understand what we believe to be the 'real' instructions he protected.

Here we are at the twentieth paragraph ("*With plants, the effect ...*") and we will follow Steiner's thoughts. We read that animals have internalised the forces of the Moon and so, compared to the plant, they are relatively emancipated from the external lunar forces. Therefore it will not work, as it does for weeds, to follow the lunar forces. It will not succeed - but the same is not the case, says Steiner, for the other planetary forces.

Even in the nineteenth paragraph the sign of Scorpio is mentioned instead of our preferred constellation. We are advised to act at a time in which Venus is in the Scorpion constellation, because they represent at the planetary and zodiacal levels the celestial bodies linked to the forces of reproduction. Venus is actually the planet that governs all reproduction in the animal kingdom, while the Scorpion - because of its link with sexuality - is the zodiacal basis. One should proceed to incinerate at this time because this is the alchemical process by which you escape from the forces of life. Perhaps it is appropriate to continue with this concept. Normally when we talk about alchemical processes, we have referred only to the three processes *Sal*, *Mercur* and *Sulfur*.

How many times have we repeated that *Sal* is the process that produces contraction, whilst *Sulfur* produces expansion and *Mercur* is the exchange that allows a balance between the two processes mentioned - and therefore is life. The expansion and contraction, however, should not be carried to the extreme because between the two there must always be an interface or a chance for dialogue. If we burn sulphur we can see that the light is not pure white but a yellowish colour, so that the light at that level must be able to recondense to form the material for further sublimation in a continuous exchange. From this circle you can exit through two doors: an upper and a lower. The higher opening happens when you can produce a perfectly white light, such as that which is obtained by burning phosphorus. That is the process that leads upward with no possibility of return and is called the phosphoric process. (Notice that Rudolf Steiner mentioned phosphorus compounds in the seventeenth paragraph as poison for the mice.) The lower door is the exit from the *sal* process through the process of incineration. The substance that represents this ash process is potassium and note that in our meetings we have always said that the fertilisation based upon potassium kills the vitality of the soil. Death is possible through an excessive condensation or through excessive expansion.

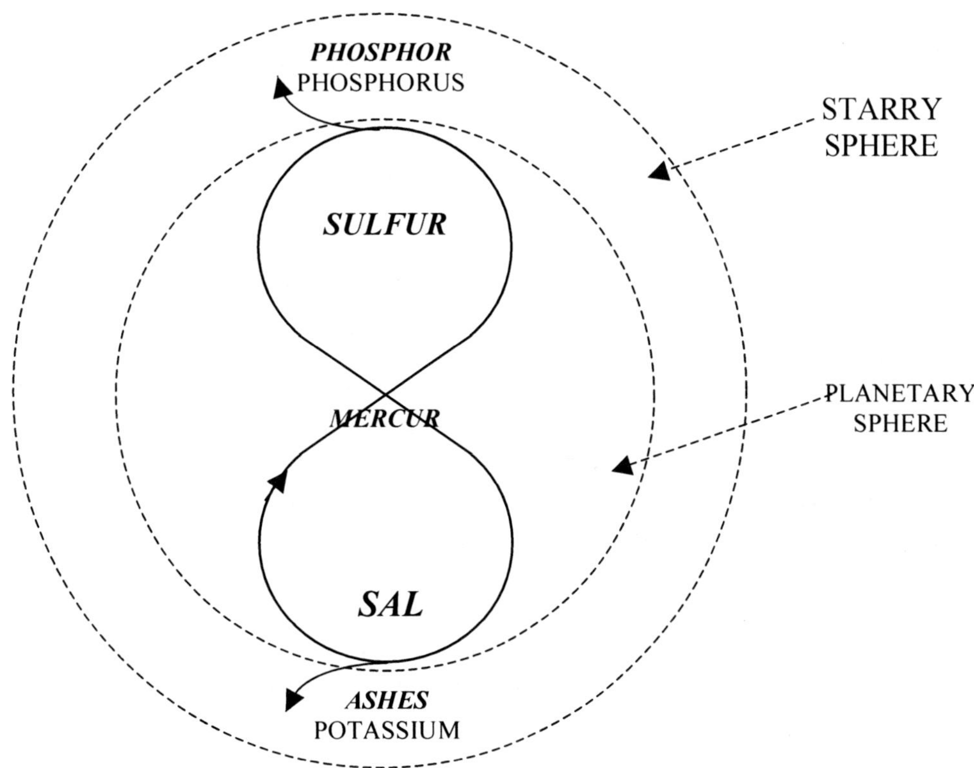

The three processes of life and the two processes of death

We can add that when we operate within the three vital processes we move in the world of planets. While working with the *phosphoric* process and with the ash this provides access to the forces of the Zodiac and the Milky Way: it is like saying that entry into the world of the stars without planetary preparation or adaptation would kill us.

Remember that the basis of chemical fertilisers is the mix of nitrogen, phosphorus and potassium. We have already seen the result where phosphorus and potassium are spread. With regard to nitrogen we can add that when it is not living it is not able to support the *mercur* processes and thus contributes to building up the processes of death even further. And if even those who practice the organic method use substances that kill the soil we should not be surprised if the layer of humus on the ground becomes increasingly thin.

This rather disastrous picture is mainly caused by human greed which pushes us to act only with the desire to draw as many resources as possible from the Earth and the Earth has learned to be afraid of greed more than anything else. Then, as frightened animals tend to gather together, in the same way the Earth closes in upon itself and becomes so compact as to become almost impenetrable, hard, dead.

Before continuing to read in detail how to incinerate the skin of the field mouse it should be noted that it is not important to have a precise technique in these operations, but it is much more important to have a good ability to think: the relationship with the spiritual worlds is a more important factor than the accuracy with which one performs a specific action. It is our opinion that in this agricultural practice, when one is not in a position to make a treatment with the right interior attitude it is preferable not to do so and to turn to the spiritual world in our thoughts, confident that this will be a more effective treatment than a treatment performed badly.

Jesus said that the Father sees what we need before we ask. Our intention therefore allows us to establish the basic connection necessary for grace to give us what we need.

Paragraphs twenty-one ("*You must burn the ...*") and twenty-two ("*It is important ...*") are straightforward. We should only remember that we must refer to the Scorpion constellation and not to the sign as we have mentioned.

At this point, we can open a parenthesis to clarify some concepts drawn from what Steiner is explaining in this lecture. The ash obtained by the method that we are looking at can be defined as 'zodiacal ashes'. Notice that Steiner indicates that one should burn the skin of the field mouse. We know that the skin is connected to the Moon, the planet that affects reproductive processes. In addition, the skin is what maintains the animals' manifest form. Thus, to burn the skin means freeing the form and therefore enabling it to return to its spiritual origins. To summarise, in the skin we have an anchor of reproduction and form. Then there is another aspect which will be repeated three times more by Rudolf Steiner - and that is the duration of the pest which is based on a cycle of four years.

We said that we have never managed to obtain good results with ash obtained as described by Steiner. Instead we achieved excellent results with another method, which produces what we could call 'planetary ashes'. The first difference between the two methods is the fact that the second, instead of using the skin, use the parts which express a creature's main characteristics. For the mice the important parts are the face and front paws. Remember what we said about the manifestation of the animal kingdom which is made up of humans who anticipated the fall on our planet when conditions were still unsuitable for spiritual development, and have remained trapped in their forms by demons linked to particular impulses of the soul, so that each animal is a manifestation of a particular aspect of the human soul. However, to burn the most characteristic parts of the creature is equivalent to releasing those entities. To work in this way it is necessary to subject the spleen of the animal to incineration because the spleen is the bearer of the forces of Saturn. Because this is the outermost planet it contains within its orbit the whole solar system and is the door through which one can enter the world of the constellations.

What we are describing is the result of very detailed studies and it has not been done just for the sake of finding a new method or to avoid pedestrian repetition of the teachings of Steiner. Moreover, we draw attention to the fact that Steiner, during the Course of Koberwitz, intervened with the rabbit overpopulation that had become a real disaster in the area. However, he did not adhere to the guidelines set down in his own lecture. First, the area over which he wished to have an impact was well over 5,000 hectares and certainly Steiner did not travel all over the property to sprinkle the pepper. Also, during the period in which Steiner was in Koberwitz, Venus was not in front of the constellation of the Scorpion. The product that we provide contains the ashes to unleash the demons, but also our 'harmoniser' product to provide protection to those who use the product.

As far as the Zodiac is concerned we do not refer to the constellation of the Scorpion because this constellation is death in relation to reproduction. (Remember that the scorpion is poisonous). When acting at the planetary level it is necessary to use the constellation that resonates with the mice, not for the reproductive aspect but for its defining or essential peculiarities, and this is the constellation of Sagittarius. Even the use of the product on the field must be done when the constellation Sagittarius is active.

As mentioned, the weakening effect works especially upon the reproductive capacity of the animal. Obviously to work as described above it is essential to identify the characteristic aspects of each animal and its 'home' constellation.

Another system that could usefully be used to replace incineration is a slow decomposition in water. This system also ensures the dissolution of form. Of course, the 'trick' is always to find the right time and developing the right attitude to undertake the operation. The ejection of pests should not be an act of transgression but a way to remove the animal after having understood the message that its presence brings. However, you must always thank him for this, and thus one will be able to recognise our limits and then to overcome them. Let us say finally that the most important thing is to entrust oneself to the higher will.

Returning to ashes, we might add that since we talked about planetary ashes, it is important to be able to recognise the planetary reference of the pest, which in the case of mice is Mercury. Ideally we would make the ashes of mice when Mercury is in the constellation of Sagittarius.

We can move on to the twenty-third paragraph ("*I am going to tread ...*") in which Rudolf Steiner brings forward the example of the sugar-beet nematode. To understand this we will draw the plant in our usual way, and show the four forces (arrows) acting around the plant. Two come from the cosmos and two rise after reflection from the Earth. It is obviously connected with the four elements and with the four Archangels.

We will place Uriel at the top left, Michael (the countenance of God) in the upper right, Gabriel (Fortress of God) at the bottom right and finally Raphael (God heals) at the bottom left.

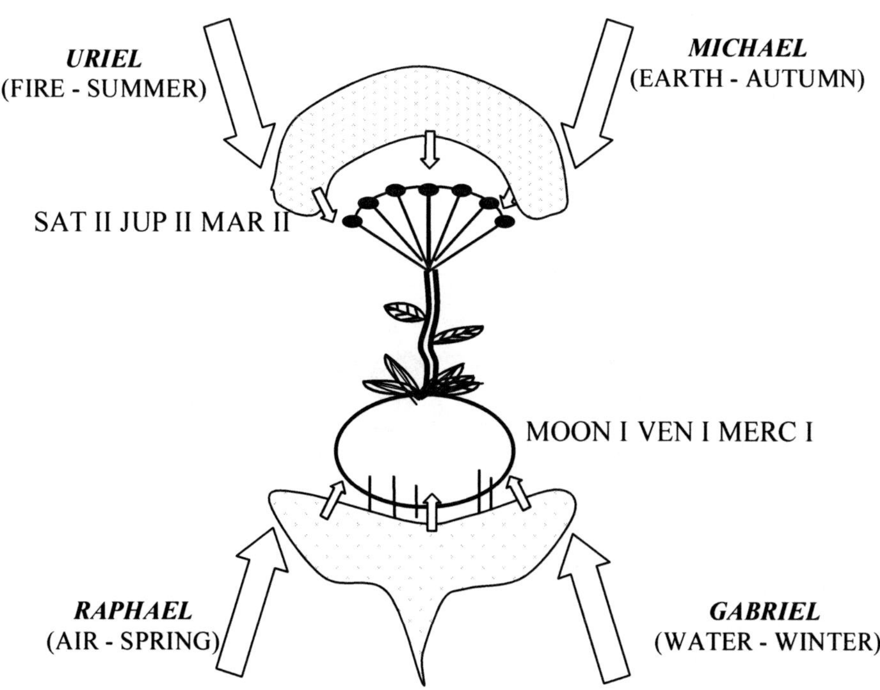

Archangelic influences upon plants

The current that Steiner indicates as descending from the cosmos is the current of

Uriel, the archangel that presides over the summer. (Remember that through Uriel one can grasp the image of the Trinity[40].) Uriel is also the archangel who enables connection with the sphere of the Zodiac and beyond. He is also the archangel who searches our hearts and judges our actions, especially in relation to the forces which descend from the sky in winter.

The influence of Uriel is extremely complex. He is located in the highest and is the one who has the strength to act most deeply. In the human, Gabriel (the creative element, the feminine within us) acts from the left side, Raphael (the rational) from the right side, Uriel from below and Michael from above. We could say that Uriel has sacrificed himself in order to be able to govern matter. If we return to the plant world, we can say that Gabriel, the current acting from below, carries the forces of reproduction and Raphael inserts an element of healing so that these forces should not dominate. Similarly Uriel brings cosmic forces and Michael is the therapist with respect to these same forces.

When Steiner speaks of two currents he is referring to that of Uriel and that of Gabriel because the other two, as we have seen, bring the therapeutic actions to these.

Then the action covered by the phrase in the middle of paragraph 23 ... " *We must be clear that this leafy middle region – which is what undergoes a change in this case – absorbs the cosmic influences from the air ...*" is the action of Uriel. On the contrary, immediately after there is a reference to forces of Gabriel, when we read: "*... but that the roots absorb the forces that come into the plant from the cosmos by way of the soil.*"

The imbalance between these two currents is the cause of disease. At this point we can learn from Rudolf Steiner the real cause of the occurrence of nematodes. These beings are linked to the fact that, for various reasons, the forces of Uriel are excessive and the forces of Gabriel are diminished. In this way the cosmic forces which should act on the upper parts of the plant, also manage to act in the ground creating an environment suitable for the nematode to live.

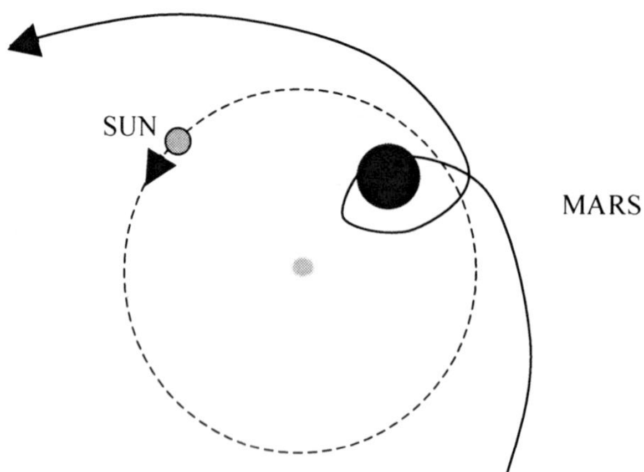

The position of Mars within the solar orbit

The forces that descend with the Urielic current are those that we recognised at the

40 For more concerning the Archangels in nature see our previous comments or read, "The Course of the Year in Four Cosmic Imaginations" by R. Steiner.

time as the secondary action of the outer planets, Mars, Jupiter and Saturn. Note that this secondary action corresponds to the retrograde motion of the planet that coincides with the apparent epicyclic step, when the planet approaches the Earth tracing out a looping path. This secondary action is called 'excarnative' in the sense that it leads the plant to flowering and fruiting, toward death.

If we ignore the forces mentioned for the moment and if we try to identify the physical conditions under which infestation is likely, we see that the nematode tends to grow in a warm climate with sandy soils, especially when there is a severe shortage of water (related to Gabriel).

Now try to place the conditions that we have just described on the drawing that we are building. Uriel corresponds to summer, autumn to Michael, Gabriel and Raphael the winter and spring respectively. So when we say we are referring to a warm summer climate and if we find ourselves on a siliceous soil we know that we are in the presence of cosmic forces, corresponding to the upper current. Lack of water also diminishes the current of Gabriel. QED.

The problem lies in the fact that the conditions that should be around the upper part of the plant are created underground, and so the insects can thrive under the soil. It should be obvious that the first therapeutic action should be to harmonise the currents we have discussed. The 'Harmoniser' preparation could solve the problem alone if the imbalance is not too extreme.

It is worthwhile just to point out that if there is excessive moisture the current which will be reinforced is that of Gabriel. These are the conditions suitable for fungi to climb above the ground and then toward the top of plant.

Steiner reiterates that the presence of the nematode is made possible by the fact that "*the cosmic forces that the nematodes need for their survival are therefore available down in the ground where the nematodes have to live.*"

The twenty fourth paragraph ("*Certain living things..*") seems to contain nothing particularly significant. It is clear to everyone that creatures "*can only live within a certain range of conditions.*" The obviousness of this statement and the example that follows makes sure that our focus does not stall at the reason why, in this context, it is precisely temperature which is used as an example. In this example a clear indication for treatment is actually provided. Steiner talks about temperature and then summer, but also the environment related to silica and then refers to the biodynamic horn-silica preparation - 501. The horn-silica forces indeed are those that enhance the relationship of the plant with the forces that come down from the cosmos. So this preparation strengthens the forces linked to Uriel but also brings an element of Christianisation and harmonisation into them through interment so that we can act without causing any adverse effect. We emphasise the process of Christianisation of the preparation. If we recall that Michael is called *the countenance of Christ*, as shown on our drawing, this receives further confirmation.

In other words we can say that the summer forces, with their excessive heat at all levels, can bring an unbalanced impulse bringing about an excessive connection with the forces that act directly from the cosmos. However, the intervention of Michael during the burial of horn-silica harmonises the impact of the summer and brings order to the flow of forces associated with it.

Paragraph twenty-five ("*For creatures that live ...*") speaks of cosmic forces with a four-year cycle and this is mentioned in connection with the development of cockchafer grubs and potato shoots, as well as a therapy against nematodes.

In paragraph twenty-six ("*With the insect...*") Steiner resumes talking of ashes, indicating the differences appropriate to dealing with the field mouse compared to an insect like the nematode. We must burn the whole insect, because, he says that: "*... a harmful root-dwelling insect like this is in its entirety a result of cosmic influences - it only needs the Earth as a substratum.*" The paragraph continues with details of the exact time, with reference to the cosmic and global framework, at which one should burn the insects. This point will be fully addressed during our next meeting. For the moment we would rather clarify the issue of the cycle of four years mentioned in paragraph twenty-five.

The force that has a quadrennial cycle is clearly that of Mars: the epicycles of Mars - the apparent looping-motion of the planet as seen from Earth - are linked to a period of four years.

We can easily see another important fact which is that Mars, when it becomes retrograde, seems to be 'infra solar'. This means that it can be found between the Earth and the Sun sphere, in the zone 'reserved' for internal planets.

This fact is very important because the world of inner planets represents Purgatory and thus the world of disease. We understand that the negative effects of Mars make themselves felt in the period in which Mars is closer to the Earth than the Sun, that is when it is retrograde.

Now, to set down a clear definition, we will define the primary action of Mars (Mars I) as being 'above the Sun' and the secondary influence (Mars II) as 'below the Sun'.

We would remind ourselves that in a previous drawing we had linked the action of Uriel with the secondary influences of the outer planets, but the only external planet capable of taking a below-the-Sun position is Mars and then it is the gateway for pests. Note also that Mars is the planet that brings to mind the old Moon phase of manifestation of the Earth, at which stage there was the Luciferic fall, and we can appreciate that the planet is also linked to the forces of Evil.

Knowing all this we should find a way so that Mars will not act like a below-the-Sun planet. Obviously we cannot move the physical orbit of either the planet or the Sun, but we can increase or decrease the strength of Mars and the Sun as if modulating their distance from the Earth. In other words, when Mars is below-the-Sun it has a dominating force on that of the Sun, because it is physically closer to the Earth. However, if we manage to strengthen the force with which the Sun acts, it would be as if the Sun were to be brought within the orbit of Mars. Mars would no longer be a predominant force as if it was in an above-the-Sun position. Well this is easy to do because we have the horn-silica whose action amplifies the forces linked to the Sun so that it will act as if it were closer to us.

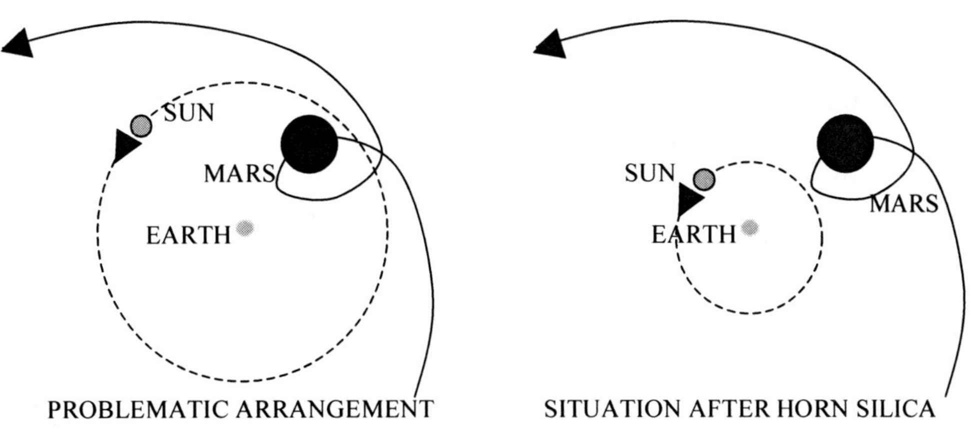

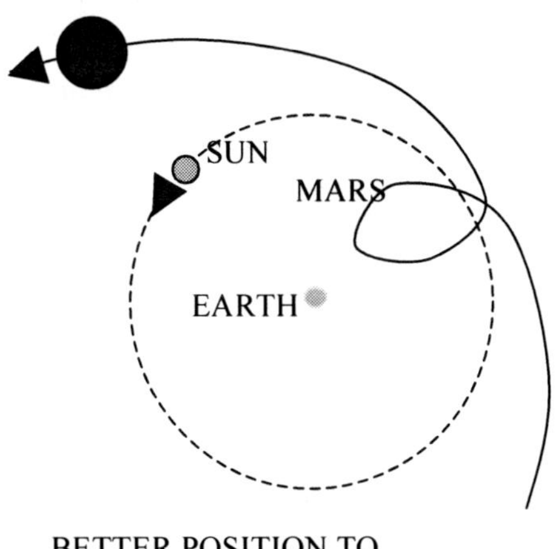

BETTER POSITION TO
APPLY THE ASHES

We have a second chance to be effective in our choice of the time when we spread the ashes. In this case we choose the moment when Mars is above-the-Sun to couple the time when Mars is 'weak' with the forces of death contained in the ashes.

This intervention against pests becomes extremely simple and results in strengthening - according to the case - either the path of Uriel or Gabriel, and in bringing the balancing power of Michael.

Meeting Twenty-one

We are still reading the sixth lecture of Koberwitz and today we will make some remarks on the last section in which Steiner explains how to counter fungal diseases. So let's start from paragraph twenty-seven ("*People have no idea ...*").

Steiner continues what he has just said about the difference in results from making the ashes when the Sun is in different constellations. He says: "*People have no idea that the Sun is actually a very specialised being*". In the following sentences he asserts that the Sun is profoundly different as it radiates from one sign rather then another in the course of a year or even a day. To really understand what these statements mean we must first clarify what the Sun is. We could ask, for instance, of what is it composed, what is its consistency, is it solid or gaseous? Actually, the Sun and its function can be understood only if we consider the star at the centre of our system as a type of hole, or rather the centre of a vortex that draws the cosmic forces towards itself. We could say that the Sun is a receiver of cosmic forces.

The first cosmic forces to be drawn into the Sun are those of the Zodiac, so when the Sun is before The Ram it attracts the forces of Aries and it brings in the forces of Taurus when the Bull is behind it and so forth. In light of this it seems appropriate to consider the profound difference between the Aries-Sun compared to the Taurus-Sun etc. The forces that the Sun picks are then transmitted to the Earth. However, given that the Sun makes a profound metamorphosis in the living forces of which we are talking, these arrive upon Earth significantly transformed.

The examples Steiner mentioned - first to the year and then to the day - is motivated by the fact that, because of the revolution of the Earth around the Sun, in the course of a year the Sun appears before the whole circle of the Zodiac, and as a result of Earth's rotation the same apparent path is made every day. So if the Sun finds itself, from the point of view of the Earth, in the constellation of Aries throughout the period from around April 21 to May 13, every day for about two hours the same constellation is in front of the observer, although clearly at different times according to the longitude upon which the observer stands

The Earth-Sun-Constellation alignments that occur in the course of the year are the dominant ones, and the daily relation is secondary. For example, at the end of April in Western Europe, the Sun will primarily be in the constellation of Aries and secondly we will find that at around 8 am it is in the constellation of Taurus.

What we are saying was deeply appreciated by alchemists, and used in order to come into resonance with specific forces from the Sun at definite times of the day and year.

Now we must proceed in our reasoning and ask ourselves what is it that descends from the stars to be captured by the Sun. Well, from the stars what comes down is light!

In the drawing that follows we will identify the Light given to us from the bright stars as 'Light I' or L1.

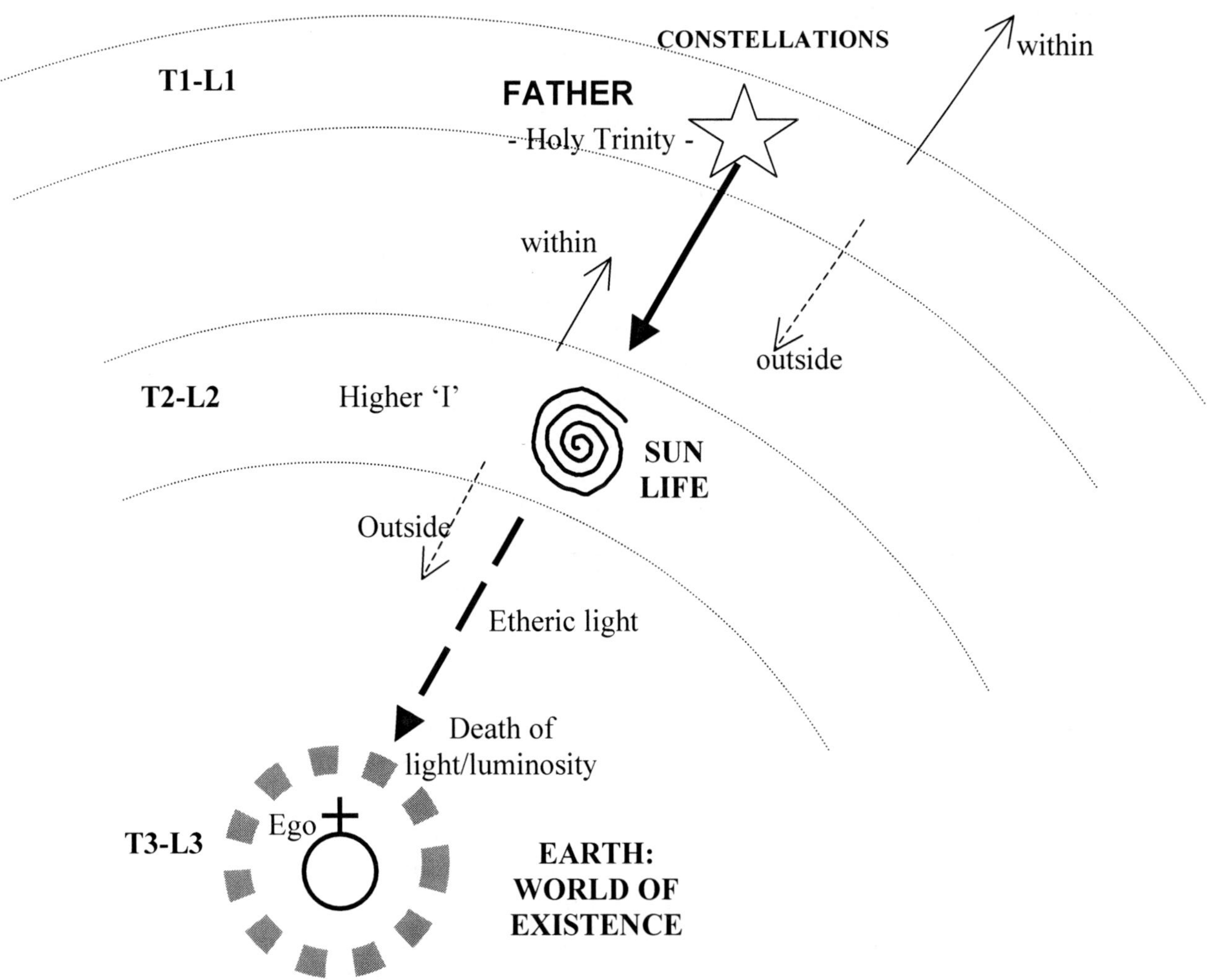

The change of light and time in relation to cosmic spheres

As we are referring to the Zodiac, we are actually talking of the Light of the Seraphim and Cherubim and this obliges us to make special considerations regarding the irradiation of this Light. In fact, every day we are regularly in contact with light sources and we can see how light - observed from the point of view of the source - will expand outwards. But if we consider the nature of the first spiritual hierarchy and the fact that the world of the spirit has no spatial or temporal limitation, the reasoning must be completely overturned.

Remember that Steiner argues that when the spiritual beings of the first Hierarchy turn their conscience inwards they *perceive* God, but when they turn their attention outwards they 'create worlds'. Now we believe it is clear that an infinite being, that by definition is not limited to the spatial and temporal, cannot produce anything outside itself but can only create something within itself. Therefore the light emanating from these entities cannot but radiate inwards, and thus towards the Sun. We must strive to understand that what from our point of view constitutes the inside of the solar system, for the beings we are talking about is an external atmosphere; that which from our point of view is outside the solar system and is therefore a still more distant periphery, for these beings is the inner world - the vision of God that passes through them.

So, recapitulating, for us, compared to the Zodiac, the solar system is inside, while the Milky way and other more distant stars are outside; the opposite is the case for the Cherubim and Seraphim for whom the solar system is an external reality and the world of the stars is within.

We recall that when we talk of the first Hierarchy we talk about the Father of the Holy Trinity. We could even add that Seraphim, which from our point of view act upon the external aspect of the Zodiac, have turned towards God, while the Cherubim have turned their activities towards the solar system. For those who wish to study more about the various spiritual hierarchies we can recommend the text entitled "The Spiritual Beings in the Cosmos and in the Kingdoms of Nature" (Anthroposophical Press).

Returning to the Sun, we said that the Sun gathers Light from the various constellations and this undergoes a metamorphosis so that the light that emanates from the Sun has different qualities than the *Light I.* In the drawing this light is indicated as *Light 2.*

Since the Sun is a star the spiritual entity whose manifestation it is (in the second Hierarchy) is subject to the same considerations as for the Seraphim and Cherubim. Therefore, the light that comes from the constellations and that from our point of view comes from without, from the point of view of the Sun can only be considered as being from its interior.

The transformed Light coming from the Sun which we can call *etheric*, reaches us in the band that surrounds the Earth which we know as the ‘atmosphere where it ‘dies’ and is transformed into luminosity. Luminosity then is the death of the etheric light of the Sun that in turn was captured from the Zodiac.

In the prologue of the Gospel of John what we have just said is expressed very clearly. The light from the constellations is called the ‘true light’. Life is the same light after metamorphosis within the Sun, and finally the same light that became Life on the Sun, becomes mere brightness in contact with the atmosphere, even if it is still called *light* by people (‘..and Life is the light of man’).

The brightness, which for people is light, is shown in the drawing as ‘light 3’ and this L 3’ is the one that allows us to know three-dimensional space. At the level of this light there is also the threefold perception of time that is divided in the past, present and future. The ‘light 2’ instead is dimensionless because it still belongs to the world of the spirit.

We can still say that on our planet, because of the three-dimensional space (light 3), we can grasp the world of form. At the solar level, however, we find the forces that allow the manifestation that we know as etheric forces or as the *sea of the ethers* (light 2). Going further out we find, in the Zodiac (light 1), the ideas (or individual principles of the species) which, passing through the sea of ethers, interweave with life, and are ready to be shaped in a specific form and bound to a seed in preparation for becoming a plant on the physical level.

If we talk about time, on the other hand, we can say that at the level of ‘time 3’ (past, present and future) there is earthly existence that begins with birth and ends with death. In terms of ‘time 3’ the ego manifests as a person’s ordinary self.

By contrast ‘time 2’ is the time of Life. By now we should all know that life takes place through many lives, which serve so that the one can face all the necessary experiences to complete ones evolutionary path that is the attainment of perfection. At 'time 2' the higher self is manifested which doesn't take on form, but shows as metamorphoses. The higher I does not live in the past, present and future, because such time does not exist at that level. We might say that the I dwells in a continual time and on our drawing we could write, at the level of the Sun, the word '*duration*'. Obviously at the level of 'time 1' and therefore at the level of the Zodiac we would enter *eternity*.

In recent days the media have repeatedly spoken of explosions occurring on the Sun, accompanied by immense eruptions that have sent a worrying amount of radiation to the Earth which disrupts satellite and radio broadcasts, and who knows the effects that this could have on humans.

From what we have now said we should reject the idea that the Sun can 'explode', but we should rather think of an implosion. Better yet we could think of an extraordinary amount of powerful cosmic forces absorbed by the Sun and sent on to Earth as Life. The disturbance caused to the electronic equipment for the transmission of radio-signals is quite logical when you consider the fact that these are machines of death, fueled by forces antithetical to life.

To understand a little more of what has happened we have to consider the fact that the Sun is now in the constellation of the Scales which is the seventh constellation starting from Aries. These first seven constellations form the so-called *light Zodiac* because its forces are embodied at the top of our bodies and bring consciousness. This part of the Zodiac is also called the *Mystic Lamb* because when its forces descend via the Sun they have the greatest potential sacrifices to give us life.

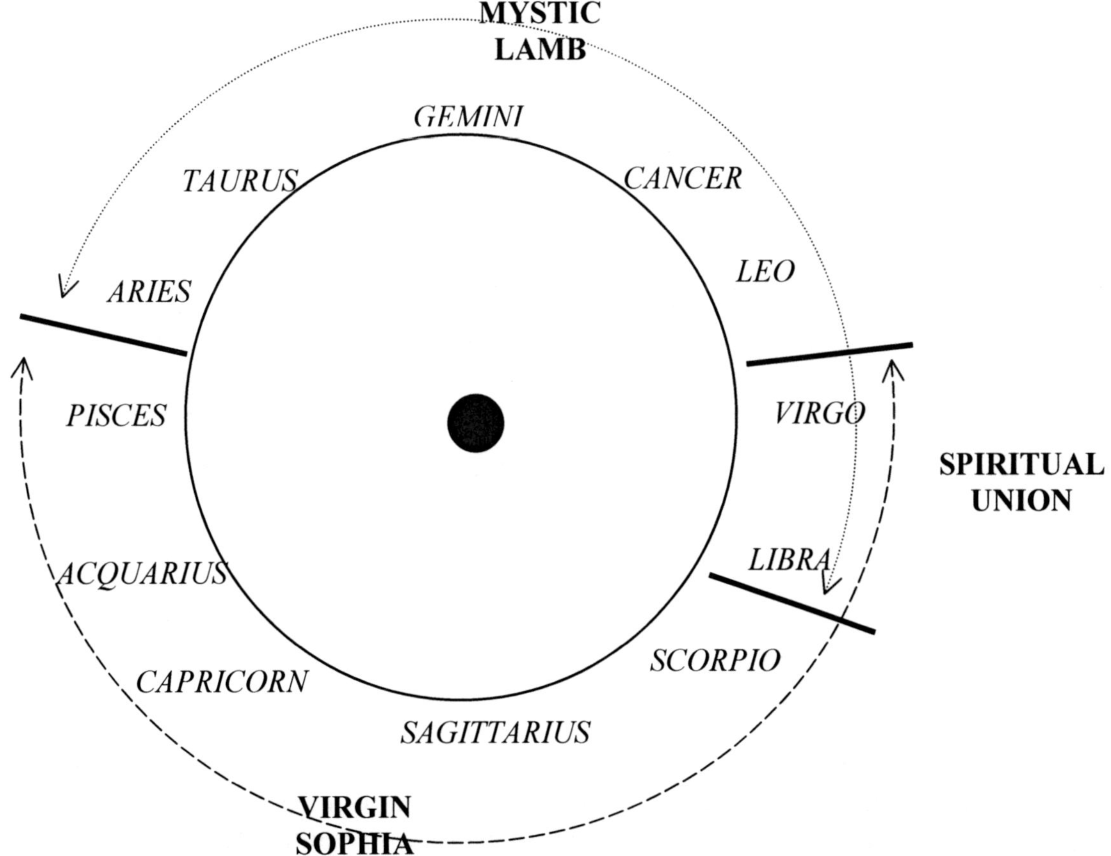

The overlap of the Virgin Sophia and the Mystic Lamb

The last seven constellations (those of the *dark Zodiac*), between the Virgin and Fishes, form a complex of forces known as *Virgin Sophia.*

The Mystic Lamb and the Virgin Sophia overlap in an arc covering the Virgin and Scales. This overlap is the marriage between the life component and that of consciousness. In other words we could also talk of union between the forces of Christ and Mary, which is presented in the Gospels in the well-known passage concerning the marriage at Cana (John 2.1).

The overlap between the Mystic Lamb and Virgin Sophia is what is referred to in the phrase: "What have I to do with you, woman?" Translated: "This is between me and you woman" (Bittleson) (Jn 2.4). Do not forget that during the baptism at the Jordan, the Being of Christ descended upon Jesus and the virgin Sophia descended upon Mary. During the wedding at Cana these two beings began working together for the good of humanity.

In light of what we have said we can better understand what Steiner says in the twenty-seventh paragraph of the sixth Koberwitz lecture.

If we accept the fact that the Sun is a vortex that '*sucks*', we can understand how the Earth itself is sucked towards the Sun and follows like a dog on a leash. The Earth's orbit around the Sun is the apparent result of a very different cycle. In fact, the Sun moves in the universe in the direction of the constellation Cygnus (representative of Christ in the Milky Way) by tracing a path similar to the one below.

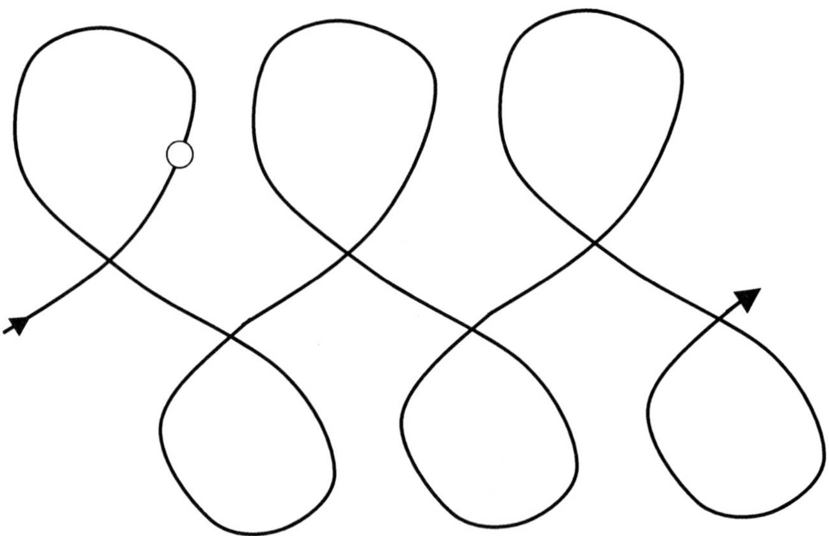

Path of the Sun through the cosmos

The Earth takes the same course at a certain distance from the Sun and the relative positions give the illusion of an ellipsoidal movement. The other planets also follow the path of the Sun, but making different lemniscatory movements. The plan outlined above is a simplification of the real movement because, in addition to moving in the way represented above, the whole system also performs a movement of rotation upon itself. The situation is therefore impossible to reproduce in two-dimensions.

Paragraph twenty-eight ("*So, if you go through ...*") reports Steiner referring back to the ashes and says that scattering them means the nematode "*gradually becomes powerless.*" Actually the destruction of form allows you to get rid of the pests from the physical plane (Space 3) and to use the sucking action of the Sun to carry the message directly to the source of the forces of life that come to the Earth. The debilitation soon follows.

It is worth pointing out that along the way another message is contained, although well hidden, namely that the word '*pepper*' used by Steiner to characterise the ashes, produces a weakening when used on our food.

In paragraph twenty-nine ("*In a remarkable way ...*") Steiner refers to what could be called spiritual astronomy[41], a discipline that has now been almost totally lost. The content of the rest of the paragraph does not present special difficulties.

Thus, we arrive at the thirtieth paragraph, where we read: "... *what is in the plant, what is in every living being, also carries within itself the germ for its own destruction.*" At first this sentence may seem strange but, if we think well, it is not really so odd because the spiritual principle of a plant species in which we can recognise the regulator of all life processes therefore also determines its death. So while we appeal to forces of the Zodiac to promote a better liaison with the Ego of the seed, we can also work with the same forces to weaken the connection with the Ego to such an extent that one can completely prevent their presence on our property. Obviously we will work to stimulate the incarnation of the plants we wish to cultivate, while we will make use of the opposite action for pests.

In paragraph thirty-one ("*It is really not so strange ...*") we now find the reference to the '*effects of time*'. Steiner reaffirms the notion expressed at the twenty-seventh paragraph about the specialisation of the Sun over the year and the day. In fact to connect to the zodiacal forces to which we want to appeal we must use the precise moment at which such forces are active, according to the mutual position of the Earth, the Sun and the specific constellation. Pre-potentised preparations, in this sense, allows us a greater freedom because they makes it possible to liaise with the various cosmic forces even when they aren't fully active from an astronomical point of view. Indeed the dynamisation of a preparation must be done at the time required (depending on the forces to which they must appeal) and be repeated three times to annul the influence of the constellation which is acting from above at that time. This also serves to establish the liaison with the desired strength and recalls the situation thanks to the choice of time that corresponded to the forces. We reaffirm that homeopathy is an instrument of freedom that allows you to connect as needed with forces that would be active only very rarely in clock time - T3. With this paragraph Steiner concludes the topic of ashes and is preparing to introduce the subject of fungal diseases.

In the second sentence of paragraph thirty-two ("*Now, we still need to ...*") we read that: "*they should not be called plant diseases.*" In the remainder of the paragraph Steiner explains this assertion. Most diseases are caused by an excessive penetration of the astral body into the physical body, due to a weakening of the protective or

---

[41] 'Spiritual Astronomy' by the Author is in translation (2008). See also Dr Steiner's third sceintific or Astronomy course

buffering function of the etheric body. Therefore the plant that has not incarnated its astral body should never become ill.

Towards the end of the paragraph Steiner confirms what we have just said: "*Now, since a plant does not contain an actual astral body, the particular way of being ill that is possible with animals and humans simply does not appear with plants.*" Steiner does not say that the plant does not have an astral body, but that it "*does not contain an actual astral body.*"

After all we can ask ourselves how can the plant communicate with the astral world without itself having an astral principle - and there is no doubt that the plant does establish a dialogue with the astral world: flowering is the proof. Well, the plant is penetrated by the astral body of the Earth.

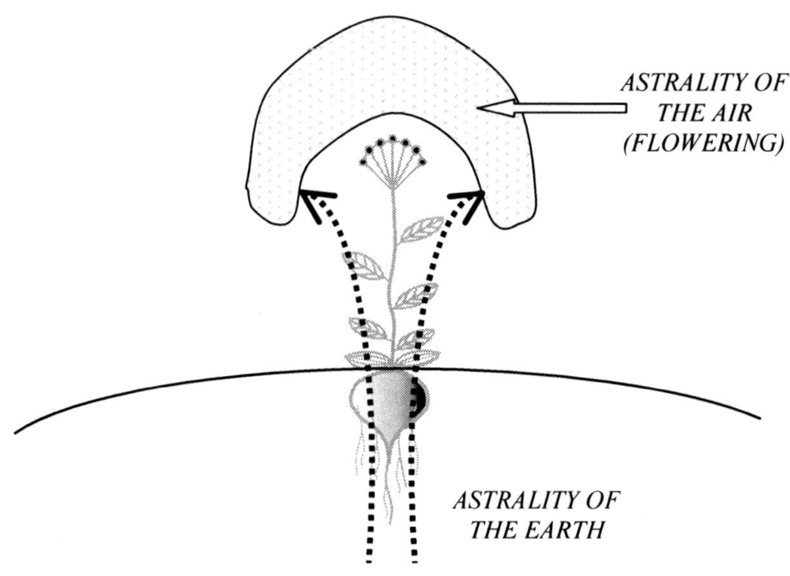

The Earth's astrality radiates through the plant

The plant then gets sick because the Earth is sick. In the light of this conclusion it seems more than justified when - echoing Steiner - we said some time ago about the fact that a correct way to fertilise can eliminate most plant diseases.

The thirty-third sentence ("*From what I have said ...*") should present no significant difficulty and we just want to point out that when Steiner says: ".. *that the soil around plants has a certain inherent vitality... although these forces are not intense enough to appear as actual plant forms, they are present with a certain intensity all around the plants*" .. this refers to the unbound Life which he connects with the forces of the full Moon. We have repeated many times that the full Moon is actually full of the Sun and thus of Life: in other words with the full Moon the forces of Zoe also abound.

In these few lines Steiner has given precise indications for understanding the world of fungi that is especially linked to the relationship between *Zoe*, the full Moon and water.

In paragraph thirty-four ("*There are several significant ...*") Steiner tells us quite clearly how to vivify the Earth and that is through the rays of the Moon that - by reflection - bring in the etheric forces of the Sun. We know that you can bring life in other ways. This can be done through manuring and we can more effectively *activate* the sea of ethers with the use of the homeodynamic preparation 'Pro humus'. But we ourselves are bearers of both lunar and solar forces and, ultimately, we ourselves can become the carriers of life forces with our properly developed consciousness. To develop the lunar forces that we are carrying it is necessary to purify ourselves of selfishness, and to develop our solar element we must be saturated by Love. So our work in the fields should not be driven by a thirst for profit, but by the ideal of assisting the Earth to evolve, having recognised that this is true, good and just. This moves the forces of Zoe and these will help us learn from our work in the fields and give the means for a dignified existence.

Paragraphs thirty-five ("*It can easily happen...* ") and thirty-six ("*Let us assume, however...*"): to understand what Steiner wishes to convey in these two paragraphs it is necessary to dwell on some ideas and to help us, as usual we will make a drawing. We have already said we can recognise four currents around the plant that pass through it. The current linked to winter is that of the Archangel Gabriel that we have shown in our drawing with an arrow upwards from the bottom right side.

Remember that Gabriel presides over the forces of reproduction and is therefore the bearer of life. Archangel Gabriel made the Annunciation to Mary. The drawing indicates the link between the forces of Gabriel and the full Moon.

The spring current is linked to Raphael. The arrow representing this influence comes from the lower left. The activity of Raphael is linked to the daily Sun. Steiner refers to these two currents early in paragraph thirty-five when describing winter and spring in relation to water.

To complete the picture, we will place the current of Uriel at the top left linked to summer, and the nighttime Sun. The current of Michael will occupy the top right of the drawing and that is associated with the autumnal forces and the new Moon.

The problem that Steiner wishes to put before us is not so much an issue of the wetness of the winter season because this is normal considering that the winter is the season tied to water. The problem is excessive moisture during the spring. In fact, if this occurs, we have too much of Gabriel's current and an insufficiency of Raphael's ('God heals'). Because of this the latter cannot bring its therapeutic action and restore balance to the environment in which the plants grow and complete their cycle up to the formation of seed. The daytime Sun, in fact, should 'dry up' the excess moisture brought by the full Moon and balance the low quality astrality that follows.

The activity of Raphael should mitigate the forces of vigour brought by the current of Gabriel to allow the metamorphosis of these forces that will determine the nutritional value under the influence of Uriel.

In the absence of the therapeutic influence of Raphael, the forces of Gabriel will go directly up the plant without undergoing any metamorphosis, so the processes of seed-formation may not take place properly. At the same time the forces that were supposed to act in the formation of seed show their activity at a lower level in the manifestation of fungi, making use of the excessive moisture that has invaded the area above the soil.

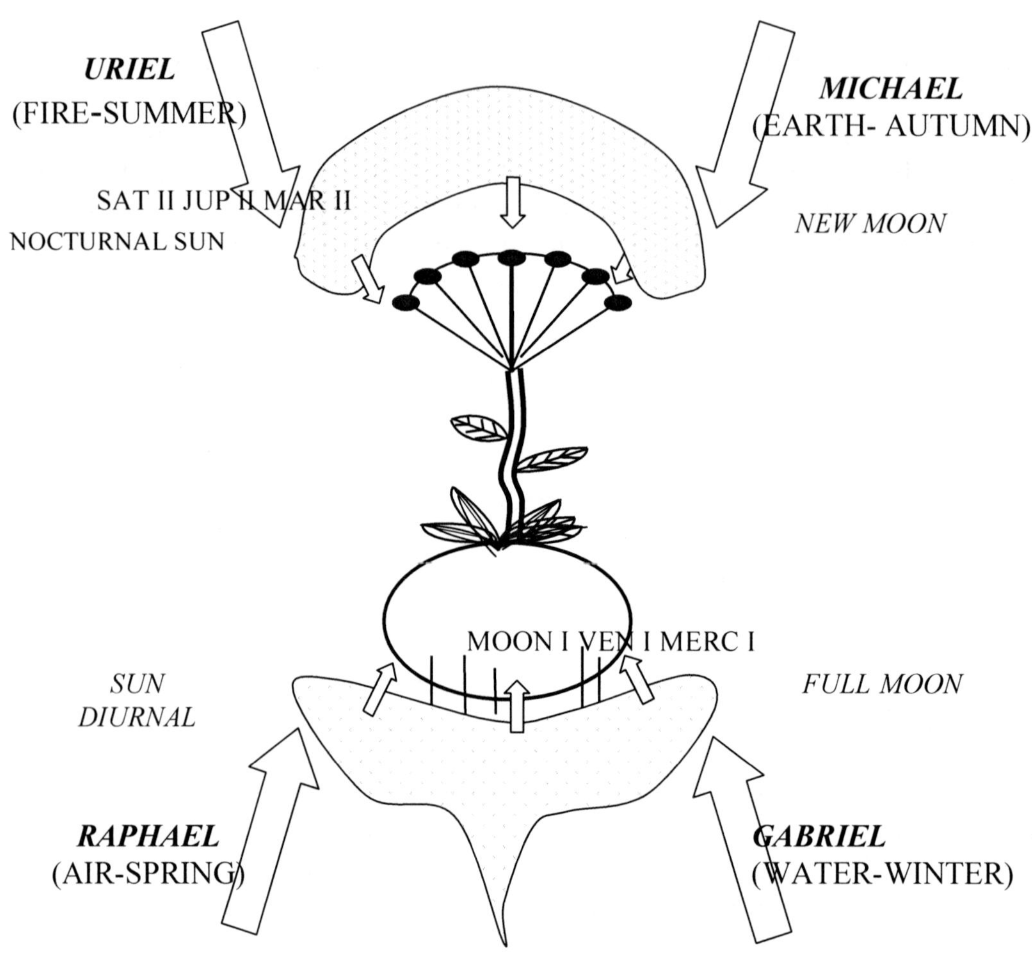

The four life-currents around the plant in relation to the seasons and Archangels

Reiterating that fungi are strongly linked to the forces of the full Moon, we could add that the type of fungi depends on influences from the other planets that can add their influence to that of the full Moon.

In the thirty-seventh ("*So what should we do now?*") and thirty-eighth paragraphs ("*This is an instance where ...*") Steiner presents '*equisetum arvense*' as a remedy for fungal diseases and briefly describes its preparation and use.

We know that the *equisetum arvense*, also known as horsetail, contains about 90% silica, is a very ancient plant, and can be regarded as the archetype of all the leafy plants. Horsetail grows on wet and sandy soils and so combines the element of silica with moisture. The horsetail has the characteristic feature of occurring at first in a form reminiscent of a mushroom (the *sporangia*) and only later manifesting in the form with which we are all familiar. Just the fact of having a juvenile form like a mushroom and of being linked to the worlds of water and of silica, perfectly places horsetail in the framework we have seen to be crucial for the development of fungal diseases.

However, our tests show that horsetail, prepared only by the indication that Rudolf Steiner gives in these two paragraphs, does not permit one to achieve a satisfactory result. We have to be a bit more shrewd. First, we start from the premise that the mushroom is the result of an astralisation of the soil ('*the plant does not contain an actual astral body*') that can only be balanced by a silica of plant origin. In this case it would be incorrect to use the forces of silica contained in the 501 because it is of mineral origin and thus is equipped to act at the level of the I. From what we have said then the use of equisetum appears to be correct.

The preparation suggested by Steiner is a decoction and has the ability to bring order in the metabolism of the soil that is linked to the current of Gabriel. We are convinced that the use of the *sporangia* of horsetail gives a greater chance of success than using the adult plant. Moreover we believe it could be used most effectively as a seed-bath preparation rather than as a spray upon the soil. Probably the most appropriate moment to collect the equisetum is about 9pm, when the most active forces in the etheric body of the Earth are those of water.

To further enhance the action of equisetum we could also make it into a real preparation. As an animal sheath we might consider the hoof of a cow. Indeed, we must take action to counteract the excess of forces that rise from the ground and the hoof is without doubt the part of the animal that is closest to the ground. The cow is also the animal that expresses the most metabolic forces. The hoof could be taken from the back of the animal because those are ones even more related to metabolism. In addition it would be preferable to take the hoof from the left side of the animal as this side is linked to the feminine part - Gabriel.

This hollowed out hoof is stuffed with the *sporangia* of horsetail, and buried from Christmas to Easter at a depth of 120 cm and in a calcareous soil. Perhaps a more viable alternative might be to prepare a decoction of horsetail in a bottle and fit it in the hoof. The decoction should be prepared for 20 minutes in order to act in metabolic processes.

We could also hypothesise that the decoction could be buried but not in a bovine hoof, but wrapped in hair from a horses tail. The horse's tail, as well as in mane, is reminiscent of the form of the equisetum. Also the forces of the Moon enter the animal through the tail, and the horse is an animal related to Fire (known to oppose Water). At this point, further strengthening of the preparation could be achieved with dynamisation.

A different kind of intervention might be made to reduce the influence of the lunar forces present in the soil. In the past we have talked about this. We have given the name of '*Mary's Mantle*' to this remedy which is made from the yarrow preparation, and clay, all prepared with another of our 'Harmoniser' products (A02) which we have called '*Preparer of the Soil*'. While not specifically acting against pests, it is able to provide adequate protection, and in all cases gives true support to specific therapeutic interventions.

The activity of A02 should be supported with the product that reinforces the connection with the I of the species (O02), and to match these two we have made a preparation '*Moon repellent*' that is identified with the code Y09.

The lecture concludes with Steiner reiterating for the umpteenth time that anyone who really wants to understand the world of plants, animals and pests, must necessarily broaden their gaze to the universe because: "*Life comes from the whole universe, not merely from what the Earth provides. Nature is a unity, with forces working in from all sides.*" The methods by which we are investigating nature today, relying on investigations of ever smaller bits of matter, leads us in a direction diametrically opposite and fatally away from the true understanding of nature.

Seventh Koberwitz Lecture

Twenty-second meeting

The seventh lecture of the Koberwitz course was held on June 15th, 1924. The translator of the present [Italian] edition entitled this lecture: "*Intimate reciprocal actions in nature: mutual relations between agriculture, fruit and livestock farming.*"

This lecture is relatively easy to understand, or at least is not as difficult as were the second and sixth lectures. Remember that the second lecture is the one that gives directions for understanding the plant in relation to the cosmos while the sixth, which we have just finished studying, is devoted to diseases. The study of the seventh lecture will allow us to discover, investigate, or intuit the basis on which the laws of the 'agricultural organism' are based. We are therefore dealing with a very important topic that Rudolf Steiner has presented in a relatively understandable way – although there are many points that benefit from clarification.

We will quickly notice that Steiner begins this lecture speaking at length about the tree, and this - logically - would suggest that he has attached a fundamental importance to it. The tree, as we will understand shortly, is actually the foundation of the agricultural organism. The forest plays the role of the I of the farm, the essential role of directing the vital processes that occur in each plant. However, this role would be devoid of importance if the fruit trees had not first captured the forces from the cosmos that the trees then distribute. During the lecture we will also learn that the forces of the universe can reach down to the very last blade of grass transmitted by the insects and birds.

The first three paragraphs of the lecture seem simply to repeat what has been said many times before, that it is a mistake to "*... look at the beings of nature - the minerals, plants and animals .. as if they stood there in isolation ... everything is in mutual interaction with everything else.*" However, if we observe Steiner's expressions with care we can grasp some nuances in his meaning that are anything but insignificant.

When Steiner talks about things that are related to each other and those things acting upon each other, he is actually highlighting two different approaches: one that considers things "*in relation to the other*" which is the logic of the astral plane and, completely different, '*one upon another*', which is the logic of power.

These two simple phrases then open unexpected scenarios, because with the first 'astral' logic Steiner is emphasising the need to be able to grasp the soul or vital-astral aspect that lives in nature, as opposed to the vital-etheric with which we are more familiar.

In other words we are urged to distinguish between life as a set of forces that finds expression in biology or what the Greeks called '*bios*', and Life that descends from the sky like an ever-flowing cosmic gift (that has an aspect of Maria) that the Greeks called '*Zoe*'. Steiner invites us to prioritise this second aspect of life and forge a fundamental and intimate familiarity with it even in the subtle communications that settle within the agricultural holding.

When we are considering the imposition of one upon another, our centre of gravity lies within the logic of the 'fall': we proud humans wanted to be as Gods over the Earth and over the beings that live there, and arrogantly impose our thoughts, our needs and our utilitarianism on everything around us. As long as we continue to adopt this approach we will perpetuate the process of the fall, excluding the possibility of a process that involves resurrection and a new sociability which follows another logic: that of mutual assistance.

On closer inspection the attitude of 'everyone for each other' is not so far from our usual way of life because few of us, for instance, have manufactured the car we drive. Someone else has made it for us. But this is inserted into a system of oppression that wants to impede the new thinking of Christ - which as we said is already operating - from being realised in full.

Obviously in our farms we must become capable of overcoming this old system so that we can fully develop the route to resurrection. This should encourage farmers to devote all their energies to the single goal of doing good 'for' the Earth and its transformation, the evolution of the lower kingdoms, and the process of vivification (humification). Production and the consequent ability to meet ones needs will, in this context, be only a secondary result for the farmer, even though perhaps the land will be more abundant than before because the Earth will return multiplied what we do 'for' her.

But, we reiterate, our goal must not be utilitarian and our acts should be determined only by working 'for' the Earth, demanding nothing in return and without even imposing our limited consciousness and will upon it.

The third paragraph ("*However, there are ...*") begins by mentioning the subtle interactions of finer forces and finer substances, and with references to the element of *heat,* and the *chemical ether* constantly operating in the *atmosphere,* and to the *life ether*. The reference to the external planets is clear, and the ethers are the memory of the four stages of the incarnation of the Earth, phases that are the origin of the living forces that still govern the life of and on our planet. These combined life forces can be called the 'forces of nature' or 'the forces of Mary'. In other words Steiner is inviting us to consider the subtle (or *spiritual*) interactions of that being which the ancients called the goddess *Natura*.

Towards the middle of the paragraph Steiner mentions animals commonly found on farms and whose usefulness is widely appreciated, but immediately after he refers to the world of insects and birds. The activity of such creatures is the means through which the forces of the universe can be received by orchards and thence passed to the woods, which in turn spread them to the entire farm. These smaller animals are the true bearers of the forces of Life.

Steiner says that people "*have no idea*" about these animals. In fact, when we consider insects and birds we do not think they could possibly be the support for the spread of the forces that we are talking about. Our thinking is shortsighted and sees only the damage that insect pests inflict upon crops and the damage done by some birds on seeds and fruits. Obviously, as Rudolf Steiner says, "*We need to shed light on these things once again, but from a spiritual-scientific perspective, in other words from a macrocosmic perspective.*"

Before moving to the next paragraph in which Steiner begins to concentrate upon trees, let us pause briefly to consider the relationship between birds and insects, in such a way that we can come closer to its essence.

One such consideration derives from the observation that most birds feed on insects. This creates a very intimate relationship between these two categories of animals, much closer than the relationship established earlier regarding their actions in respect of the vital forces that must pervade the agricultural organism. What is established is a dialogue of Life that goes beyond mere nutrition, which is real but relatively insignificant from a more macroscopic point of view.

In paragraph four ("*In looking at a fruit tree ...*") we think it is important to note that Steiner introduces this entire section devoted to trees by talking about fruit trees. This is a clear signal from Steiner that in the agricultural organism's macroscopic processes of nutrition the relationship with the cosmos occurs primarily through the fruit trees which, thanks to the insects, seize the diffuse Life that we have often called free astrality or *Zoe*. Steiner was not explicit about this but he wants us to realise it ourselves so that this can be properly understood and internalised in order to settle harmoniously into our existing store of knowledge.

In the second sentence of this brief paragraph Steiner, emphasises the need to "*tell already from* [a fruit tree's] *outer appearance that it is something totally different from any type of herbaceous plant. We must determine the exact nature of the difference, otherwise we will never understand the role played by fruit within the household of nature.*" We believe that Steiner did not casually (as if!) use the term '*household of nature'*. Anyone with any familiarity with economics knows that the term 'economy' means household management. Well, in the management of the household of nature, which consists in treasuring that which descends from the cosmos (Zoe), the fruit tree functions as the first essential link in the chain. The fruit is the part of the tree that serves the nutrition of the higher realms of nature - animals and humans. Moreover the fruit is the part where the tree has especially developed the qualitative elements that are particularly connected with the forces of the cosmos. This is very important because we know that for farmers to attain and maintain an appropriate level of consciousness they must have a high-quality diet: this consciousness is essential for a correct approach to nature and the practice of agriculture as an act of love for the Earth, plants and animals.

The fifth paragraph ("*Let us look at a tree ...*") resumes the consideration of the tree that we have found in Steiner's fourth lecture. Of a tree, " *...the only parts that we can consider plant-like are the thin green stems, which bear leaves, flowers and fruits*". In this way of considering things, the tree trunk is nothing more than raised soil that - just because it is elevated - is a little more alive than the soil on which annual plants live.

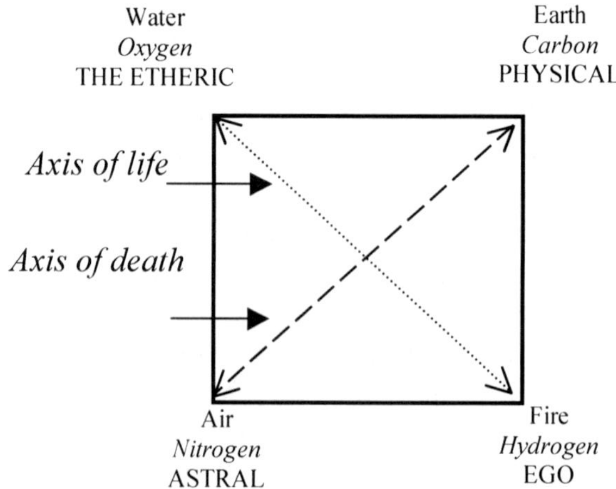

On the other hand, the trunk of the plant is the most dead, because it is mineralised. If we refer to the well-known square of the elements, the trunk can be seen as Earth (it is raised earth) carried into the air (the atmosphere). There we find the diagonal that we called the *'axis of death'*. We know that the chemical element linked to the Earth element is carbon and that linked to the Air element is nitrogen. Carbon and nitrogen combine to give the strongly poisonous cyanides. Remember that the other diagonal joining oxygen (water) and hydrogen (Fire) is called the *'axis of life'* because oxygen and hydrogen produce water - the basis of life.

Remember that the fourth lecture characterised the tree by its astral richness in the foliage and etheric poverty in the trunk and roots. Basically we are saying that although a tree that may live for a thousand years it is much less vital than an annual plant that sometimes lives for less than a season - and this seems a contradiction. In reality, what we have said is true if we are referring to biological life (bios) which we often identify with vigour, but if we are referring to unincarnated Life (Zoe) we must completely reverse the reasoning because the tree is extremely rich in cosmic forces (of which it is the receiver) while the herbaceous plant has relatively few. The abundance of biological life translates into a turbulent acceleration of all the processes of life and hence a rapid development of all manifestation.

Steiner presents the foliage of the tree, formed by the leaves and flowers, like a parasite upon the trunk and branches and asks: "*.... are these plants, these more or less parasitic growths on the tree, really rooted?*". In other words, if the plants really grow on the trunk, where they are rooted?

The answer comes to us in the two succeeding paragraphs, the sixth ("*In the roots..*") and seventh ("*What I have sketched ..*"). The plants that live on the tree have in fact lost their roots as commonly understood. But a 'root' is really only that which we can understand. And here Steiner brings a hypothetical example of herbaceous plants that grow so close together that they form a tangle of roots as if to constitute a single body. Indeed we can imagine that this tangle has a will and decides to organise in a way that "*the sap of the different plants would start to flow together down below*".

We wish to point out that in this circumstance Steiner refers to the spiritual logic (*each one for each other*) and this is motivated by the fact that the roots are primarily connected to the group I of the plant.

The seventh paragraph ("*What I have sketched ...*") starts with the explanation of a sketch made by Steiner and reproduced on page 140 [of the 1993 edition of the Creeger & Gardner English translation]. This gives further clarification to the concept of the communal root. The sentence ends with a reminder of the logic that we just mentioned, referring to the spiritual aspect of the tree - that is to how much divine Life is manifested through it.

None of this is found in annual plants but occurs in perennial plants precisely because their trunk has a relatively dead component. We often come across situations where what is higher manifests itself in what is lower. In this way the spiritual forces of the zodiac are manifest in crystals that belong to the world of minerals - apparently the lowest realm. Note how the word crystal has a strong similarity with the word Christ. Moreover, Christ himself announced that he did not come for the healthy and righteous, but for the poor, for sinners and the sick. But this being 'lower' applies only to the logic of the material world because, when viewed in the logic of the spirit, the material condition is a sacrifice and corresponds to the condensation of spiritual laws into darkness. So, mysteriously, in the depths of darkness one finds the crystals. We can consider these as the thoughts of God manifested as light solidified. The Earth may contemplate God because it has crystals.

The seventh paragraph concludes the explanation of the sketch and clarifies that the plant grows on the tree in a similar way to the plants mentioned in the example. They have sacrificed their roots - understood in an orthodox physical sense - and have combined them in a more "*etheric*" way as the cambium. This cambium can then be considered as the communal root of all the annual plants that form the canopy of the tree. This can be understood as being of an etheric nature because the cambium does not take part in any transportation of minerals or nutrients that are carried by sap.

Let us pause a moment to clarify what we are saying using a drawing of the tree above the soil. Directly under the bark and exterior to the cambium the sap is descending (*phloem*), whilst the ascending sap is more internal (*xylem*) than the cambium. Going down the trunk we arrive at the collar area, where we know that the arrangement described above undergoes an inversion.

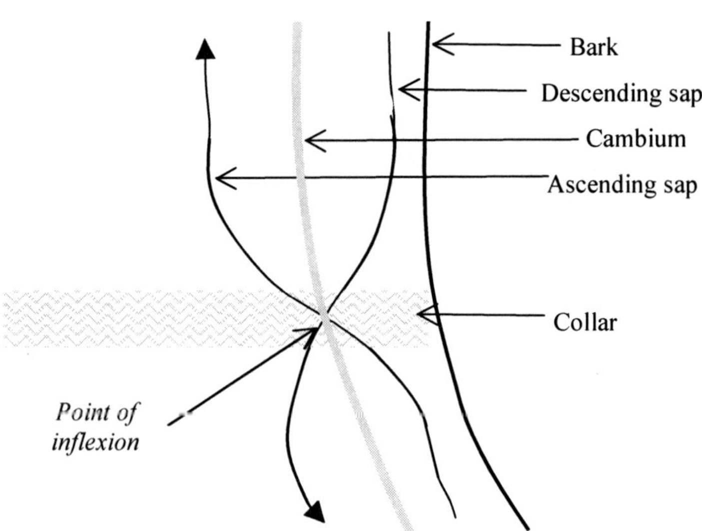

At the point where there is an inversion of the lymphatic vessels there is a change of curvature that is called an *inflection* in geometry. Precisely because of this reversal of curvature, the part of the tree that is situated above the collar has a convexity towards the outside and a concavity towards the inside, while under the collar there is the exact opposite.

We know that a convex form radiates etheric forces while a concavity accumulates astral forces. Knowing this, the tree above the collar will have a richness of etheric forces within, while the astral forces are received from outside. Alternatively beneath the collar the inside will receive the astral forces and the outside radiates etheric forces. If we turn our attention to the function of the individual vessels we can detect that the ascending ether-rich sap is outside the cambium in the root, while it runs

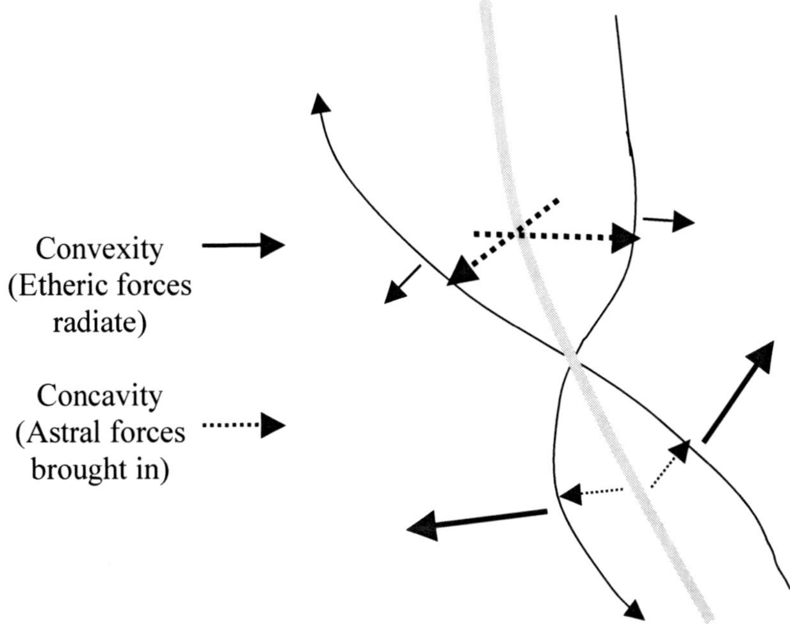

more centrally than the cambium in the trunk. The descending sap, however, is richer in astrality and runs in the trunk more superficially than the cambium, but crossing the collar it goes within. It follows that below the collar the tree is more ethereal (always in relation to mineral) and above the collar the tree is more astral. At the point of inflection a dialogue occurs between the etheric and astral forces and the harmony of this dialogue is the basis of the trees' health. Using different terminology we could say that health is the result of a harmonious dialogue between vital and 'soul' processes.

All this should lead us to conclude that between the tree bark and the Earth's crust we should find a reversal of forces, and this could lead us to deny what we have always said about the continuity between soil and the bark of trees. We have always argued that to identify the ideal environment for a tree it would be enough to match the characteristic markings of the tree-bark to the patterns in the soil when it cracks. We will see later that this contradiction will be resolved.

We are at the eighth paragraph ("*Thus, in the tree ...*"). The way in which Steiner speaks about the tree in this paragraph provides us with the opportunity to represent the tree with the following sketch, which essentially consists of a lemniscus that is smaller at the lower etheric part, and larger at the higher astral part, which becomes the foliage with flowering buds and fruit. The sketch also includes the inflection points that we just discussed. We get a representation that highlights two points of inversion: in addition to the one located in the area of the collar, now we also see one between the trunk and foliage.

The diagram introduces this other point of inversion of the forces, and this gives us the opportunity to mention the importance of establishing a fruit tree's canopy with the proper structure. Indeed, the way in which we establish the top of the tree is anything but indifferent with regard to the distribution of forces and their area of influence. In this regard, note that a structure with twin branches is the bearer of etheric-lunar forces. If we bring to mind that the top of the tree should be open to astrality we can understand how this approach is not ideal. The part where the etheric component should dominate the tree (always consistent with the mineral aspect) is the

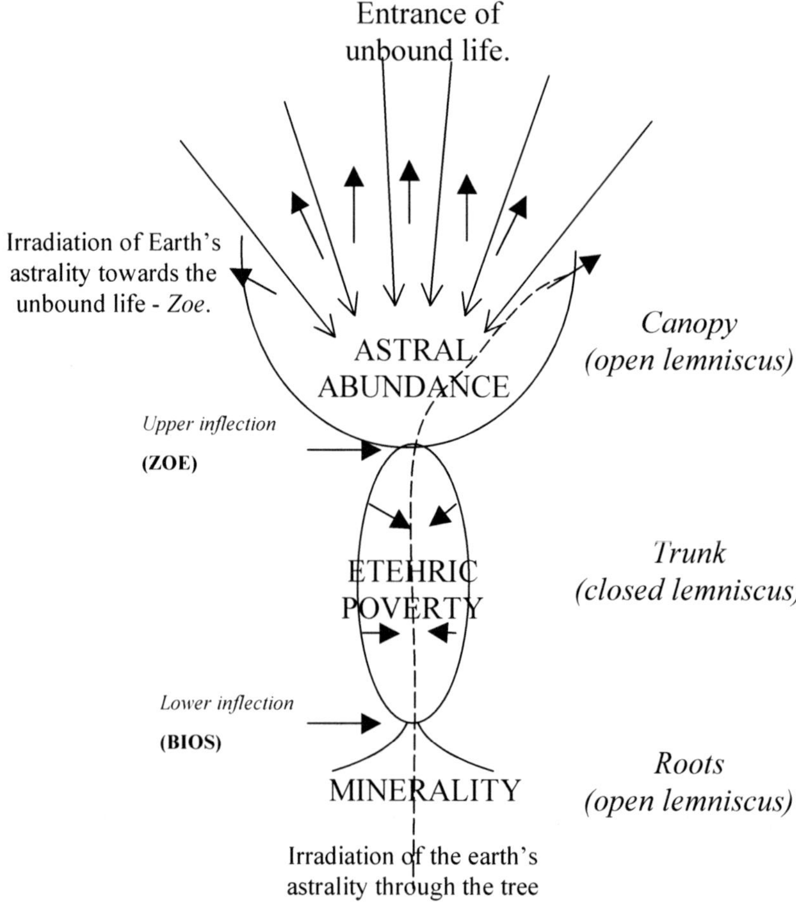

trunk. Therefore we believe that it would be much more advisable to establish a three branched structure, both because it is asymmetrical, and because the number three is more resonant with astral forces that descend from higher worlds. In order to further increase the asymmetry of the three branches one should let them emerge from various points, distributed irregularly up the trunk.

However, the attachment of branches is a point at which the exchange occurs between ethericity and astrality so it is also a key point for the health of the tree. It is no coincidence that the attachment of branches is one of the points where there are imbalances and at which it is common to find deposits of pest eggs etc. Maintaining a state of health must also respect the normal distribution of the tree's forces. For this it is not correct to introduce forms that induce the forces to act outside their normal field of influence.

The other crucial point for the health of the plant is, of course, the collar and it is essential that this is not under the soil surface. Precisely because of the importance of these two points, tree pastes - used to promote the biological health of the trees - must be spread from the area of the collar including the roots close to the surface as far as the intersection of the first main branches from the trunk. We could even argue that if we had only a little paste available and had to choose the points on which to apply it, we should cover only the points in question in view of their importance for the health of the plant.

In the ninth paragraph ("*That is the macrocosmic ...*") Steiner clarifies that, "... *with respect to the air and outer warmth, what grows up there on the tree is something totally different from the herbaceous plants that grow out of the soil*". We must not forget this sentence because we always have the presence of mind necessary to realise that when we see a growing herb we are faced with conditions that are completely different from those that we meet on a fruit tree. Therefore the understanding of any disease must take account of such profound differences. Only with these conditions can one then introduce a correct therapeutic intervention.

Reinforcing the foregoing, we may wish to dwell on the first part of the next sentence that reads: "*A whole different world of plants exists up there. They have a much more intimate relationship to the surrounding astrality, which is given off in the air and warmth...* ". We are accustomed to thinking of the tree capturing forces from the free astrality but here we read of an astrality that is expelled into the air and warmth. The reasoning leads us to say that if the astrality is given off it must emerge from within the plant - and this seems strange because we know that the plant does not have an incarnated astral body. Well, the astrality mentioned by Steiner is the astrality of the Earth. This cannot manifest in the trunk because of the etheric forces that dominate there, but they pass through to expand in the foliage and to radiate outside. This dispersion of astrality by the tree allows the free astrality to establish a dialogue with the astrality of the tree canopy and then have the chance to act and accumulate.

If we broaden our view we can deduce from what we have explained that the astral body of the Earth converses with the free astrality through the trees. This allows the unbound Life to become embodied life with the collaboration of the insects that mediate the encounter.

The sentence goes on to say: " ... *so that the air and warmth can become mineralised in accordance with the needs of animals and human beings.*" Evidently the air and warmth mentioned are those that are found outside the plant and it is difficult to understand that they have a degree of minerality. But the mineral belongs to the physical world and we know that the physical material world has a particular resonance with the astral world. Now we can finally think that the minerals brought with the plant's flow of etheric forces, arrive in the foliage and undergo a process of sublimation. In this process they are transformed into *substances* (substances and no longer matter) similar to those found in the constellations (of which they are the earthly representatives). It is a process very similar to that caused by homeopathic potentisation. From the same sentence we learn that humans and animals need this mineral. Steiner evidently refers here to cosmic nutrition[42]. From this we can understand that if there were no plants our cosmic nutrition would not be possible.

This is the main reason why the indiscriminate cutting of trees is so damaging - not for their production of oxygen as is commonly claimed because oceanic algae produce 90% of atmospheric oxygen. But the fact is that trees make cosmic nutrition possible for our animals and us. Recall in this regard that cosmic nutrition occurs through the sense organs - the most mineral part of us - through light, heat and sound. Thanks to these organs we perceive the 'homeopathic' substances that are finely dispersed in the air, which we take in and use, condensing them at the level of our "collar" (the area of atlas bone) to form our body lower down.

We believe that a relevant experience is that often, after a long walk in a forest, one is not particularly hungry, and this is precisely due to the fact that in the woods

---

[42] See 'Earthly and Cosmic Nutrition' by the author. This is not yet (2008) available in English.

one can take in much subtle nourishment.

If we shift the discussion to humans the two corresponding points of inversion are below the head and in the region of the sexual organs. These are the locations from which our two famous adversaries act. As in plants both points are vulnerable to attack by 'pests'. It is no coincidence that Michael defeated Lucifer in the head with the sword and Ahriman at the base of the spine with the spear. To confirm this we can add that in hatha-yoga there are two *asanas* considered essential for the maintenance of health - the turtle and the deer and these positions harmonise the perineal area and the base of the skull. The turtle asana, according to the Eastern tradition, also reduces the need for standard nutrition.

Another confirmation comes from the practice of some farmers. It shows that raising pigs by putting them in an area full of trees encourages exceptional development, especially in the aspect most affected by cosmic nutrition which is the formation of lard.

Returning to the lines that we just read, we can make a different observation. We said that minerals are sublimated by the plant and then dispersed in the air and in the warmth, but while on the one hand the process of sublimation releases a higher substance, on the other hand it always produces a coarser byproduct. This happens, for example, when a wax candle produces light but also smoke. The trees' gross product of sublimation is the condensation of the trunk. The annual plant does not produce the trunk because it did not disperse the astrality that the tree possesses. We might add that a tree's ability to sublimate minerals is reflected in the quantity of minerals deposited in the timber making wood harder and more durable. Since few timbers are harder than oak we can understand, in relation to the earlier example, why pigs derive so much benefit from *pannage* - being put under the oaks to eat acorns.

The oak, moreover, is linked to Mars, the planet that brings to mind the ancient Moon. The 'fall' in which Lucifer approached humanity occurred in this phase. The pig, for its part, is the animal that most embodies the *fall*, and is therefore considered by many religions as an 'unclean' animal. In this regard we can bring to mind the episode in which those possessed by demons are healed as told in the Gospel of Luke (8:32). The demons that had entered the possessed ask Jesus, who was driving them out, if they could join the herd of swine that were feeding on the mountain. The human, the oak and the pig represent the *fall* in three different kingdoms. Besides the pig has the blood that comes closest to that of humans and it is no coincidence that technical medicine thought about the pig when seeking organs suitable for transplantation into humans.

Returning to the main argument of the lecture, we believe it is sufficiently clear why one needs to include fruit trees in a farm.

The tenth paragraph ("*This is actually the easiest area ...*") begins with the statement that if we observe what happens in the world of nature, and use this opportunity to formulate the right thoughts and correct ways to act, it allows us to progress along the evolutionary path as far as 'passing the threshold'.

Certainly it is possible to sharpen ones senses to grasp the subtle aspects of the influence of plants upon the environment, especially with the sense of smell. In this way we could distinguish the wealth of astrality around trees and the relative poverty surrounding herbaceous plants. Today unfortunately the world of smells has become parched, demonstrating that we have not only fallen but after the fall, rather than rise, we have delved further into the world of matter, exacerbated by the production of artificial flavourings and scents.

Healthy farming methods can recover the scents of blossoms and this is a sign that 'cosmic fertilisation' has been reactivated. Consequently we can gradually reduce crude fertilisation.

The next paragraph, the eleventh ("*In order to see ...*"), suggests some questions to consider. What happens in the tree trunk? What does the cambium do? In the next paragraph Steiner poses further questions about the role of foliage.

The first sentence of the twelfth paragraph ("*You see, the tree makes ...*") talks about the cosmic nutrition we mentioned earlier. Then Steiner says that the herbaceous element that makes up the canopy should, because it is leafy, have a strong vitality. But "*the cambium acts to damp down this vitality to a more mineral level.*" That is the role of the cambium in making astral riches around the foliage in contrast to the etheric poverty in the trunk.

The reasoning is pursued all the way to the root that, being so distinctly mineral, absorbs vitality from the land making the land on which the trees grow poorer in the etheric.

From what we have read and commented upon we believe we have a sufficiently clear and detailed statement emerging from our course that the environment surrounding the tree presents a wealth of astrality around the foliage and an etheric poverty in the soil.

Twenty-third meeting

Before continuing to read and remark upon the seventh lecture we should continue to develop our picture of the tree since trees are of such fundamental importance in agriculture. We know that nature can be seen as the result of three types of processes: those that are contractive, those that are expansive, and those rhythmic mediating processes. These processes are also recognisable in the soil where the expansive processes can be found in the upper layer on the surface of the Earth, the contractive one in the deeper levels (the layer of rock) and, of course, in the intermediate layers we find the *mercur* processes. The plant world can itself be seen as an expression of the process of expansion (which includes diversity) of the upper layer of soil. When a soil contains many pollutants it reacts by producing weeds that must be considered as a way to reduce the various excesses. In this sense, the presence of weeds is the manifestation of a process of removal of negative forces. Because this is a response to purely natural requirements it should not be seen as a negative event, but only as a signal that there are undesirable imbalances in the soil.

Of course the layers of soil, which find expression in the three processes mentioned, do not have equal thickness because the *sulfur* process occurs in the thin layer of top-soil, contrasting with the *sal* process that affects the mineral layers of the bed rock. The *mercur* layer is the area where we find the sub-soil that, typically, might be 10-50 cm thick.

We have read that a tree trunk can be considered as raised soil so we should find the threefoldness that exists in the ground reflected in the trunk: the plants of the outer layer should express a *sulfur*-like metabolic activity and be rich in diversity because it is precisely from this layer that we see the development of buds from which arise the branches, leaves, flowers and fruits. In the innermost part we find the wood as an expression of a process of condensation. Wood is clearly the hardest part of the plant. The *mercur* element, corresponding to the soil humus, is the cambium.

If we want to deepen the considerations that we have formulated we must broaden our vision to the cosmos and reflect on the fact that the Earth, in the process of creation, represents the phase of maximum condensation in which all forces from the cosmos have been materialised.

In previous meetings we were able to study how each planet has two different influences. We have noted a primary or incarnating, and a secondary or excarnating influence. The Earth then is the product of the incarnating forces resulting from the planets. In the lower layers of the soil, those in which we recognised the *sal* process, we will find the activity of Saturn, while the surface is the site of the action of the Moon. In the middle layer we find the humus or *black gold* of alchemists that manifests the influence of the Sun.

Taking into account the fact that the trunk is raised soil, we should also find an expression of the planetary influences in the trunk. But we have also shown that in the transition from the region of the tree that is below the surface to that part which is above there is an area (the collar) in which the forces undergo an inversion, similar to what we saw happen to the flow of sap. In this case, however, the reversal does not affect the position of vessels on the physical level, but this is a reversal of the influence of the planets that, as they ascend, become excarnating. The plant, from this point of view, is not a simple swelling of the Earth's surface but an inversion. To get some idea of this we could think of the turning inside-out of a glove.

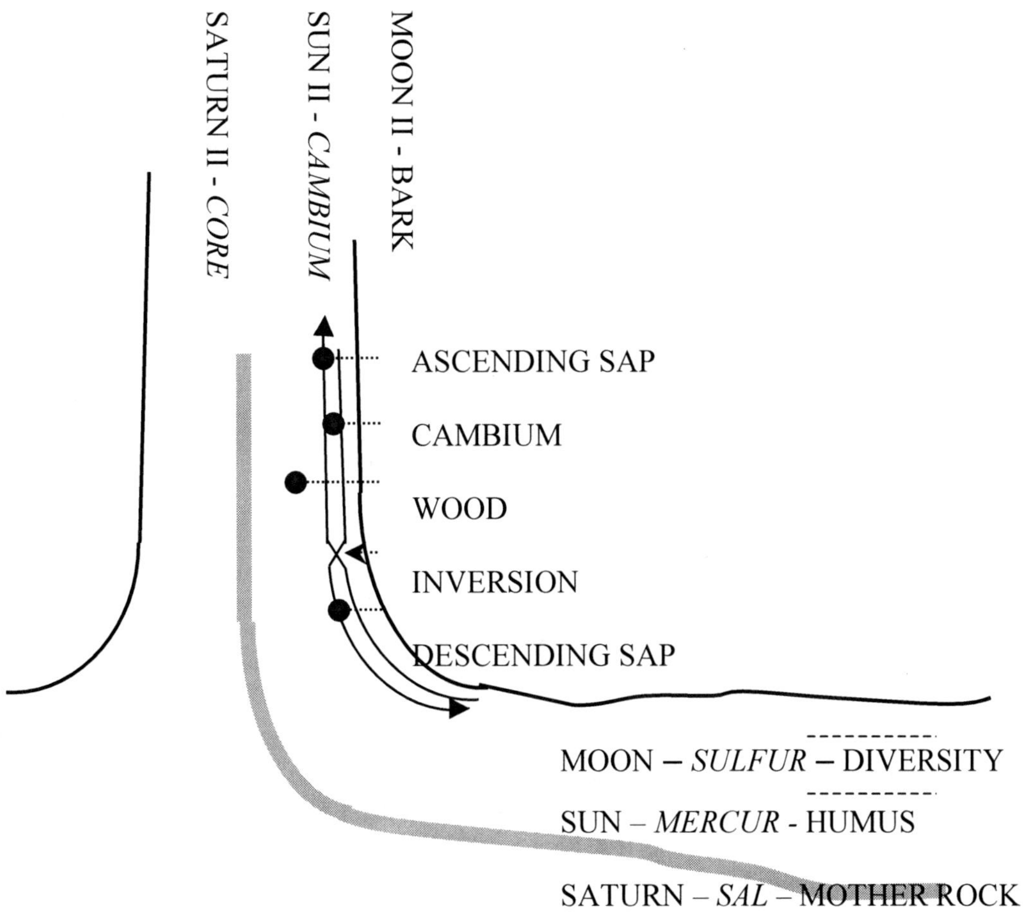

Relationship between the layers of soil and tree

In the innermost core of the plant we should be able to recognise the secondary influence of Saturn, as in the upper layer we should recognise the influence of the Moon. In fact the Moon is expressed in skin (bark) and cell division (desquamation). Obviously in the middle we should always find the Sun.

What we have said, however, must be confirmed by a perception, and not remain an abstract thought. By changing our point of view, that is by looking at a tree trunk in section, we will find that the influence of Saturn is recognisable by the presence of the rings that surround the various layers of annual wood and these correspond to the rings of Saturn.

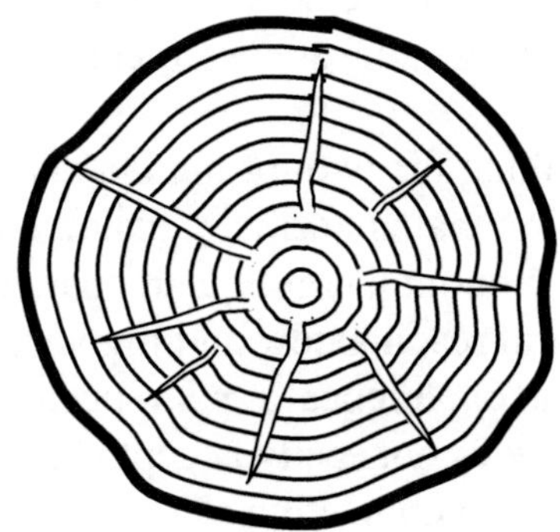

Section of a trunk with medullary rays

Outside of Saturn we should find a sign of Jupiter. But the mass of trunk wood itself is the expression of Jupiter. The planet Jupiter brings the forces of Light and we know this radiates from the centre outwards. The perception of this is in the medullary rays. What we are saying is also found in the fruit: if an apple is cut horizontally and not vertically, we find a five-pointed star.

Sections of an apple cut transversely

In beech trees the same light forces descend a bit further toward the forces of the Alchemical ether and manifest in the characteristic parallel veins of fibres rather than in the radiating cracks.

Still moving out from the centre of the trunk, we find the channels of the ascending sap and the force that drives the sap upward is a Mars force. Further out we find the cambium linked to the Sun and then the channels that take the descending sap, linked to Mercury, with the nodes and evaginations of the branches. Venus, with its characteristic protective role, finds expression in the cortex. Finally the Moon is in the part of the bark that often comes loose - like our skin - such as cork.

So, recapitulating, in the ground are the primary planetary influences whilst in the tree we find the secondary planetary influences.

Progressing up to the tree canopy, beyond the second point of inversion we discussed earlier, we revert to the planets' primary influences, although they manifest a different quality than when acting in the soil. The secondary influence really begins to make itself felt when the fruit begins to mature.

In the plant we can recognise two principle stages of metamorphosis: the first in the transition from the soil to the trunk, the second in the transition from the trunk to the branches. There is also a third in the transition from the growth phase of the fruit to the ripening stage. Moreover we must bear in mind that a metamorphosis is only possible when a new force, acting from the outside, enters to disrupt the balance that had previously held sway. This enables the achievement of a new equilibrium at a higher level.

The same thing also applies along the path of human evolution: every step forward requires a profound transformation that is normally called *conversion.* Without recognition of our mistakes and without a consequent change of direction one cannot overcome the present situation and reach a new and higher equilibrium. Each conversion, however, requires the experience of pain, because the part of us that will be converted is the soul and this does not easily accept change. It prefers to remain in its old habitual form and our I has no choice but to provoke the transformation.

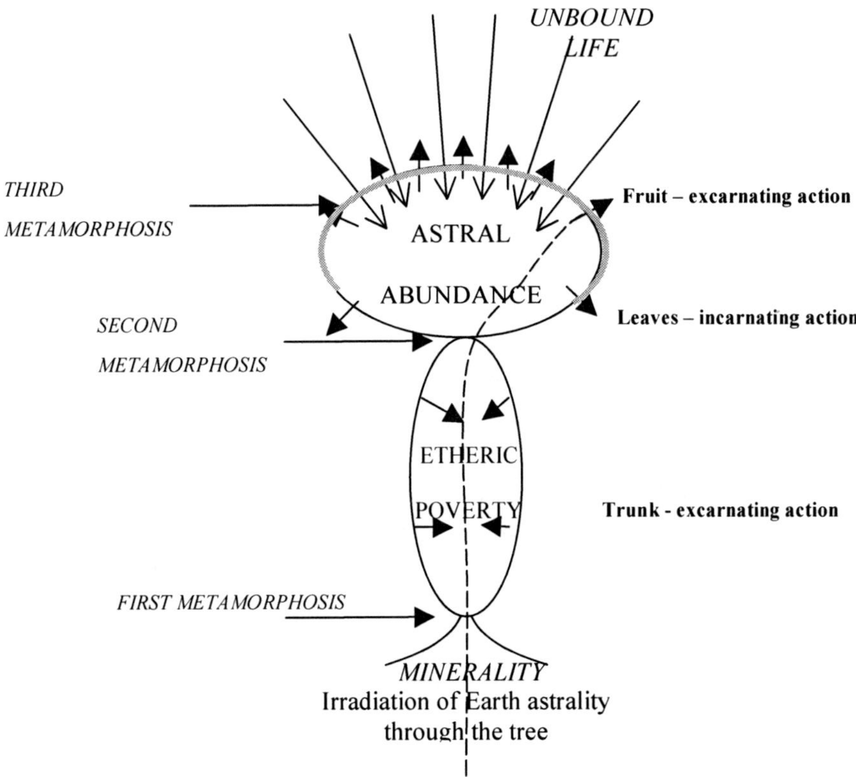

We can never hope to become beings of Light without taking on changes. Similarly we cannot imagine that the soil can become a tree and fruit without undergoing conversion or metamorphoses.

From these brief considerations one can understand that the being of the tree is much more articulated than we commonly think. We could say that it is the archetype of processes that affect all of nature's living beings. In effect, the development of a child until maturity, even on the purely biological level, crosses two moments of metamorphosis: the first in the period between birth and the exchange of teeth, which coincides with the change to an individualised etheric body - and the second during puberty which involves a change for the astral body. The same thing happens in animals.

It would be very interesting to study how the different compositions of soils, leading to different balances in the distribution of various planetary forces, affect changes in the processes of plants and thus - in their own way - make themselves manifest. We could substantiate how an imbalance of Mars forces at the soil level (reserves of water), that direct the ascending sap after the first transformation, can influence the upper part of the plant and then, after the second transformation, affect the flowering impulse.

Obviously we must learn to look at the tree, its parts and the processes that take place there, but in a completely different way. This is absolutely necessary because, having begun to understand the importance of the tree in the economy of the agricultural individuality, we can no longer afford superficial evaluations based only on modern orthodox knowledge. In this way the wood will not be appreciated only as the timber of a tree, but will be appreciated as the transformation of the forces of Jupiter, of Light. We can also understand that the strength with which the buds manifest originates from the core of the tree. The bud is the beginning of a new life and is thus connected with warmth ether and the Saturn forces that occur in the tree's centre. We can even understand why a tree hit by a frost which has destroyed the new buds, activates the so-called 'wandering buds'[43] which live in a fluid form in the sap and we can also understand the effect of the homeodynamic product 'Pro-flowering'.

We talked about 'wandering buds' and we believe it is worthwhile to point out that these buds are not in a physical material state, because they are alive and therefore live in a liquid condition such as the sap (water is the basis of life). These gems are 'daughters' of Saturn, or the 'heart wood', and this allows you to dramatically revalue the function of the heart as an essential part of the tree that is normally of practically zero popular interest. The heartwood must be taken into consideration and since we said that the core has a strong relationship with the deepest layers of soil, we must also admit that it is very important to strive for knowledge of these layers because of their effect upon the development of buds.

Since we're talking about buds and we know that sometimes in sylviculture deep incisions are made into the trunk as far as the cambium to stimulate the 'wandering buds', we might ask ourselves whether this type of incision can influence the manifestation of buds compared to those that come through the wood. So if we consider that the fruiting buds are formed mainly on the horizontal branches and timber in the vertical branches, we can similarly assume that a horizontal incision favours the manifestation of fruiting buds and a vertical incision favours timber buds.

After this necessary diversion we can resume reading the Koberwitz text. In the thirteenth paragraph ("*A phenomenon like this ...*") Dr Steiner speaks about the insect in relation to the forces of the tree. The adult insect nourishes itself upon the astrality that rises through the tree and reveals its richness in the canopy. The larva lives under the soil where there is a situation of etheric poverty. We should already be aware that there is a resonance between the stages of development of insects and those of the plant: the insect is born from an egg that can be considered as analogous to the plant seed. The following caterpillar stage corresponds to the leaf stage, so much so that the

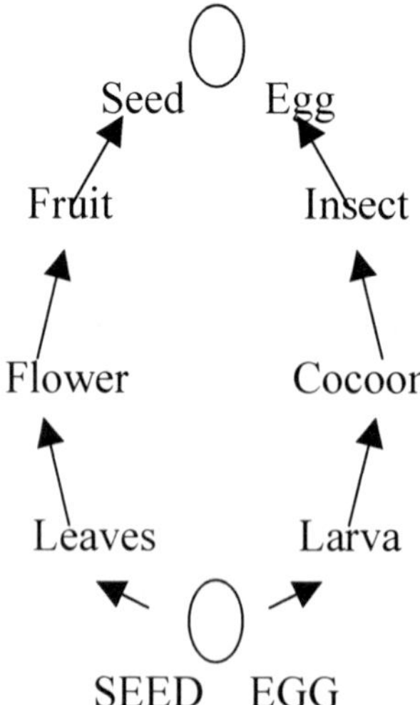

The parallel development of plants and insects

[43] '*Gemme vaganti'*, in Italian.

leaf is the typical food at this stage of the insect's life cycle. We see that plant and insect have a parallel development and dance together through their stages of development. We could even suggest that the caterpillar inches along the tree to get the leaves, in the same way that the leaves grow up all around the branch, following a systematic order. Then the light and colourful butterfly corresponds to the flower on which it alights, and finally the mature adult imago corresponds to the fruit that produces the seed, as the insect produces the egg.

At the end of the third sentence of the paragraph Steiner mentions human karma. At first this may seem unimportant, but if we think for a moment, one cannot escape the relationship between human karma and the Earth. In fact, every action produces consequences on several levels: on a personal level as individual karma, on the people to which we belong, as general human karma, and even upon the karma of the Earth. Man should face the problem of how to begin to remedy this last type of karma. When we said that a farmer should be concerned to work for the good of the soil before being concerned for their income, we referred to this.

The paragraph continues with the observation that insects, as I said earlier, could not live without trees but we could add that the trees could not live without insects. This allows us to understand what we have repeatedly stated in the past about our attitudes to insects: if we consider insects to be harmful we actually miss the important role they play in the economy of nature. What makes them appear harmful are disharmonious man-made conditions. This causes unbalanced populations of insects in the world. In healthy conditions a huge variety of insects regulate each other and also regulate the exchange of the free astrality of the tree, as we said in the previous meeting.

Returning to speak of insects and the consequent relationship with the trees and with the free astrality we noted earlier, we reaffirm that the astrality poured out by the trees is the basis for our cosmic nutrition. That is one crucial aspect of the tree that is so central to the life of nature and man.

Let's now discuss the fourteenth paragraph ("*This is a further ...*") in which we find an interesting comment: "*Within every plant there is a certain tendency to become tree-like.*" We could begin to ask "*who*" in the plant wants to become a tree? To resolve this question let us represent the evolution of nature schematically. Our scheme can be constructed from a single point, from which the paths split taking two distinct directions.

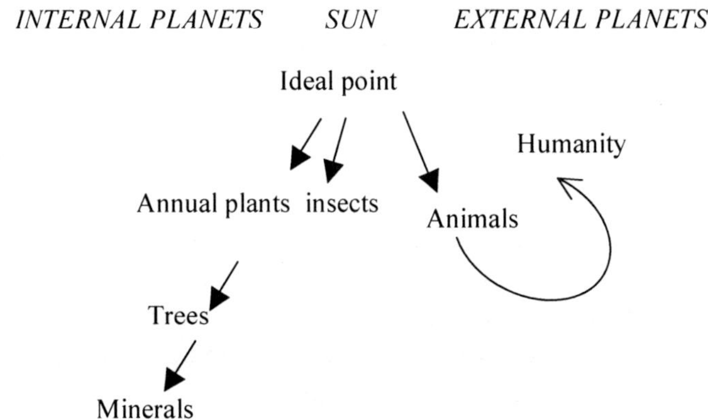

Following the first path we meet the plants, then the trees and then finally the minerals. The second direction leads instead to the animals and then to humans: the arrow which shows the human lifted back up represents the strength thanks to which man reverses direction and may redeem all creation bringing it back to a higher level. The first way is a path of the fall and along which each plant would become tree-like. All can understand that this process should not be encouraged.

The evolutionary line of plants develops primarily due to the forces of the internal planets. On the contrary, the other evolutionary line refers primarily to the forces of the external planets. In this context humans, whose evolutionary line refers to the outer planets as far as Saturn, can ascend up to the Zodiac to draw the spiritual forces that allow the redemption of nature.

From all this we can deduce that the point from which the development began is/was the Sun, which we have seen to be the representative of the Zodiac in the world of planets.

We know that the animal kingdom can be divided into two broad categories: that of warm-blooded animals and that of cold-blooded animals. Animals that are upon the evolutionary line leading to humans are particularly those with warm blood and are those that are also called *higher* animals. The world of insects, which are cold-blooded, can be represented in our scheme by a second arrow drawn beside that which refers to the internal planets. We should not be surprised by the fact that, in this context, we place the insect with the interior planets, even though we have always said (and still maintain) that the origin of insects is to be found on old Saturn. Here we are using a different point of view that considers evolution - not origin. This allows us to contrast insects with higher animals: the insects do not have warmth and are of the animal kingdom representing the sense organs, which are the most mineral.

Let us once again take this opportunity to urge a certain caution in regard to such schema. They undoubtedly have a role in developing concepts but we must also be ready to drop them as soon as necessary and not be caged by them. The schematic is a form, but the understanding of reality involves the adoption of many points of view that we need to embrace. We must be prepared to break the forms that we build for the understanding of individual particulars when they are no longer useful.

The end of the paragraph considers those insects that remain at the larval state and then continue to live at ground level. We are dealing with the so-called nematodes or, more generally, of soil-insects. These insects do not feel the impulse to rise to the astral world or to the canopy and so remain bound to the astrality of the Earth.

The fifteenth paragraph ("*As we shall now see ...*") presents us with a third type of 'larva' which is not bound to the astrality of the Earth but, for Love, denied itself the rise toward the astral of the plant in order to stay put in the ground. This is the earthworm. It is the regulator of the vitality of the land and the worm succeeds in this thanks to the strong limestone component of its intestinal glands. This particular type of limestone can be characterised as astral/living because its living aspect gives vitality while it absorbs the astral.

The activity of the earthworm beneath the soil surface corresponds to the bees that regulate the astrality in the air. Both of these animals are driven by the power of love, but the action of the bee takes place in harmony with the forces of the Sun, while the earthworm works with those of the Moon. It is no coincidence that the reproduction of the earthworm happens in the full Moon night, when the *annelid* emerge from the ground. The element of sacrifice that unites the two insects is found in the bee's renunciation of sexuality, and in the earthworm living beneath the surface of the earth and thus away from the benefit of high spiritual entities which live below the surface of the Moon and preside over humanity's evolution. The Romans recognised the importance of the earthworm by calling him the farmer.

We could say that the bee is the servant of astrality (Air and Fire elements), the earthworm of ethericity (Earth and Water elements). For completeness we could also say that the cow synthesises the actions of all four elements.

So we come to the sixteenth paragraph. In the first sentence Steiner says: "*So, down in the ground we find earthworms and other creatures vaguely reminiscent of larvae.*" We want to note the precision with which Steiner has used the word "*reminiscent*". We know that earthworms are closely related to world of gnomes. (The gnomes are characterised as beings of the head. Earthworms, being headless, are partnered in the world of elemental beings, by the gnomes that 'donate' their head, while earthworms give their intestines to the gnomes.) Gnomes are the bearers of the memory of nature. We know that in German there is a single word for both *memory* and *recording,* and in the case of gnomes it would be more correct to speak of recording: these beings do not have an I - the spiritual part that enables true memory.

In the second sentence Steiner says: "*And indeed in certain soils - which can be recognised just by looking - it would be good to encourage the earthworms.*" He's talking about land that does not have a 'memory' of its origins, of the thirteen holy nights, or of the actions of archangels, etc.

Do not forget that earthworms form the humus by uniting the dead organic matter with clay. Clay - as was indicated in the sixteenth paragraph of the second lecture - is the remedy for a "*soil that does not carry these influences upward during the winter as it should*". Therefore, the earthworm plays a very similar role to clay, but here the earthworm is mentioned because the issue in this lecture is animal health and the agricultural organism, while in the second lecture Steiner spoke of planets and forces. Hence the importance of the earthworm is not only in its ability to form humus, but to bring the cosmic memory to the Earth and to the plants.

The metamorphosis of the Prologue of the Gospel of John made by Steiner begins with the phrase: "*In the beginning was the memory and the memory was God and the memory was with God.*"

Today's conscious human, who works 'freely' to establish a direct relationship with the divine, must first revive the memory of his own divine origins. Today, with the advent of the etheric Christ, we finally have the forces to conquer this memory and the realisation of this can take place in our hearts. We must reach '*ri-cordare*' (with a heart) and not '*ram-mentare*' (with the mind)[44]. Today it is possible to establish a direct relationship, which does not need the mediation of any man or institution - something that was not possible until a few decades ago.

As always Steiner is leading us along a route as a path of initiation. In fact he begins speaking of the earthworm (which embodies a lunar memory), then we talk about the butterflies, still later of the birds. In this way he leads us through three stages of memory, ending up at the Sun. Again, we have the opportunity to appreciate that what we are reading is a book about agriculture but it is intended to give us the means to help the kingdoms of nature through a path of spiritual evolution. At this point we may not be surprised to read in the last sentence of that paragraph that the benefit of the earthworms is: "... *not only for the vegetation, but also, as we shall see, for the animals.*"

---

44 Italian: *Cuore* = heart. *Mente* = mind

In the seventeenth paragraph ("*Now again, there is ...* ") Steiner introduces us to knowledge of the world of birds and tells us that there is a close collaboration between insects and birds whose combined role is to gather the astral forces of the cosmos and spread them throughout the agricultural organism. They should not fail to be present where they are needed.

Towards the end of the paragraph is a phrase that must have seemed just a hypothetical possibility to those people who were listening to Dr Steiner's lectures in 1924. It seems to have been presented only to highlight the importance of the presence of insects and birds in nature, but surely they would not have imagined that after a very few decades this situation would actually occur. Now the presence of birds in our fields has been drastically reduced, not to mention that of beneficial insects, and unfortunately we must also recognise that - as prophetically announced by Rudolf Steiner - we are seeing "*a certain stunting of vegetation*" understood as the diminishing nutritional value of plant products. Every day we are confronted with beautiful and lush products that lack taste and aroma. The most serious is that the fruits of the fields, in addition to having lost the ability to satisfy our senses of taste and smell, have also lost the ability to provide adequate human food, especially in its finer aspects.

The loss of nutritional value is also made evident by the reduced shelf life of plant products. Once apples lasted all winter and were not kept in a refrigerator but in the kitchen which was the hottest place in the house. Today this is no longer possible. We have said that the fact that the fruit does not keep is an indication of the loss of quality, and we know that quality is bound to heat and time. The birth of time can be dated back to old Saturn, the first stage of manifestation of the Earth, and whose remnants in the animal world are the insects and birds. It should not be surprising if the drastic reduction of the animals mentioned is attributed as the main cause for the loss of nutritional value of foods.

We might add that on old Saturn the seed of our sense organs was also sown, so we can say that insects represent the sense organs of the animal kingdom and as mediators with the world of astrality, even the vegetable kingdom.

The sterilisation of our stables and the disappearance of insects also determines that it is impossible for domesticated animals to enjoy a strong relationship with the forces of the cosmos. The same thing applies to the world of birds. Once there were swallows and their nests in the byres. This is no longer permitted, unless in some quaint old byres which are considered not to move with the times and that meet with many problems with the authorities for 'substandard hygiene' when animals are kept.

The paragraph concludes with emphasis on the connection between the plant world and those of insects and birds and the exhortation to farmers to create the conditions to enable a moderate presence of all these animals.

Paragraph eighteen ("*We need to keep these ...*") reiterates that insects are the means for the "*right astralisation of the air*" and then it is shown that this astrality "... *is in interaction with the wooded areas*", so that areas could then direct and distribute that astrality in the right way to all of the plants as required. Then Steiner adds: "... *just as in our body certain forces direct our blood in the right way*". The forces that direct our blood in the right way are those of the I, so the woods are equated to the ego of the agricultural organism. Therefore, we can understand that just as a cell of the body dies if it is not blessed by our blood - and thus forgotten by the I - so any plant is going to get sick and die if they do not receive the forces of the forests. The presence of woods is essential to the practice of a fundamentally sound agriculture.

Nowadays, on most agricultural holdings, the forest no longer exists and this can only have negative consequences on the health of plants which will always be weaker, lacking in vital forces, and increasingly ill. This does not mean that our plains must be brought under conditions in which they were a few centuries ago when, for example, the Po Valley was completely covered by forests. The issue cannot be considered only from the biological point of view but also from a spiritual perspective.

The last sentence of the paragraph says: "*The effect of a forest is felt over a very large area, and in areas with no woods this function must be performed by something else. We must realise that in areas where fields and meadows alternate with woods, the vegetation is subject to quite different laws than in the vast treeless regions.*"

The "*something else*" that Steiner refers to, precisely because it is mentioned towards the end of Course, must refer to something that has already been mentioned. Thinking that the central aspect of the Course must be recognised as the biodynamic preparations, they should also be able to replace the wood. In our view these are: *fladen* (not named in the Course, but later devised by the Thun's and also known as 'cow pat pit' or 'barrel compost'), horn-manure (500) and the yarrow and nettle preparations (502 and 504). These preparations (*fladen* excluded), potentised and united into a single homeodynamic product, constitute the so-called 'Pro-forest' and the 'Activator' for the preparatory stage of the use of the 'diffuser'[45]. About 3 ml of this Activator is equivalent to three hectares of forest and serves to ensure that the forces of the other preparations that we would like to distribute reach the plants in the most distant fields of the property.

We conclude by noting that the expression "*vast treeless regions*" today no longer has the same weight as in those times. We believe that the influence of a forest has been drastically diminished - maybe even to a few tens of metres. In our day this suggests that the "*something else*" becomes urgently necessary.

Returning to the question of whether it is useful or not to reestablish the woods again, we see that an increase in the forested area has a very strong effect on the vigour of the plants which, being linked to the more vegetative pole, is also linked to the hereditary line of plants and thus the biological life. Limiting the area under forest enhances the quality of the plants, tied more to individualisation and the life that comes from the cosmos.

Moreover, we cannot ignore the change over time in the relationship between the plant, with its nutritional aspects, and man. Human nutrition has changed profoundly in recent decades and not always just because the available funds allow access to foods that were prohibitively expensive not long ago, but also because human work is very different and requires different nutrients. Think of sugar. A century ago the consumption of sugar was limited to just over one kilogram *per capita* per year. Today every person consumes 30 kilograms of sugar a year, so there is a tendency to say that this is a sign of faulty nutrition. In fact, we should know that sugar supports the work of thought and there is no doubt that we think more than once was the case - probably the thoughts produced in a recent year were not produced during a lifetime of someone from the beginning of last century. So an increase in the consumption of sugar is justified.

This allows us to maintain that now we have to maximise the nutritional value of fruits and this is possible only if the presence of the forest is contained and supported with different systems.

[45] A device used by some homeodynamics agriculturalists to spread the homeodynamic preparations over wider areas.

Twenty-fourth meeting

Let's continue to comment upon the seventh lecture, which we know can be characterised as a lecture on the 'agricultural organism' dealing with the relationships between the various organs that make up the whole. Steiner presents the tree as a fundamental agricultural component and for this reason we have tried to understand trees better and consider them from various different points of view. These considerations are also vital for the issues that we are about to study as they prepare us for an unusual but equally interesting line of reasoning concerning ponds.

Let's start reading from the nineteenth paragraph ("*You can tell ...*"). Remember that at this point in the lecture Steiner is talking about the forest and had just said that if we were in a situation where there is not enough woodland, it would be useful to replace it with "*something else*". We have suggested that the biodynamic preparations are this something else.

During the meeting we concluded that preparations that are able to replace the forces of the forest are the horn-manure preparation called '500', and the 'compost preparations' made of nettle and of yarrow. We have also said that these three components are combined to make the product known by the name 'Activator' for those who use the homeodynamic 'Diffuser'. The diffuser is able to radiate the influence of homeodynamic products to cover as much as 400 hectares[46], and it is the influence of trees that is carried by this preparation. If the diffusion were from a sufficiently large forest (at least 1 hectare) the *activator* preparation would be superfluous. From this point of view the forest should ideally occupy the centre of a farm so that its influence can reach all over the farm, although from a practical point of view it may well be more appropriate to plan the woods away from the centre.

The paragraph starts with the observation that in regions where there is a large amount of natural forest the growth of herbs and the grass of surrounding farms is already promoted. Common sense, says Steiner, should warn us not to eliminate the woods and if one were to notice a lack of vigour in plants it would be possible to remedy this by increasing the percentage of existing forest. Our opinion is that around 7% of an area should be under trees, increasing up to 10% in drought-prone areas.

Obviously the proportion of forest could be decreased if plants exhibit excessive vigour to the detriment of their ability to produce good quality seeds. We may recall here that the forest takes the function that is performed by the I in the agricultural organism. We know that the I presides over the organisation of the whole human organism, coordinating the activity of all the various components in order to maintain the whole in dynamic equilibrium. The I has a dual function: one part is involved in and inhabits the lower bodies (physical, etheric & astral), the other addresses the spiritual world from which it is formed. To achieve the first of the two functions the I uses the so-called 'ego organisation' which presides over and, in the broadest sense, conducts and coordinates all the soul and vital processes, the most subtle of which influences are those that most affect the physical body. The ego or I organisation, in other words, allows the I to act on the various bodies. To clarify the concept further one could consider the I as a king who has power over all that exists in his kingdom, but who cannot conduct all his governmental affairs personally for all his subjects in the complex range of activities that they perform. However, to deal with all the aspects of the life of his kingdom, he uses a number of ministers who will each have

[46] A more recent 'Diffuser' has been made which covers up to 2500 Ha

their ministries, a system of public administration, and law enforcement officials who respect the wishes of the king. All the apparatus through which the government implements the king's requirements is known as the 'I organisation'. In the agricultural organism all this is the role of the forest.

In the human we know that the Ego plays a number of different roles depending on whether it is acting upon the physical, ethereal or astral and the same thing is true for the influence of the forest. At this point, though, we have to investigate further, perhaps continuing to use the parallel with humans.

We can start by asking how the influence of the I is manifest in the physical body. Well, its primary manifestation is our upright stance. This enables us to understand that plants which are unable to maintain verticality - such as the grape and kiwi vines, the cucurbits, many of the pulses and others - have a physicality showing that the spiritual element acts weakly because of a weak ego organisation. Obviously these plants, when they are used for food, carry such weakness to us, although this is not always immediately perceptible. Conversely, plants with a clear verticality - we might think of the conifers - have a very strong I.

The ego organisation also plays the important role of organising the system of warmth in the physical body. It is thanks to the I that our body temperature is normally close to 37° C. When there is a need for a greater intensity of our I, for example to activate a process of healing in the event of illness, our temperature rises.

In agriculture this influence allows the formation of special microclimates. In fact in the vicinity of a forest one can enjoy a climate characterised by a reduced difference in temperature between day and night compared to a completely exposed area. In the summer the forest mitigates the typical excess of heat of the summer season. It is no coincidence that the areas of the earth where there are the greatest daily temperature ranges are the deserts, where there is a total absence of trees.

In the human the I is responsible *inter alia* for the formation of the arch of the foot. This formation of the foot is typical of the human (no animals have it) and represents a 'space of freedom' in touch with the physical world. Anthroposophy says that the arch of which we are talking is the result of the separation that occurs between the Zodiac constellations of Aries and Pisces. So today the constellation of Aries corresponds to the head (with the dome) and Pisces to the foot (with the arch). The I in the physical body is also expressed in the proportion between fingers and the palm of the hands.

Having said this with reference to humans we might now ask ourselves what part of the plant corresponds to the arch of the foot. Recalling that the plant corresponds to an inverted human, we can see the flower at the top of the plant, which carries out the function of acceptance of the cosmic I and, therefore, the connection with the Sun - not so much in the sense of heliotropism but more in terms of coordinating with the group I. The correspondence with the proportions of the hand becomes, in the plant, the ability to fully express the archetype, as the hands allow a person to demonstrate their talents and fulfill their karma.

Conifer forest can express what we have just described and therefore develop the ability to manifest the I on the physical level. This type of forest shows an imposing physicality, but at the same time brings forces of warmth as is witnessed by the resin and the fact of being evergreen.

Let us now consider the effect of the I in the human etheric body. The fundamental effect of this influence is the state of health induced by the perfect coordination of all vital functions. The I in the etheric body also governs the smooth functioning of the immune system. A consequence of this is a person's physical strength and stamina, and the ability to find reserves of energy when necessary - even when feeling tired. In the plant this translates into what we have been calling vigour and allows the formation of strong, rich and abundant plants.

The I's work upon the etheric body is also manifested in good humour and equanimity. Remember that the '*humours*' of Greek philosophy (yellow bile, black bile, blood and mucus) correspond to the four temperaments, which in turn correspond to the four aspects of the etheric body. So good spirits can be seen as the result of harmony in the etheric body.

The corresponding plant characteristic can be found in the active ingredients, predominantly in medicinal plants. What we have often said is not a coincidence - that medicinal plants grow well near a forest. However, they can be self-sustaining in the absence of trees just because they have a strongly active I in the etheric body.

The equivalent to optimism or the fund of good humour in plants can also be seen as the ability to connect to the life-Germs[47], or the so-called *etheric ring* surrounding the Earth[48]. This is a new world of elemental beings towards which we are trying to open the plant with the help of seed baths or with treatments during the 13 holy nights.

Now we can take a further step and specify what type of forest is most suitable to support the I's influence at this level. Well, it is the deciduous or hardwood forest, because the leaf is the part of the plant where the etheric is most expressed.

Let us now consider the effect of the I on the astral body. The first is the control over impulses of sympathy and antipathy. In the plant the soul component is undertaken by elemental beings - beings of an astral and etheric nature. More specifically we can say that the gnomes and undines carry out their work driven by antipathy whilst - on the contrary - sylphs and salamanders are driven by impulses of sympathy. Therefore the group I, dominating impulses of sympathy and antipathy, has the mastery of these elemental beings governing the plant and can guide their action as needed. In our times an understanding and control of elemental beings is extremely important because humanity now no longer recognises them as we cultivate the land and pollute the water and air. This induces the elemental beings to retreat leaving room for those related to negative entities. In addition to 'control' of the elementary beings it also allows and facilitates the action of the 'new' elemental beings linked to Christ.

Another aspect of the I on the astral body is to give a direction to our actions, mitigating - for example - the instinctive reaction to assume an immediate stance whenever criticism comes from the way we operate or are. In this way one can take from these criticisms any element of truth that they might contain and work upon this for our improvement.

In a plant this can mean an appropriate reaction to the treatments. Do not forget that the influence of biodynamic preparations (and even more for homeodynamic preparations) is extremely robust for the plant and is not among the actions for which the plant is primed. Therefore the plant would initially react with antipathy. However,

[47] The zone of the vital germs is found beyond the Zodiac, between it and the Milky Way; from it originate the germs of the four natural kingdoms – see 'Theosophy' by Dr Steiner.

[48] The etheric ring is part of the new etheric body of the Earth, resulting from the sacrifice of Golgotha, and within which the Christ is currently manifest.

we can help the plant to give a proper direction to the influence, according to the I's requirements, in response to the message received.

Courage is one manifestation of the I on the astral body. In a plant this can be translated into the ability to resist climatic stress, pruning, water depravation and extremes of heat etc. Another result can be called 'loyalty', a quality that - in the plant - can be translated as resistance to GMO pollen, as a form of loyalty to their individual principle of their species.

An attribute resulting from the I's influence upon the astral is the ability to manage time wisely. This means the plants can complete their life cycle, and tailor their vital processes to changing climatic conditions.

The type of forest that supports this action is made up of plants capable of producing conspicuous colour and scent - a highly astralised forest.

The last influence to be taken into consideration is that of the I within itself - within the I. This shows up as the ability to reflect upon oneself and is effectively expressed in Steiner's phrase: "*In the thought that thinks itself one finds the self-sustaining I.*" The plants' capacity for such self-awareness can be seen in the Group I recognising the plants to which it is connected. Of course the I cannot even recognise the plant that has been subject to unnatural manipulation by people. The group I does not only read what the plants show, but also - upon their death - the experiences that the plant underwent in its lifetime.

Another result of the I on the I is clear thinking, which in plants can be seen in the extent of manifestation of the idea in the form of the plant. It should be obvious that all plants which emerged from cross-hybrid seeds or what was cross-grafted may not be a clear manifestation of the idea, and this is reflected also in terms of food, because food involves bringing into your body an element of disturbance to the clear thinking.

The I working in the I also brings humans a moral quality, namely those related to the laws of the good, the true and the right. Well, in the plant, these qualities correspond to the resistance to pests. So the presence of a wood capable of moral forces should also help protect plants from pests.

The highest activity of the I within itself is the ability to forgive. On the farm this can be regarded as the pardon of Mother Earth for our greedy and exploitative misdeeds. The plants suitable to support influences like those just discussed are related to the Sun - like olives, lime, ash, citrus trees, and partially solar shrubs such as dog rose, whitethorn, arbutus, etc.

Having said that, trees are few and far between in our modern countryside and so the influences described are extremely weak. We must strive with everything in our power to remedy the shortage of these impulses, which, as we said, do not necessarily mean the re-establishment of forests, not least because this would require such a long time before its positive influence is realised.

Let's now read the twentieth paragraph ("*Now we can continue ...*"). Note that there is a mistake on the last line. It is not "*larvae and insects*", but "*the larvae of insects*". In this paragraph Steiner unexpectedly mentions limestone again, by which, in relationship with earthworms and grubs, the etheric is "*lead off*".

Let us take the opportunity to know more about limestone, at least with regard to the three basic types of limestone. The first is the bearer of lunar forces. It is a mineral limestone which finds its highest expression in slaked lime and partly also in calcified seaweed. This type of limestone has the ability to depress the etheric.

Then there is a limestone that we could characterise as earthly, which is organic in nature and is found in eggshell, the oyster, in earthworms, in snail shell, and partly in quicklime that desires the etheric. These types of limestone are able to support the etheric.

The last type of limestone we will consider now has the ability to connect with the forces of the Sun. This allows limestone to revive the etheric and allows connection with the etheric ring surrounding the Earth. In the animal world this limestone can be found in the shell of the crab. There are moments in the life of this crustacean that its carapace would prevent its continued growth if it did not use a process by which the shell is dissolved, absorbed within the animal and, after a phase of rapid growth, reconstituted as a new shell. In the intermediate phase the crustacean is particularly tender and so is specially appreciated in some seaside kitchens. This process of dissolving and recondensing is reminiscent of a process comparable to resurrection. This limestone brings with it the propensity to dissolve and so it is used as a remedy for kidney stones in anthroposophical medicine.

Other types of limestone that bring forces for connecting with the Sun are calcified seaweed, which we have already found as one of the dead limestones, and - of course - preparation 500, the horn-manure. We believe that in preparing the spray products, both 500 and 501 (the horn-silica), a transmutation takes place during potentisation in which part of the limestone becomes silica and part of the silica turns into limestone.

As we have defined the first two types of limestone as 'mineral' and 'earthly', we can identify the last as a limestone of 'synthesis'. The 500 we know is the highest synthesis between the realms of nature, but this kind of synthesis is also found in calcified seaweed. These are made up of very small plants living in symbiosis with the tiniest crustaceans. In the case of calcified seaweed, the type of influence that it brings depends on the process in which they are incorporated, so if they are included in the compost heap the resurrection aspect predominates, but when used just out of the bag without being included in a process of transformation it brings forth its dead aspect.

In paragraph 21 ("*You see, if we take ...*") Steiner moves on to cite the "*instinctive clairvoyance*" once common in people but which is inevitably lost today. Steiner says that the blame for this loss lies in the materialism that has inspired all aspects of life and leads people to forget that there is not only the physical aspect of things in nature, but there are also more subtle aspects that don't serve the intellect or cleverness: these serve wisdom. Knowledge of nature through wisdom is possible when the process of comprehension is not only guided by intelligence but also by feeling, a process that we could define as 'thinking with the heart'. Unless this understanding becomes fully realised knowledge becomes useless - even the study of anthroposophy - because it is reduced to a kind of intellectual dogma that cannot help to reveal "*the way things in nature really are*". What we have just said about the three types of limestone makes no sense with the modern scientific intellect because, from that point of view, any physical type of limestone is the same because it is always expressed by the same chemical formula. But this does not take account of the inherent qualities of these profoundly different limestones.

This different approach to nature brings a different understanding and diagnosis and, therefore, a different set of therapeutic interventions. The cause of a disease, from the physical-material point of view, is sought in the presence of something that came from outside and has caused an imbalance manifesting in certain symptoms. It then follows that increasingly powerful and sophisticated interventions are required to discover ever smaller microorganisms in order to combat and eliminate them, and this is the allopathic approach. However, if we approach the same problem without the microscope, but with the '*macroscope*' (to use an increasingly common expression in our association) to try to grasp the reality not so much in its particulars but in its more general aspects, perhaps we can understand that the main cause of the disease is not found in the existence of a particular microorganism, which under different conditions will not produce any harm to humans. Instead we may find the foundation of the problem is the wider imbalance that can be addressed not with opposition, but with harmonisation. Moreover we believe we can say that if humanity came into the world naked and was placed in the midst of nature in this condition, then nature is the necessary and sufficient condition so that man can understand everything, even in its most intimate quality, and find the way back to God.

Thus, we arrive at paragraph twenty-two ("*As our perception becomes ...*"), wherein Steiner begins to establish the relationship between the world of plants and animals. The paragraph begins by saying: "*... that the bird world can become harmful if there are no coniferous forests nearby to make good use of what the birds accomplish*". This phrase, as indeed those that follow in the subsequent paragraphs, only presents a description of a particular relationship without providing sufficient means to understand why they exist. Of course we will not be content with descriptions alone and will try to discover some deeper connections.

Note that in this first sentence a specific reference to conifer forest is made for the first time. Steiner refers to an avian influence that has a positive effect in the presence of coniferous forest but which has a negative effect in the absence of such a forest. We must ask ourselves what this influence is and why the presence of the forest changes the effect so radically. The answer to these questions can be sought by deepening our understanding of birds. We will start by saying that birds represent the evolutionary stage of old Saturn when everything was warmth. The remnant of Saturn in people is the head and the entire bird is a head[49]. What we see as the physical head of the bird is really the head of a head! This link with the first phase manifestation of the Earth helps us grasp the existence of a relationship with conifers that are also linked with Saturn.

At this point we must think that the bird, being an animal, is the bearer of astral forces, particularly so in an animal that has such affinity with the air. As we saw, the forest, as well as strengthening the forces of vigour, is the bearer of the I and therefore, balancing between excesses, mitigates the impact of the astral forces brought by the birds.

Now we add a further consideration into our reasoning and see that in the plant the forces of Saturn are expressed strongly in the seed, so you can establish a link between the world of birds, and the conifer seed. This realisation, if governed by the ego organisation, is balanced and acts positively on the world of seeds. If the forest is lacking there is an excessive movement towards astrality which carries the seeds 'beyond itself', reducing the vitality.

Notwithstanding that all birds are bound to Saturn and are therefore linked to warmth, we have to acknowledge that over time some birds have developed in an atypical way. There are birds that have lost the ability to fly and have become terrestrial. These are the so-called *ratite* that today are without a sternal keel and include the ostrich, rhea, cassowary, emu, and kiwi, etc.

Then there are *impennis*, which are marine birds whose wings were transformed for the purposes of swimming. The birds that have kept the hull and the ability to fly

[49] See 'Man as Symphony of the Creative Word' by Dr Steiner

are called *carenati*.

Among those classified as *carenati* are the birds of the Earth that, while able to fly, prefer to stay on the ground and walk or run. These include chickens, turkeys, quail, peacocks, and crows.

Birds that are classified as being related to the Moon include the web-footed birds such as the goose, duck and swan. The birds of Mercury are sparrows, magpie, swallows, blackcaps, wrens, great tit, and larks for being small, fast and playful. The birds that show a special bond with Venus are those that take care of the young, such as pigeons and doves. Tied to the Sun are all songbirds, such as the nightingale, the canary and the blackbird. The raptors and hawks, for their aggression, are connected to Mars. Those linked to the wisdom of Jupiter are birds like the eagle and stork. Finally Saturn is linked to the spotted and little owl because of their nocturnal habits and because of their link to woodland.

These types of birds bear the forces of the forest plants supporting the seven life processes, which in turn are already connected with biodynamic preparations.

Saturn supports the process of respiration, Jupiter supports the generation of heat, Mars supports nutrition, the Sun supports individualisation, Venus supports preservation, Mercury supports the growth, and finally the Moon supports reproduction.

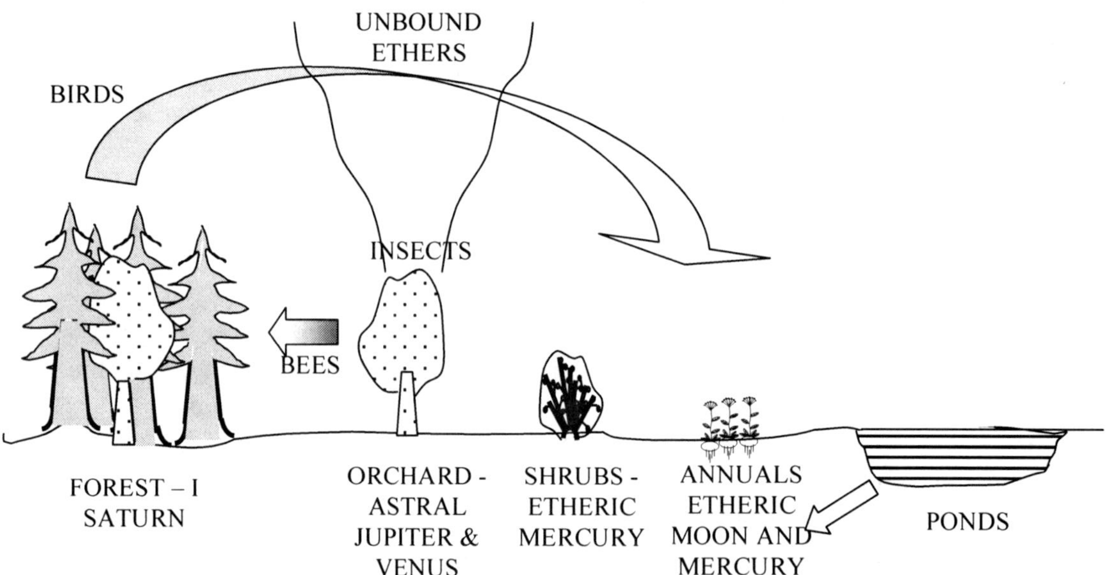

Relationship between orchards, trees, shrubs, seeds and ponds

With what we have said we can conclude this discussion of the agricultural individuality which we can sum up as follows:

- Orchards collect the unbound life (Zoe), thanks to the world of insects. This is a concentration of life that the bee will carry to the woods.
- The forest, which we saw acts as the I organisation, distributes these forces to all annual plants through the world of birds, which we have already ascribed to seven taxonomic classes.

We briefly linked the birds with life processes and the world of the preparations.

When we described the role of preparations in the plant we used the diagram that is reproduced below to help us. When we made this diagram no one asked who is 'behind' the arrows that we drew. Now we know that they are the seven vital processes sustained by the world of birds. For the record we can specify that the 500 and 501 should be seen as a polarity of the same force: that of the Sun (respectively the 'night' Sun and the day 'Sun').

Having said that, we are also better able to appreciate the genius of Rudolf Steiner who noticed that the world of birds would suffer considerable changes, the forests were virtually disappearing, the orchards were being denatured, and so he has tried to give farmers the means to replace the forces that would we would very soon come to miss - the biodynamic preparations,

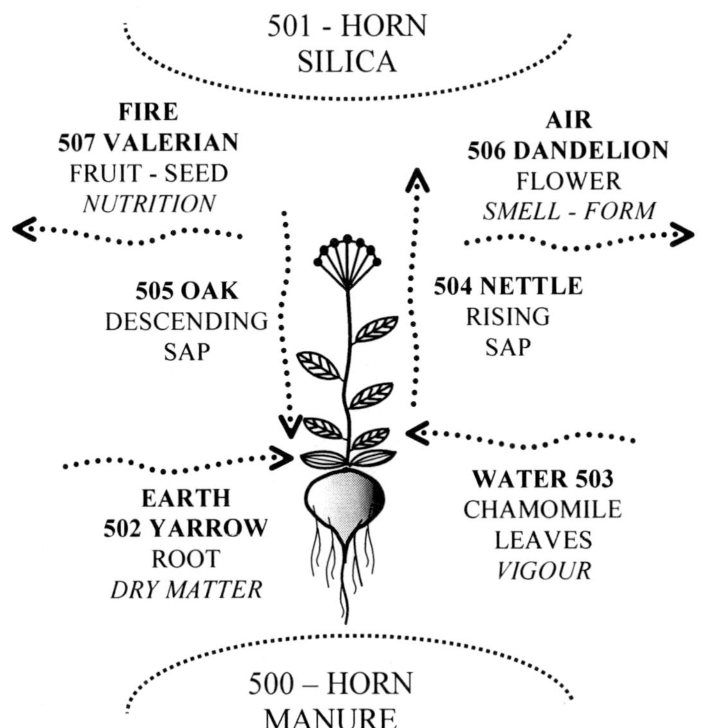

The plant and the seven vital forces

It is up to us to rebuild balanced and natural environments, and we also must try to bring - in a natural way - the forces necessary, using only preparations to compensate for the particular imbalances brought about by forces outside of our farms. Now we have all the information we need to do just this.

In the remainder of the paragraph Steiner submits another link to the Koberwitz delegates. He says that this link between mammals and shrubs is a "*delicate and subtle*" one. Shrubs are usually connected with Mercury for their strong ability to develop regrowth, but also with the forces of Mars for their hardiness. The relationship between the shrubs and cows must be sought in connection with the latter's own connection with Mercury, the planet of healing. It is no coincidence that the central process in a threefoldness is called *Mercur*. The cow is the *mercur* element in the rural animal world (between the bee which carries the solar forces, and the earthworm that bears the earthly-lunar forces), and this is why it is strongly recommended on a farm.

Another link, this time at the level of the forces of Mars, is reflected in the fact that beef is linked to the forces of the red planet, which we have seen is also influential on the hardiness of shrubs. Dr Steiner adds that the shrub has the capacity ... "*to have a wonderfully regulatory effect on everything else they* [mammals] *eat*".

The following paragraph is the twenty-third ("*If we trace ...*") in which a link is made to the lunar forces. In fact we are talking about fungi and how they affect plants and - in as much as they affect bacteria - also the lower animals. The world of mushrooms is then placed in connection with the pond.

In recent paragraphs Dr Steiner has established the relations between the animal and vegetable worlds at the level of the two Fire planets (Saturn and Mercury) and two Water planets (Mars and the Moon). In the two Fire planets we can see the forces of the origin of Life and healing, while the two Water planets are connected to the biological aspects of life.

Continuing to comment on the twenty-third paragraph we encounter a phrase that acquires its proper meaning only if properly integrated. This is the sentence that reads: "*You will then experience the remarkable fact that if you have even a small area where mushrooms are growing, their relationship to the bacteria and other parasitic creatures will keep those creatures away from everything else. The mushrooms have a much stronger relationship to these creatures than the other plants do.*"

Let's begin with clarity that fungus has the task of breaking down organic matter and the task of the pond is to radiate the lunar forces suited to support fungi and bacteria in this irreplaceable role in the soil. Then the worms unite the remains with the mineral component in a colloidal form and this is the soil humus.

It is time to understand how the pond acts. (We refer of course to those climatic conditions in which we usually operate - the temperate climate.) Immediately it is important to be clear that in order to act as such a pond, the water must be at least two and a half meters deep.

We can also say that the function of the pond is to activate the fungi and, if properly organised, these active fungi will act under the soil surface to decompose the organic substance. The problem arises when - for a variety of reasons including environmental reasons - the pond brings excessive lunar forces that activate the fungi that grow above the ground - a pathological condition in the plant world. To better understand all this we can assist ourselves with a diagram.

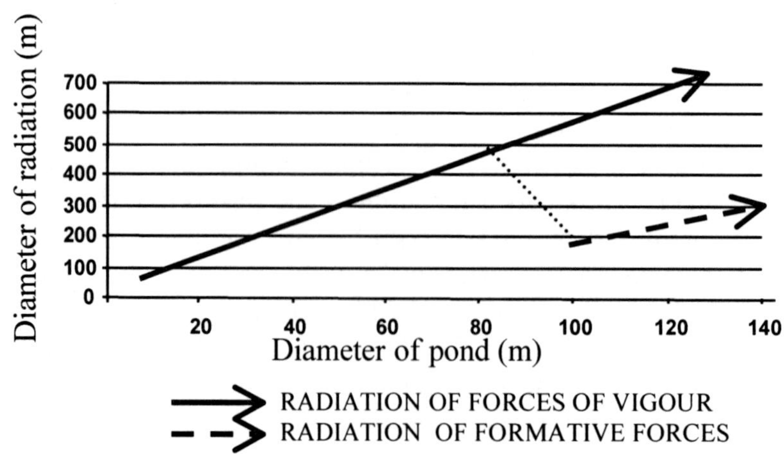

The capacity to spread their influence is directly proportional to the area of the pond up to a certain size. Beyond this the influence spreads over a proportionally reduced region. More precisely, the forces of vigour extend from the pond for a radius approximately ten times that of the diameter of the expanse of water for a diameter from 50 to 100 meters. For larger ponds something strange happens at around 100 metres diameter. Beyond this the radiated forces change in quality and transform themselves into a bridge between the plants and the formative forces. The area covered by this is no longer ten times the diameter of the pond, but twice the diameter.

To help us understand, let's make an example of another kind. We can refer to a man and say that until he reaches the age of 21 years his food is primarily diverted toward his physical growth. Having lived beyond that age, the man's interest no longer simply revolves around his stomach but he also begins to develop his inner world that is more connected to the spirit. Similarly, a pond up to a certain size acts only to stimulate vigour, but when it becomes a little larger it begins to promote a dimension of quality that helps plants to create a link with the planetary forces and the zodiac.

Another schema may help us with this. We can trace a circle at the centre of our plan representing a pond of 100 meters diameter. The central part of the pond is 50 meters in diameter and radiates forces of vigour for 500 metres, but a pond with a diameter of 100 meters radiates formative forces for 200 metres.

Given that we want to establish our farm on sound principals, we may now ask where the garden will be situated and where we will establish the orchard in a landscape similar to that depicted in the drawing. Well, the orchard should be placed where there are more forces of quality and the garden where there are stronger forces of vigour. Therefore, the orchard will be planted in the inner circle (200 m diameter) and the garden in the outer circle (500 m diameter). Obviously our reasoning applies only to large ponds over 2.5m in depth. Otherwise, they only produce forces of vigour.

If we return to the diagram above we can say that the area influenced only with the forces of vigour (1:10 ratio between diameter and area effected) is that area in which there is a greater connection with the interior planets, while the outer area (in the ratio of 1:2) represents the connection with the exterior planets and the Zodiac.

The centre line of the diagram represents the Sun and the inversion in towards the line (point of inversion) can be assumed to be the first change of space and time. This is compatible with the fact that in the potentisation of our homeodynamic products, the various times we use to be in resonance with the planets are affected by this reversal. It requires an increasing time to connect a product with the interior planets up to the Sun, but then these times fall before growing again approaching the periphery of our system.

Also in this centre, talking about space and time, we said that, coming to the Sun, three-dimensional time - consisting of past, present and future - turns into *duration*, while spatially we enter into the etheric Light that generates formative forces.

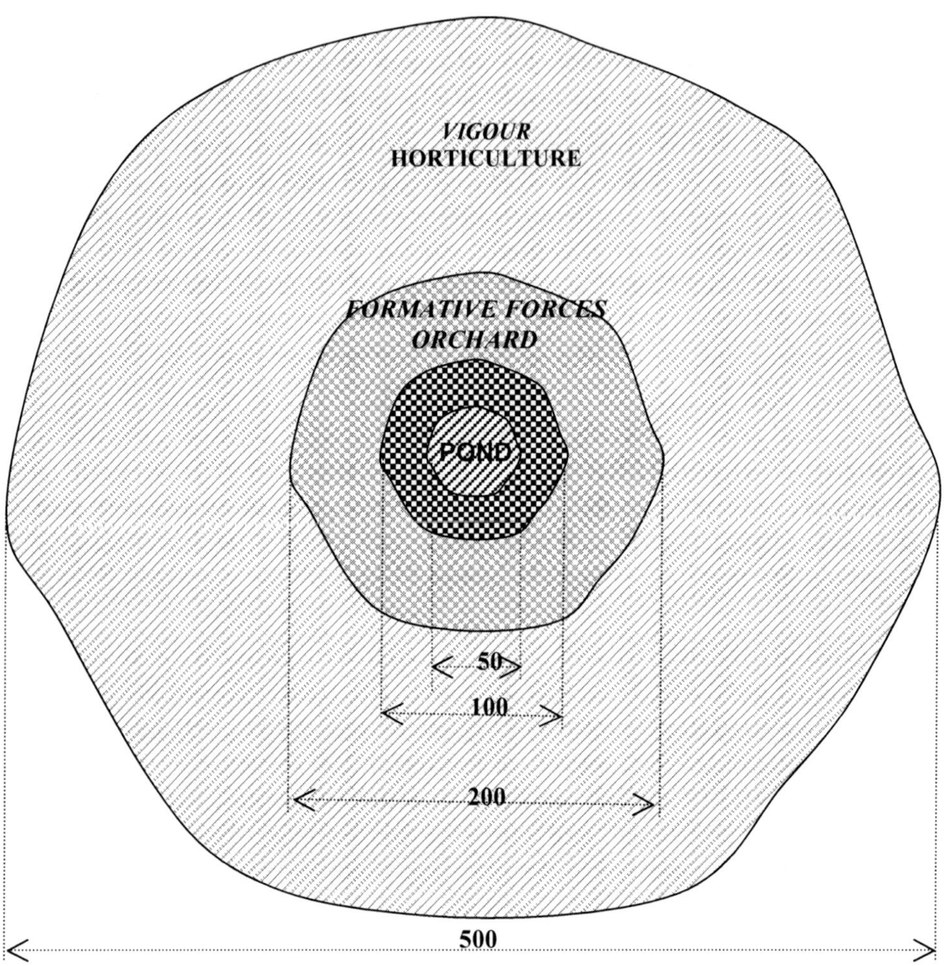

Of course, at first glance, no one would suspect that a pond can have the functions mentioned, although we all know that water of the pond is composed of hydrogen and oxygen, and we know that hydrogen is the chemical element linked to heat and cosmic life (Zoe), while oxygen is more connected to the biological life (Bios). Therefore the being of the pond can indeed be the bearer of all the forces and functions about which we have talked.

Twenty-fifth meeting

We are preparing to complete our study of the seventh lecture and we will see how, towards the evening of June 15th 1924, Steiner introduces the topic that will receive fuller discussion on June 16th, in the final lecture. This is the subject of animals, considered especially in their relationship with the plant. So let us continue reading from paragraph twenty-four ("*The correct balance ...*").

In this paragraph Steiner lists the components of the agricultural organism and then devotes a few sentences to emphasise how counterproductive it would be not to create this organism, "*even if this means a slight reduction in your tillable acreage.*" Otherwise " ... *the resulting loss in quality will far outweigh the advantage of being able to cultivate a larger area at the expense of the other things.*" In other words we could say that even economically it is not sensible to increase the amount of arable land at the expense of the presence of the various organs which should compose the organism, because the imbalances that will result would be such that to try to remedy them would cost much more than would be recuperated from the short term increased production.

The last sentence [of the Italian translation] contains a term that we have already found in previous lectures. It is the word *nexus*[50], or complex of relationships. We believe it is worth spending a few words to put more focus on the meaning of that term in this context. We just said that a farm estate should contain certain components. We should try to understand the '*who*' of each of these, or rather the spiritual element which is their inner essence and which - in the Catholic culture predominant in Italy - is called the Holy Spirit. The term Holy, in fact, means 'enshrined' or - more understandably – 'distributed in all things". The method that we use to investigate this spiritual component, which makes something different from any other, is based on the study of forms.

In attempting to clarify a rather unusual concept for modern culture, we have often said that the form can be considered as the 'signature' of the spiritual element that is manifested in the material. Through understanding the form we can go back to the *idea* that is embodied in what we are examining - mineral, vegetable or animal. Understanding the essence of any object is a faculty linked to the development of imaginative consciousness. To try to help those who wish to start to grapple with such exercises we can suggest that one should become familiar with the search for the essential. There is a publication entitled "Therapeutic Thought," which is available to all members of l'Albero della Vita[51]. This is the result of the work of a study group that has engaged in such research.

Coming back to Rudolf Steiner and the agricultural organism we can say, therefore, that if we consider the various organs of the organism, for example the orchard and woods, and try to grasp their '*who is*', we are really addressing our efforts to grasp the Holy Spirit which is dispersed and expressed in these organs. But if we try to take a step further to try to grasp the *relationship* between the forest and orchard, we ascend to a level that embraces both spiritual essences in words and we can get to grasp the Son through nature.

[50] '*Nessi*' - in Italian. In English: a *nexus*, web, or net of interconnections, relationships and interactions.

[51] An association dedicated to homeodynamic work based in northern Italy.

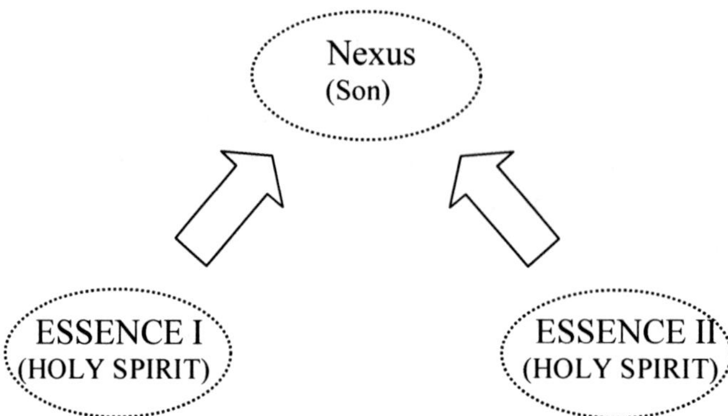

We realise that we are proposing work that requires a lot of application, but we understand how important it is if we think that the Son is the 'healer' *par excellence* and that failure to understand the nexus can be a source of errors that - although committed unwittingly - are no less serious for that.

If, as we said, the first step may be expressed with the words 'read the book of forms', the second step that Steiner suggests there may be called 'reading the Book of Life.'

This is a growth also in the quality of consciousness: if the understanding of 'who is' the plant requires imaginative awareness which leads us to the threshold of the Sun, to grasp the nexus requires inspirative consciousness, which allows us to reach the presence of the Father through the Son. The journey does not end here, because we should also assist the Earth to vegetalise, or, if we prefer, to complete its process of etherisation, and this will enable us to enter the sphere of the Father. The discourse that we have just concluded shows how a simple word, which can pass completely unnoticed, was wisely used by Steiner in a phrase, and can open incredible vistas.

So we have arrived at the twenty-fifth paragraph ("*Now there is still time ...*"). Steiner once again broadens the horizon and invites us to consider the links between different kingdoms, in this case between the animal and plant worlds. Earlier we talked about 'reading the book of Life' and we said that this requires the development of inspirative consciousness, but we could express ourselves in a different way, perhaps closer to those who deal with the subject through their hearts. So we could say that this means we approach nature with love, because life is nothing but Love. This word is often now demeaned with only its lowest connotation but even at these levels it is easy to grasp the link between Life and Love.

Perhaps in the way in which we normally express ourselves we mainly restrict the word to thoughts of the head and rarely consider the fact that it is possible to think with the heart. But if we think well both with the mind and the heart we can find two ways to the same essence. With the mind we grasp thoughts, which, although apparently the result of our brain activity, in fact come from the spiritual world and our brain acts only as a kind of receptor. In our hearts on the other hand we can grasp the 'inner thought', but what are these inner thoughts if not the voice of the world of the Spirit?

Let's now examine the beginning of the twenty-fifth paragraph. Rudolf Steiner says: "*Now there is still time to consider certain points that will bring home to us the fundamental relationship between plants and animals. What is an animal, actually?*

And what actually is the plant world?"

We should not miss the fact that Steiner refers in one part to the plants as a whole and in the other to a single animal. This is because animals and plant are at different stages on the path towards individualisation. The animal, unlike the plant, has the ability to bear within itself the forces that are condensed in the internal organs. The plant does not have such organs within itself so, as we have often said, the plants' organs still reside in unbound Life from the cosmos. We know that only the human individualises completely because we bear within ourselves our spiritual component too. The animal has its own soul incarnated, completing a process that, not surprisingly, is called animalisation. For the plant we can only speak of vegetalisation, that is life without consciousness. When a person is in a coma, that is when they are not aware of themselves or even of their emotions, they are said to be in a vegetative state.

In the next sentence we again find a reference to the interactions between animals and plants and the necessity of understanding this nexus. We will now try to clarify these relationships on the basis of knowledge, but we cannot forget that in nature there is a vital link - on the nutritional level - because each animal needs food to which it is adapted. The problem does not arise if the animal is free, but when we house them in a stable it can no longer devote itself to the search for food. It is the stockman who assumes the responsibility to provide for the animal and the problem of selection becomes real. Indeed, if people have not really understood the links between animals and plants we run the risk of not providing suitable food. Indeed, if attention drops from the low level of food adequate to maintain the animal to the need to make savings, even more unnatural things may happen that can fall so far as to force the herbivorous animals to eat feed of animal origin. On the other hand, if the nexus is the basis of health and healing, not respecting the nexus becomes the certain source of disease. Clearly if these links are not known it is virtually impossible to act in a manner consistent with them.

The interrelations are not only between animals and plants, but also among all the kingdoms of nature. During previous meetings we asked 'who' are the chemical elements such as calcium, silica, nitrogen, and others. Of course if we do not care about such questions, and do not try to understand the links between these entities and the plant world, we are set in a position to make mistakes similar to those mentioned in relation to the animal. If we do not really understand the links there is no difference between spreading living or chemical nitrogen on our fields, or trying to contain the weeds through understanding the imbalances that the weeds reveal rather than to assault them with all kinds of herbicides.

Towards the end of the paragraph is the word 'proper'[52] in the sentence, "*with an understanding of the proper relationship between plant and animal.*" This term underlines that only one who is able to understand the nexus can adopt *right* behaviour in daily work practices such as feeding, that is 'right' in as much as it satisfies the laws of the spirit world. The relationship with nature should be guided by love, because if there is this awareness nature herself begins to whisper her secrets and to guide us in our actions so that we do not commit errors.

The twenty-fifth paragraph ends with the question: "What are the animals?" Paragraph twenty-six ("*You can certainly study ...*") begins to answer this question. Steiner starts by saying that if we want to really understand animals we must consider

[52] 'Giusto' in Italian, meaning *proper, just, right,* or *appropriate.*

them in relation to their surroundings. Let us return to the concept of the nexus between the environment (unbound Life - Zoe) and animals. We have already alluded to the fact that the animal has the ability to relate powerfully with the forces of Zoe that enter the animal to form the internal organs which can be considered as the destination for these forces.

The animal can be seen essentially as a receiver of these forces and this particular activity is carried out mainly by the nerve-sense system that has the brain and vertebral column as its main constituents. This system has two specific receptors in the horns and tail. Unbound Life has a predominantly solar aspect, which the animal receives with horns, and a lunar aspect received through the tail.

Of course we are talking about horns because instinctively we refer to cattle because they are the most common animals on farms. However, we know that there are animals that do not have horns and we understand that they have different frontal receiving organs. For example the horse has its mane, the ass has it ears, others their canine teeth - perhaps transformed into tusks. It would also be interesting to try to understand the relation to Zoe through the observation of the colour of the animal's coat. For example, we may ask why the zebra has a striped black and white coat, while the horse's is sometimes white, sometimes bay, sometimes black or brown and so on.[53]

Meanwhile consider the fact that black and white are the 'colours' of the human because, as the ego bearer, man represents the union between matter and spirit and therefore between Light and Darkness. Rudolf Steiner, in his book "The essence of Colour", tells us that black and white are two imaginary colours. More specifically white is the soul image of the spirit, and black the spiritual image of death. Man, in turn, is the image of the spiritual world and thus, through dying (black), revives (white). We can then add the observation that the horse is the last animal 'expelled' from humanity and that this occurred concurrently with the last faculty that was

[53] See "Man and Mammal", Wolfgang Schad.

gained, which is thought. We can then hypothesise that the horse represents thinking.

The pleasure of riding can be traced back to the pleasure of dominating this icon of thought. The knight, in ancient times, was an initiate and so riding also represented the power to control his thoughts. Because of this, riding can be used as a type of therapy.

In the spectrum of horses the most developed is the zebra. Moreover, of the full range of their capacities the African people have placed less importance on the ability to think compared to Europeans. It is probably compensatory that a superior impetus towards the faculty of thinking had been brought there as reflected in the fact that the horse took a step further in Africa.

Let us indulge a bit more in our digression so we can also see other things about the horse. For example, we can consider the fact that not all horses are used for the same purposes. For example there are working horses and racing horses, and this distinction does not arise only from human choice at the moment they are being used, but from two different modes-of-being of the animal. It is interesting to note that in the two cases mentioned the gestation time is different. If we consider the work horse the gestation time is 342 days, or four synodic cycles of the Moon (27.3 x4 = 118 days) plus the time of one orbit of Venus (224 days). The racehorse instead has a gestation of 333 days, which corresponds to four sidereal cycles of the Moon (109.2 days) added to a cycle of Venus. The Koran reveals that the work of transforming the work horse into a racing horse was undertaken by the prophet Muhammad and this can be considered the last work of spiritual genetics made by an initiate based only upon the knowledge of the forces involved in determining the characteristics of living beings - and not using a material technique upon cells. It is clear that the horse has a special connection with the Moon, sidereal or synodic as appropriate, and with Venus.

We know that human birth is linked to Moon cycles. A normal pregnancy lasts 10 of these periods. In this context it is interesting to note that in this time there are two moments of particular resonance with the forces of the cosmos, which correspond to the completion of the seventh month and the ninth month. The end of the eighth month is not considered a positive time. It is preferable to be born after seven months rather than eight. The eighth month is equivalent to the Fall for the foetus.

Returning to the teachings of Rudolf Steiner, we now hope it is even clearer that to understand animals we need to see them in their relationship with the forces of the cosmos. If we do not, it is practically impossible to understand the 'who' of the animal and, more importantly, it will be impossible to grasp the links with other realms of nature and so act in a healing manner.

The increasingly sophisticated methods currently used to investigate the animals mean that: "*You can certainly study animals by dissecting them and viewing the delightful forms of their skeletons which you can study in the way I pointed out earlier. You can certainly also study their muscles and their nerves, but all this will not help you understand the real significance of animals within the household of nature. This can only be understood by looking at the aspect of an animal's environment with which it interrelates most directly and immediately.*"

At the end of the paragraph the animal is presented as a receiver of forces passing via "*air and warmth*". Steiner is evidently referring to the ethers of Light and Warmth. In the air especially there is also humidity, but the animal does not nourish itself with water from the air and therefore should take it in another way, that is physically through the mouth. The animal, however, has two options to obtain water: the first is to drink and the second is through eating grass that we know is as much as

90% water. Then the animal will also eat minerals, part of which come from the food eaten and another part from direct salt intake. The type of physical feeding that is widely understood, is only a supplemental nutrition to that which comes directly from free forces, what we have often called *cosmic nutrition.*

This capacity also marks a sharp boundary between the animal and plant kingdoms, because the plants are unable to connect so directly with Zoe. Hence the importance that our farmers learn to understand clearly the relationship between the two worlds and are therefore able to choose the type of forage suited to the type of animals they choose to husband. Even if we only consider cattle we know that some animals are bred for meat production and others for the production of milk, but from what we have just said, it should be intuited that the two types of animals need different fodder.

Later in this twenty-sixth paragraph Steiner confirms how an animal feeds itself. "*In its nerve-sense system and part of its respiratory system an animal assimilates directly from its environment everything that comes via air and warmth.*" It is then stated that the warmth serves to form the skeleton, while the air forms the muscles, because in the air and warmth the forces of the Moon and Sun are active, which are closely linked with cattle.

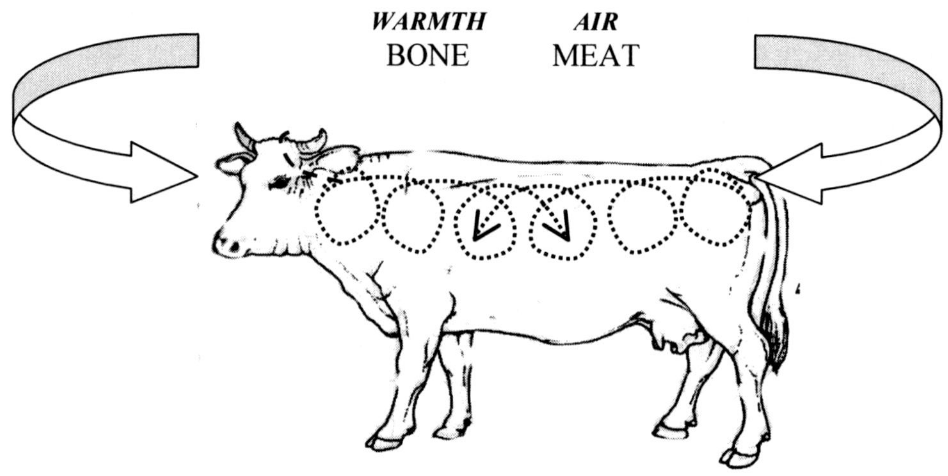

What we said about the condensation of Light and Heat in the flesh and bones respectively also applies to humans, and this should enable us to understand a little better some diseases such as osteoporosis that is certainly linked to a lack of warmth. At the time we saw that Light and Heat belong to the upper pole of the plant, linked with silica, so the treatment of these diseases might be to prepare a decoction of horsetail and to act from the periphery of the body, namely through the hands and feet. This contrasts strongly with the principles of therapy on a chemical rationale, whose practitioners diagnose a lack of calcium which would require administration of the calcium. We know, however, that in the living world calcium is obtained by the transmutation of silica.

As we anticipated a few moments ago, the twenty-seventh paragraph ("*On the other hand...")* discusses the inability of an animal to process directly what comes from Earth and from Water. For this the animal must have a body capable of developing what is assimilated through the mouth.

In this way the two types of sustenance that we have called the *cosmic* and *earthly* nutrition streams were outlined. The topic will be developed further in the eighth lecture, but here we will mention the fact that through the cosmic nutrition stream the animal takes in forces to form the bone and muscle, but curiously not the main organ through which this cosmic nutrition stream is captured, which is the brain - the centre of the nerve-sense system. This organ is the result of what comes into animals through the terrestrial nutrition stream and is linked to the Water and Earth elements. Now you can understand the importance of high-quality forage, which is not there to fatten up the animal, but is needed to form an organ to collect and condense the cosmic nutrition.

What we are saying clearly applies to humans, with the far from negligible difference, however, that the internal human organs are not comparable to those of animals. The latter represent only end points for the planetary forces, while for humans hey are also vital spiritual centres.

As we said, the plant is constituted of water and dry mineral substance that completes the animal's diet. The paragraph concludes with the observation that "*Of course, the assimilation I have just mentioned is more an assimilation of forces than of actual substances.*" and paragraph twenty-eight ("*In contrast, let us now ...*") deals with the latent question: "*What is a plant*?" and continues to consider the part of the nexus not yet covered. ".. *a plant lives an unmediated existence in Earth and Water, just as the animal lives in air and warmth.*"

What Steiner says in the plant is "*a kind of breathing*", is what happens in the leaf, while "*something remotely akin to a sensory system*" is the root. These constitute what we have called the *pole of vigour*, which we know is linked to the Water and Earth elements. In our sketch (see later), where we had written that which corresponds to the influence of horns as 'silica', now we write, 'calcium'. This allows us to highlight a fact that at first sight we would expect to be represented in the opposite way, namely that the bone system owes its formation to the forces of silica, while the brain owes its existence to the forces of calcium. We can also infer that calcium governs the vegetative system in the plant, which presides over reproduction: this is clearly the case because you can succeed with plant cuttings from a root or a leaf, but never from a flower.

Paragraph twenty-nine ("*Now, after seeing ...*") says that after thinking about what we just read, and then following a purely rational logic, we could infer that: "... *a plant must assimilate air and warmth within itself, just like an animal does earth and water.* " The use of the conditional tense, however, allows us to anticipate that things are not quite as they seem. ... "*But that is not the case.*" The next sentence warns us that we will not to be able to investigate spiritual truths only using analogies and inferences. It is fine to investigate the purely outer physical world in that way. At our present level of evolution humanity can no longer be satisfied to investigate nature as it did in the Middle Ages. At that time the alchemists had developed the "science of nature", but now we must turn to the knowledge of things from the point of view of the spirit. Steiner called his science, 'spiritual science' for a good reason.

The third sentence of the paragraph is translated improperly [in Italian]. The correct translation should be as follows[54]: "*Contrary to what happens when the animal receives the Earth and Water elements, the plant does not accept the Air and Heat, but secretes them to live together in the ground.*" This means that the animal receives the unbound Life (Zoe) directly from the skies and carries them within, while the plant lives primarily in the unbound Life of the Earth, of which Steiner spoke in the third lecture when describing the roots of the plant that are interwoven, increasingly subtly, to form the humus.

The next sentence also deserves a correction! The phrase "*Air and warmth do not go into the plant, at least not to any great depth; instead of being consumed by the plant, they are given off - and this giving off process is the important thing" (Creeger Gardner edition)* " should be translated "*... do not achieve substantial interpenetration, but go out and are eliminated from the plant rather than consumed...*"[55] Therefore the plant lives primarily in the root and leaf stage, receiving the life of the Earth, then turns to the Air and Warmth as it goes to flower. However, it is not saturated by the external air and warmth. This is so much the case that we always represented the astral of the plant outside of the plant itself.

To facilitate the understanding of what we are saying, we will reproduce the representation of the plant that we usually use.

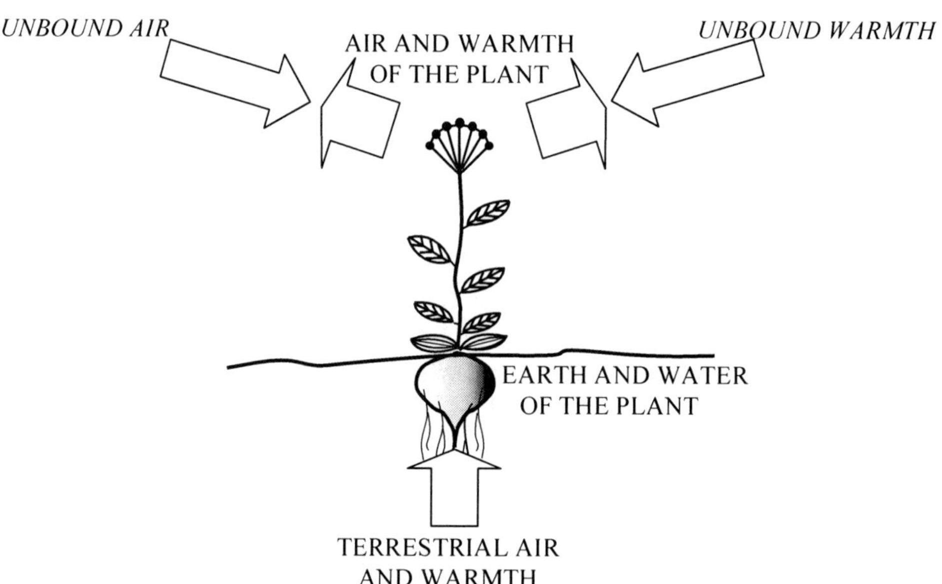

[54] "*Al contrario di quanto fa l'animale quando accoglie gli elementi Terra ed Acqua, la pianta non accoglie dunque l'Aria e il Calore, ma li elimina vivendoli assieme al terreno*" is the Italian version. The Gardner/Creeger edition says "*In actual fact, while animals take in earth and water and assimilate them inside themselves, plants give off air and warmth, inasmuch as they experience them together with the soil.*" George Adams wrote: "*The fact is this: whereas the animal consumes the earthy and watery material and assimilates them internally, the plant does not consume but, on the contrary, secretes – gives off – the air and warmth which it experiences in conjunction with the earthly soil.*"

[55] "*...non la compenetrano molto, ma si espandono; vengono eliminati invece di essere assorbiti ..*" – in the Italian edition. Enzo prefers "*... non la compenetrano sostanzialmente di molto, ma vanno fuori e vengono eliminati dalla pianta anziché consumati...*"

The plant lives in the eliminated Air and Heat - the Air and Heat that are of the Earth. Through this elimination a sympathetic relationship is created with the cosmic Air and Heat that penetrate it from the periphery, nourishing it. What the plant gives in this process of elimination is the perfume and pollens.

Now we might understand the action of orchard trees better that perform the function of capturing the free astrality. These have the characteristic of producing flowers before leaves, proving they have learned to feed directly from the cosmos itself and bringing in the free Life to the farm.

The thirtieth paragraph ("*As organisms plants ...*") sentence begins by emphasising the importance of the process of elimination described in the previous paragraph. The plant can receive only from the outside if it has the capacity to eliminate. Again, the emphasis is placed on the difference between the plant and animals, as the plant is a gift of love, a virginal component of nature open to the cosmos. This contrasts sharply with the animal who retains everything inside of himself in a selfish gesture.

A particularly significant sentence of this paragraph states: "*A plant lives by giving off air in exactly the same sense that an animal lives by taking in nourishment.*" For a plant to give off Air and Heat is to say that they produce and donate their fruit, and the plant lives through this gift. That is why it is so important that people harvest that fruit with the right attitude of gratitude. Reaping the fruits with feelings of greed and with the soul focused only on personal gain, does not please the elemental beings who bring the plant to complete its life cycle. It does not encourage them to continue to be fruitful. Do not forget that the activity of elemental beings is essential at all stages of the life of the plant. Their activity is also vital in giving the fruit its nutritional value (which is linked to the radiating of Warmth, as flowering is linked to the giving off of Air.)

The paragraph concludes stressing that "*Plants give, and live by giving.*" Only in this light can we understand the plants as they are, what they give and what they receive. By understanding that what it receives is the basis for our life, we can also understand how important it is that we not only manure the plant but also support the growth of the plant towards thriving beauty, but above all to support its ability to give Air and Warmth.

With these lines Steiner also enables a better understanding of the importance of some of the biodynamic preparations, particularly 501 (the horn-silica), 506 (Dandelion) and 507 (Valerian), which are elements of the homeodynamic preparation 'Pro fruit'. Especially one can gain a deeper appreciation of their role in the improvement of the qualitative aspects of the plant. Do not forget that the consciousness of the farmer is fundamental to the practice of agriculture directed primarily to support the evolution of plants and animals, so he must do everything he can, in the knowledge that what he can not reach God will complete with His grace. But it is necessary that no effort is spared. Let us not forget that Jesus said, "Knock and it will be opened". Who goes to knock does so because he has faith that on the other side someone will listen and he shall open the door at the first touch. It is only necessary to knock with trust.

We think that the remaining paragraphs do not require any particular comments. However, before reading and remarking upon the eighth and last lecture, we should still share some considerations about animals to further clarify the function of some of its parts. For example, we may ask what role do the hooves have. Meanwhile we can note that not all animals have this part, and also that in different animals that do possess them the hooves are in different forms. For some the hoof is the end of one finger, in others two fingers and indeed in others three or more.

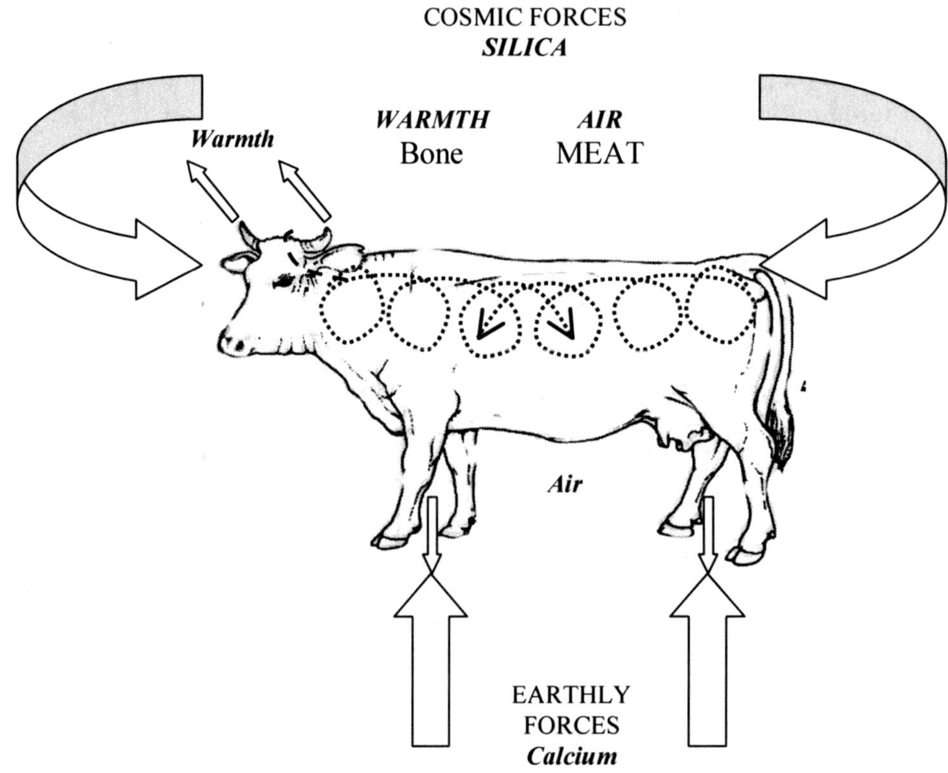

Earlier, we saw how the cow absorbs the forces of free astrality through its horns, and in the comparisons of these forces the hoof can be seen like a route for elimination. But the hoof also serves to absorb the forces of Earth and Water that serve to complete the animal even if they are secondary compared to the previous forces. Their importance is increased in countries where there are excessive free astral forces, as for example in the tropical desert. It is no coincidence that when preparation 500 (horn-manure) was made in north Africa by inserting the manure in the hoof of the camel instead of in a horn, the result was very good.

Regarding the elimination of absorbed forces, we can say that Heat forces are discharged mainly through the horns, and Air forces are discharged through the hooves. This will highlight another important function of the animal. As we have just said, the animal can eliminate excess astrality that, if not eliminated, will contribute to the cause of pests in plants. In light of that statement we can understand how useful is the practice, almost impossible nowadays, of ploughing the fields with draft animals. Each hoof beat on the ground brings the impetus for expulsion of the forces which manifest as pests.

With this last thought we believe we can also consider that we have completed our remarks on Dr Steiner's seventh Koberwitz lecture.

Eighth Koberwitz Lecture

Twenty-Sixth Meeting

At last we are reading and commenting upon Steiner's eighth and final lecture. Compared to the preceding lectures this is actually relatively easy to understand. In part this is because we have already done a great deal of work that has enabled us to approach a different vision of the world: the one we probably had before is commonly regarded as the only valid model. I would like to express some satisfaction that such a large group of people has persisted with this work for twenty-six meetings despite the mental effort it requires. I also realise the significant physical effort some of you have had to make to get to Monselice from so many different and far-flung regions of Italy.

Whoever translated the text into Italian has given this lecture, held on 16th June 1924, the title: "*The nature of fodder.*" In this lecture, Rudolf Steiner draws together some of what he has sown in the seven previous lectures. However, as we shall see, he has added something new into the now-fertile ground and this can be clarified without too much difficulty.

The first paragraph begins with the premise that what is being commented upon can be completed at different levels, depending on the needs that arise during the lecture. We believe we can begin in the same way, because in this lecture we will touch upon the topic of food. This could be limited to the animal world but can also be extended to the subject of food for human consumption. We cannot cover the problem in the same detail as the special course held a few years ago in Padua. Nevertheless, the subject lends itself to considerable investigation, in numerous details and in great depth as the need arises.

The first paragraph emphasises the fact that in the subject we are addressing, although there are common rules of a general nature, the way one tackles each case is "*subject to a great deal of individualisation and personal discretion.*" It is a topic that requires a deep knowledge of spiritual scientific laws to determine the precise relationship between the animal and its forage, and this does not bypass the requirement of understanding the world of the plant. Only in this way will it be possible to choose, as appropriate, food more suitable for the meat animal, or to increase milk yield, or for breeding creatures.

Of course this applies to more animals than cattle so we will have to acquire a good level of knowledge of all other animals we are called to husband. It is not an easy task because the ability to recognise the relationship that we talked about in our last meeting requires more than biological knowledge because, on that basis alone, the risk of committing errors is really high. Bring to mind that in our last meeting we emphasised that the animal builds its body by turning the Heat and Light absorbed from the outside into flesh and bone, while the plant lives outside Light and Heat. The polarity between animal and plant is so clearly evident we can even think of the plant as the natural completion of the animal.

Having said that, another set of questions arise which are linked to the way the plant becomes food for the animal. The plant can be used fresh or dried, can be used in whole or in part, or the foliage or the flower or fruit of the plant can be used. These are all choices that require knowledge of the difference - in terms of forces - that the various parts of the fresh or preserved plant are able to bring.

It is a similar problem to that asked about eating raw, cooked, or dried fruit. If we address the problem as a physico-chemical issue it is clear that there will be little difference between the three, apart from some difference in the quantity of water. But if we try to grasp the difference in terms of forces, the various methods are very distinct.

A fruit which has been dried satisfactorily, namely with the use of the Sun, does not lose significant vitality. Life is contained in the seed and it does not undergo significant changes during drying. What is heavily amended is the ability to support consciousness, because when fruit is dried by the forces of the Sun it is further 'matured' and increases in those qualities related to the forces of Light and Heat. Let us not forget that it is said that the food of the saints is 'bread and dried figs'.

The choice between fresh or preserved fodder is similar. Fresh fodder, because of its higher moisture content, is better able to support biological vitality in animals, while hay, like dried fruit, supports the processes of consciousness. In animals this results in a better ability to grasp the multifaceted impulses arising from the individual principle of the species. These impulses are commonly called 'instincts'. One of the best indicators of connection with and government by the individual principle of the species is a higher level of health indicated by a harmonious function of all the vital processes in an animal. Anyone who has had some experience of farming knows that an animal fed with hay produces a smaller volume of milk, but this is offset by its higher quality. What is more, it supports higher consciousness in those humans who consume the milk. Of course, if the farmer reckons only upon the quantity of production with its link to a greater financial profit, he will have no interest in the distinctions in nutrition that we are considering.

In this paragraph one can also heed a call from Dr Steiner to avoid the logic of those who wait passively to receive all the lessons that could develop us. Everything must be considered as something like a seed from which it is possible to get a new plant, but which cannot make this new plant alone. Received information must become something extremely intimately involved in us. This can then form the basis of an opening for a perception of the world around us that is illuminated by what comes from the spiritual world.

This leads us to note that issues related to forage for livestock are raised without even having understood the real parameters of the issue such as those we have just mentioned. This observation is also the content of the second paragraph ("*Just think how little insight ..*"), where Steiner confirms that without such understanding it is not even possible to understand what *animal nutrition* really means.

In the third paragraph ("*As I have already pointed out ...*") the basic terms of the subject are better defined. The current orthodox understanding of food takes into account only matter. According to this one must ingest a certain amount of protein, vitamins, carbohydrates, fats, etc in order to restore what the body is assumed to have consumed. (Now we are talking of the animal, but the problem is quite similar for humans.) A discipline has even been invented to examine the problem from this point of view and this has been called a science. However, almost every day we are bombarded by information - often contradictory - about the 'correct' way to eat. We are educated by slogans and phrases that have become so common that they are senseless such as: 'We are what we eat'!

Whoever has approached the problem of nutrition with some spiritual-scientific knowledge knows that things are really completely different. Steiner said: "*Basically, people imagine that first the food is there outside, then the animal eats it, and then what the animal can use is deposited inside the body and what it cannot use is excreted.*" This way of seeing things is very limited because the food that we eat first undergoes a process of destruction of its form, so that the 'subtle' nutrients are freed from the material component that served only as a support. These elements, during their passage through the stomach and intestines, undergo a genuine process of dematerialisation, which is often defined as 'etherisation'. In this condition they rise up to the top of the body where, above the atlas bone at the base of the skull, they undergo a process of rematerialisation, thus sustaining our brain. In this sense only our brain is directly linked to what we eat.

Through the senses the brain, as the centre of our neuro-sensorial system, 'observes' the unbound etheric forces that surrounds us like a heavenly gift and which, upon being condensed, give substance to our body from the neck down. The importance of a high quality diet is the ability to build a brain capable of adequately providing this '*cosmic nutrition'*.

The idea that food directly provides the energy needed to live and carry out our activities is equally unfounded. This understanding flows directly from the mechanistic conception that digestion works like incinerators that destroy the material and in so doing release energy. In fact our strength and vitality are not more or less effectively extracted from foods, but are the result of a process resulting from the need for our digestion to destroy those things we eat. For instance the structural form of the plants we eat is needed for the living plant but are inappropriate to continue within us. The destruction of this vital organisation strongly commits our etheric body, which is the bearer of our life forces. Our etheric body has been individualised and organised in a specific way and cannot tolerate foreign living forces. The etheric body, however, is governed by laws that are complementary and opposite to those of the physical body, so the etheric body is strengthened by this commitment. The more vital and living the food that is ingested the more the etheric body is called to engage, and so much more will it be reinforced. We all understand that eating meat is, from this point of view, the least adequate to support our vitality when compared to that of plants.

Precisely because the effort of destroying plants' greater vitality is so important and so challenging, we understand how people who are debilitated from exhausting physical activities or disease, might decide to eat raw vegetables. However, it is advisable to cook vegetables in these cases to depress the very vitality of foods making them more easily digestible by the stressed etheric body. One can then build oneself up again during and after the convalescence.

Food can be considered of the Earth element. When this is properly chewed it is reduced to pulp and this can now be considered as being of the Water element. In the stomach, as a result of the acids, the part that will be fully digested is transformed to Air. Finally it reaches the intestine as Fire. We know that Fire and the warmth which is bound up with it are unable to be retained by walls and therefore, in our intestine, what is left of the food can easily go through the intestinal walls and transform itself into Warmth ether. During the materialising path it is enriched becoming first Light ether, then the Alchemical ether and at the end Life ether. Because we know that Life ether is tied to the Earth element, the account is correct because our head is the most hard and dead part of our body: the most earthly.

As we mentioned, the vital aspect of food is dealt with by our own etheric or life body that demolishes the food's own organisation or structure, resulting in active forces within us such as movement, a capability to support all our vital processes which result in circulation of blood, as well as allowing us to move in the outer world.

Only the portion of food that cannot be used is finally expelled from the body. Clearly, as we progress this will be treated in greater detail.

The fourth paragraph ("*People talk about* ... "): here Steiner reiterates his insistence that a crucial error is made when comparing nutrition to combustion. Clearly this comes about because of a widespread and increasingly materialistic mentality. "*Combustion is a process in non-living nature.*" Combustion in the mineral world involves the destruction of the same physical substance and the cessation of all vital processes. But unlike the combination with oxygen that is happening inside animals and humans, it does not welcome in any new Life and integrate it within the vital processes of the organism.

We believe it is important to note that the paragraph ends with the word '*sentient*' referring to the process of the internal transformation of food. The term refers to a process of consciousness and suggests that this is relevant to the type of food that we should ingest. For example, there is no doubt that the ingestion of foods rich in sulphur has the effect of darkening our consciousness and that, if we understand our evolutionary process, we must learn to make a limited use of such foods. Sulphur is a strong stimulator of the expansionary processes (which are not surprisingly called *sulfur*) and these exercise their first influence in our intestine and thus cause swelling. Later a similar effect occurs in the brain, obviously not on the physical level but on the level of stimulated functions and level of consciousness. Moreover, we know that there is a close relationship between the intestine and the brain, as the latter represents a transformation of the former. We must bear in mind that some of what we eat stimulates our lower soul faculties - the sentient-soul in fact - and so we must learn to behave accordingly.

Some sulphurous foods, such as garlic or onion, should not even be considered as food, but as remedies to be taken only when there is an actual need and then in just enough volume to stimulate the internal processes.

The fifth paragraph ("*When people talk* ... ") should present no significant problems and does not add to what has already been said, so there is no need for further comment. We will only repeat what Steiner says in the penultimate sentence: "*It is a typically modern phenomena for people to do things that are totally at odds with what is happening in nature. That is why we need to spend a bit of time looking at what's really involved here.*"

Steiner is suggesting that the only master is nature and that perhaps we should stop reading books and start reading the books of nature, which are, as we have already taken the opportunity to say: the book of forms, the book of life and the book of consciousness. These are often needed together, perhaps stimulated by a question that requires knowledge written in these books. When we ask the question of 'who is' a particular plant, we can formulate a response only if we have some knowledge of the book of forms. When we try to plan companion planting to help plants, or to understand how the life processes can be stimulated by the forces brought from other plants, we approach the book of life. The same is true with the present work on the relationship between crops and animals. The further step is to ask whether the way we use the knowledge derived from these two books is right or not.

Of course, we cannot buy or borrow these books. We read them within us when our busy interior is calm and we can hear what nature continually whispers and teaches us.

In paragraph six ("*Let us consider something ...*") Steiner reconsiders the plant and reiterates that this has a physical body and an etheric body, but not an astral body. The second sentence reads: "*The plants themselves don't reach the stage of astrality, but it hovers around them*". Particularly interesting is the third sentence of this paragraph where we read that ... "*when a plant enters into a particular connection with the astrality, as is the case in fruit formation, something nourishing is produced which can then support the astrality that is in the animal or human organism.*" Steiner has spoken about the connection that the plant may make with the free astrality, ie with the unbound Life, which is reflected in the formation of flower and fruit. Here he emphasises the fact that this connection bears fruit in the ability to sustain the astrality of animal and man. We believe we can better grasp the concept and clarify that astrality in animals and humans is the basic stuff of consciousness - the soul. The need is emphasised for these connections to have the opportunity to act fully to enhance the nutritional value. We can, therefore, understand how detrimental it is to harvest fruit that is not fully mature. It is not so much a matter of taste because we know that some even prefer the taste of unripe fruit, but it highlights the impossibility of unripe fruit to stimulate our highest consciousness. Instead we understand that eating ripe fruit supports soul processes and induces the soul, adequately reinforced, to be manifest. We can also say that eating unripe fruit is a sign of a person's reluctance to move in the world of feelings.

Even more useful for supporting the consciousness is cooked fruit, because during cooking the fruit undergoes further maturation. Similarly sun-dried fruit is suitable to sustain a high level of consciousness. Of course when we talk about 'fruit' we refer not only to literal fruit such as apples, pears, peaches, and so forth, but also the fruit of horticultural plants. To clarify further, let us say that, in this context, the fruit is that part of the plant that bears seeds.

The paragraph concludes with the observation that in order to understand fodder it is also important to consider the issue of the capacity of the plant to support the animals' astrality, obviously after having focused on the being of the animals concerned.

The seventh paragraph ("*There is not such ...* ") begins to extend the exploration of the animal and right away, in the first sentence, there is the observation that in the animal's body there is no such clear three-foldness as in humans. In animals the nerve-sense system and the digestive-limb systems are particularly well developed, while the rhythmic is rather strongly influenced by the other two systems and not so well defined as its own entity. We might wonder if there is a process that originates in the metabolic-limb system and that reaches right through into the nerve-sense sphere. It is the process of formation of horns. It is no coincidence that pregnancy causes the formation of a distinctive ring on the cows' horns. Also the process of formation of stag antlers, as we have said when we discussed the yarrow preparation, is closely related to the pregnancy of the female. Conversely, bee poison is an example of something starting from the nerve-sense system that reaches into the sphere of metabolism. The poison is linked to the higher consciousness of the bee.

Bee-sting poison is the result of a transformation of oxalic acid that the bee has carried into its metabolism. It turns into formic acid in the thorax, and poison in the head. In the head the poison is also linked to a bee's ability to see in a 'subtle' way through the three 'extra' eyes on its forehead. Furthermore, the colony is held together by its own distinctive flowing astral poison which expands through transpiration, and the same poison - as an expression of the colony - is also responsible for swarming.

The new queen's poison has different characteristics from the old queen and therefore brings a different 'astral smell'. This will be binding for the new family but it is also incompatible with the old family. In order not to surrender their identity the bees leave the hive.

We also suggested that the weakest element in the animal world is that of rhythm. But there is an animal that is the representative of rhythm - the earthworm. For this reason we usually say that the three creatures that should be present in every farm are bees, earthworms and cattle.

Of the higher animals the animal that has the most defined rhythmic portion are the cats. This is why the lion is considered the king of the forest. But in general we can say that if the animal has a particular weakness in the sphere of rhythm, it is necessary to take something from outside that might support this sphere. Animals find this in their relationship with their farmer which then assumes its rightful importance.

We can add a further consideration that will allow us to understand another aspect of the animal better. In fact, as each being is threefold (and we have seen that animals are weak in the central or rhythmic sphere), so is each of these three parts in itself threefold. This is particularly clear in regard to the head of the animal where we can recognise a lower part (mandible), which represents the sphere of metabolism and movement (through which the food enters that is worked upon in the stomach and metabolism), a central rhythmic part linked to the nostrils (through which the air enters that goes into the rhythmically working lungs), and finally, on the top, the seat of the brain, the epitome of the nerve-sense system that is, so to speak, the head of the head. It is interesting to note that the rhythmic aspect of the head, the area of the nose, is particularly vulnerable and prominent as if to compensate for the weakness of the primary rhythmic sphere of the animal.

Unfortunately, these things normally go completely unnoticed so we do not see even very striking things. For example, there is no doubt that the elephant's trunk is a noteworthy feature, but few understand the real reason why the elephant has such an extraordinary feature. It may be regarded as a curiosity but people do not try to understand the reason for such a strange phenomenon. The problem lies in the fact that the animal is considered only from the physical point of view. The reality would be clear if we were able to perceive the dimensions of the elephant's etheric body because the etheric head exceeds the size of the physical head and reaches as far as the trunk. (Moreover, remembering that the etheric body is the seat of memory, an etheric body that is so great supports the proverbial memory of this animal.)

It's not sensible to be too discouraged, because it isn't useful, but we must recognise that we know very little of the subtle aspects of nature, including the etheric. The odd thing is that our generation should be the one to encounter Christ in the etheric plane, but it seems legitimate to ask how we can possibly recognise the magnitude of Christ in a reality in which we move with uncertain steps. We hope that at least this awareness will stimulate us to a greater commitment. We should not be satisfied with the limited but predominant modern view of the world. Unfortunately, we are mostly educated for basic utilitarian goals so, for example, the various types of timber differ not so much for the forces that they can bring, but only for how they may be exploited in a purely material sense. The same is true for animals of which we are taught, for example, their specific and precise capacity to transform fodder into milk. However, we are told nothing about why the cow transforms fodder into milk at a lesser rate than the goat or why sheep, a close relative of the goat, at an even lower rate. No one is concerned to understand what kind of relationship exists between the goat's hair and its milk, and between the fleece of the sheep and its milk. Such questions should not leave us immobile, but should lead us to recognise our ignorance and lead to a greater commitment, in an attempt to become worthy of being called farmers.

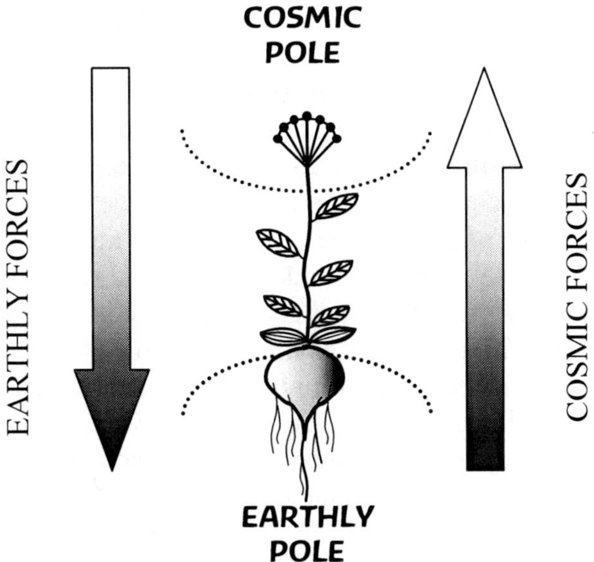

The eighth paragraph begins digging deeper into the processes of nutrition. Already in the first sentence we find a reference to what we anticipated: "*Now, everything in the way of substance in the head system consists entirely of earthly matter.*" Earthly matter is the food that was swallowed and is called *terrestrial* as opposed to those subtle forces that enter us through the senses, which are called *cosmic*.

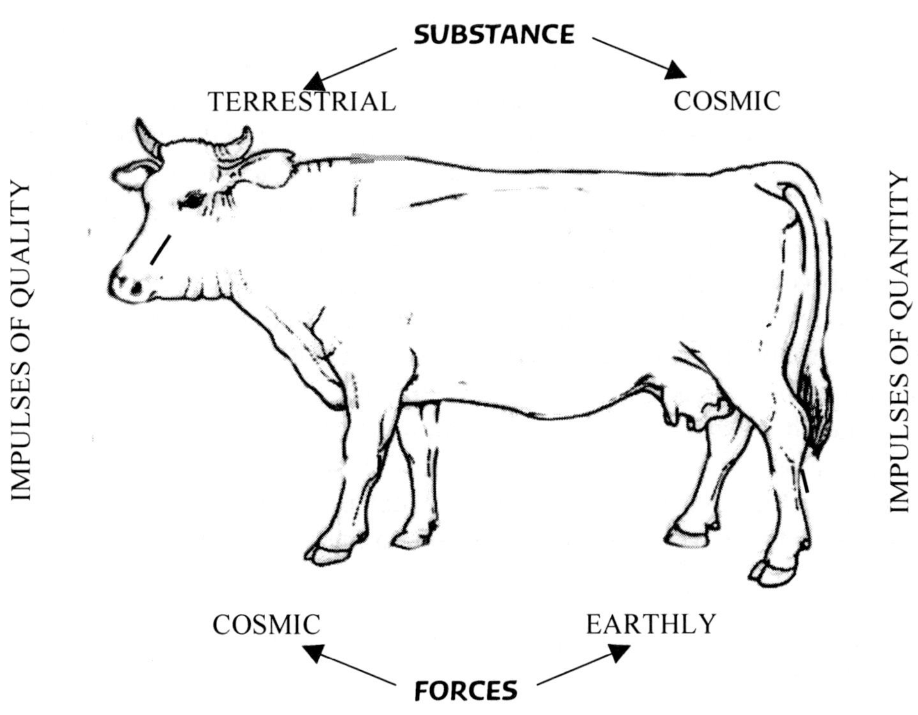

It is curious that our brain carries out more advanced and more spiritual functions such as the processing of thoughts, undertaken within an organ made of *earthly* matter, while the rest of the body, formed of *cosmic* substance, carries out such earthly tasks. This type of inversion between substance and quality of forces should not be a novelty for us because we also found this in plants where the lower part is made of earthly material and cosmic forces, while the upper part is cosmic material shaped by earthly forces.

Pointing out that there is a complete reversal between man and plants, we see that the cosmic forces govern forms guided by a principle of unity, while the earthly forces express diversity, multiplicity or fragmentation. The structure of Man's head is subject to the unifying principle but as we move toward the sphere of the will, which is more subject to the force of gravity, there is a growing multiplicity. The arms and legs start with one long bone, and then divide into two and reach the hands and feet ending with five fingers formed by many bones of ever smaller dimensions.

For the plant the same thing happens as the unity of the root and trunk contrasts with the division of the branches, increasingly accentuated as they ramify towards the canopy, reaching the multitude of leaves. So in the root we can recognise earthly substance in cosmic forms, and in the flowers we see cosmic substance in earthly forms. The root can be considered to be the daughter of humus, whereas the flowers and fruits by contrast are predominantly children of the cosmic forces. When we use the 'Fruit Plus' to improve the quality of fruits, we are merely emphasising the cosmic nutrition. A typical root that corresponds to what was just said is the radish, and the flower is the inflorescence of the *umbelliferae*.

This contrast between substance and form or function and force, is not typical of anthroposophical thought alone, because it belongs to the alchemical world. But even earlier we find the same contrast in Chinese culture. In the alchemical world the contrast is between the *sal* and *sulfur* processes. If we adopt this model to look at the physical head, it is hard and dominated by the earthly physical processes of death, representing the *sal* principal. But the head is the bearer of functions related to *sulfur*: with thoughts one can grasp aspects of God, or one can easily and in an instant move thousands of kilometres. By contrast, the physical members expressing division (*sulfur*), cannot easily help us escape our earth-bound reality and allow only limited movement through the physical world (*sal*). The Chinese express the dynamic in terms of Yin (corresponding to *sulfur*) and Yang (corresponding to *sal*), but despite the different terminology they are essentially expressions of the same considerations.

In the third sentence of paragraph eight there is a reference to the embryo that "*must be organised in such a way that the head acquires its substances from the Earth.*" If we are referring to the plant world the correspondence is to seed chaoticisation. When we give the homeodynamic 'seed baths'[56] we are merely guiding the embryonic organisation of the plant. Once the plant is established it is harder to direct development, but in the embryonic phase one can create conditions so that the group I can manifest the characteristics that normally remain unexpressed. This enables us to offer an effective influence upon the growth of plants.

This way should not be regarded as seed manipulation, because we are merely unblocking the imprinting of the plant's 'group I' so that the plant can manifest more of its own forces. Otherwise this can be difficult because of the various forms of pollution that have to be overcome in its approach to the seed. As the embryo receives

[56] This is a homeodynamic spray that is applied to seeds bringing a specific message to the etheric body that then informs the growth of the future plants.

nourishment from the mother's blood, in the same way during its early life the plant receives nourishment from Mother Earth through the soil, which corresponds to the blood of the Earth.

In the rest of the paragraph we meet considerations that we have already anticipated about the substance that fills out all that is below our neck. In particular one sentence refers to the function that is stimulated in us by overcoming the etheric aspect of vegetarian food, causing the reaction of our etheric body that supports all the forces of movement. "... *what the animal eats is simply there to develop its forces of movement, so that the cosmic factor can be driven into its metabolic limb system - right into its hooves for instance. These parts of the body are filled with cosmic substantiality.*" Then Steiner adds the ninth paragraph: "*With respect to the forces, however, the reverse is true. In the head, we have to do with cosmic forces since the cosmos is perceived with the senses, which are located primarily in the head. In the metabolic-limb system, on the other hand, we have to do with earthly forces; with cosmic substances, but earthly forces. Just think how we are constantly engaged with earthly heaviness when we walk, how in fact everything we do with our limbs is connected with the Earth*".

We believe that in light of what we have said the contents of these phrases can be understood well enough. In these phrases we continually hear Dr Steiner talk of substances and forces. We must now acknowledge that forces are practically never taken into consideration in the world today. This is the case even in movements such as organic agriculture that intends, in various areas of human endeavour, to avoid damaging nature and man. However, their attention is focused only on the use of substances that, on the physical level, are not harmful to the environment. They do not engage with the problem of the forces that they bring with them. Such an approach, despite its laudable intentions, cannot provide a sufficient response to the problems that we come across when involved in agriculture, farming, housing, power and so on.

Without adequate knowledge of the world of the spirit, and hence of the subtle forces involved in life, it is impossible to tackle any problem without taking the risk of incomplete solutions or even solutions inappropriate to support life, even if they may seem appropriate on the physical-material plane. We can say that the organic world pays attention to the biology of nature, but to implement an ecology that we might define as spiritual, we must also take account of the world of forces, which manifest in the world of forms, since the form is how the spirit manifests itself.

Twenty-seventh Meeting

The Italian translator of the eighth lecture of the Koberwitz agriculture course gave it the title, "*The nature of fodder*". As we have already seen the lecture is not only about fodder, but also about the world of the animal and nutrition in general. However, if one of the functions of the animal is to be a conveyor of forces, especially through its manure, the problem of choosing the correct way to feed them becomes crucial and the title rightly stresses the central point of the lecture. Indeed it is easily comprehensible that only when an animal is fed properly will it give manure of good quality which, in turn, can bring the forces of Life that sustain the life of the soil and of the plants that grow there.

We certainly will not open up the whole problem of fertilisation once more because it was largely handled by Dr Steiner in the fourth and fifth lectures. But it may be helpful to recall a sentence from the second paragraph of the second lecture, in which we read: "*From the perspective of an ideal farm, any fertilisers and so forth that are brought in from outside would indeed have to be regarded as remedies for a sickened farm.*" This sentence clearly indicates that it is not enough that the manure is of a high quality because it was obtained from well-fed animals, but that it should be produced on the same property by following the principles of sound agriculture. In fact when the animal eats forage, as we will be able to consider, it extracts only a part of the plant. This is especially the case with regard to the etheric component and what connects to the astral and spiritual worlds. So what remains in the manure as life and the capacity to connect to the world of the 'Group I' is particularly suitable for plants and the animals which feed upon them if the animal is fed plants that originated in the same place, and that grew up with the forces of that same agricultural holding. In this way you can create a virtuous cycle so that the animal is able to produce manure suitable for improving plants, which in turn feed the animal better, which can produce manure of even better quality, which itself does not fail to make further improvements on the plants, and so forth.

It is virtually impossible to achieve this today because there are rarely suitable conditions for each farm to have its own animals. Animals, for reasons of short-term profit, are concentrated in a few large companies that are designed only with maximising production and profit in mind. For this reason the problem of nutrition imposes itself in all its dramatic and direct consequences upon the animals, but also with implications for the health of those who eat the meat or other products from such animals.

In these conditions it is often more likely that growers will not use manure at all and will seek a totally different method of bringing fertility. It is possible to use plants that can greatly facilitate the formation of humus, such as certain types of Mexican beans, or even to adopt a totally different approach and use methods that bring fertilising forces without the need of any physical support. The indications of Rudolf Steiner will help us to focus even if we have departed from the ideal way to work. At least we may avoid further mistakes. Unfortunately, the cost of maintenance and care of animals has become so high that it can be unsustainable for small farms. Therefore, it is necessary both to accept the reality and to recognise the problems that come with departing from the ideal; understanding can help us minimise further errors.

The first sentence of the tenth paragraph ("*If you are going to use cattle ...* ") is almost certainly unclear to inexperienced readers. Nevertheless, a whole area of knowledge - that we believe it is appropriate to expand upon - is summed up in a few lines. Of the nourishment that comes from the animals' fodder (and this applies to human food too) the vast majority of the coarse part only serves as a material vehicle that is eliminated in the faeces. This coarse vehicle carries a number of more noble components (etheric, astral and spiritual) and these undergo a process of liberation at different levels during the process of digestion. On the other hand the physical food undergoes a process of disintegration that progresses from the Earth element (in the mouth) to the Fire element (in the intestines) and is then conveyed in an etheric form upward toward the head. Here, beyond the threshold of the neck, these components are condensed back to become the brain substance. Steiner says: "*In this respect it is the head, and not the big toe, that is directly dependent on the stomach.*" The head is therefore the direct result of foods that enter animals (and humans) through the mouth, after being transformed by the digestive organs. Since it is physical food we say that it consists of earthly substance, and that substance is therefore directed to the head and brain.

The sentence quoted above continues as follows: "*And you must understand that the head can assimilate the nourishment it receives from the body, only if it can also obtain the forces from the cosmos.*" In other words, what we have defined as earthly substance can be condensed to form the brain only if the head can receive the relevant forces of the cosmos.

That is why the next paragraph ("*That is why ...*") indicates the need for animals to have access to graze freely and thus to enter into a perceptual or sensory contact with the surrounding world.

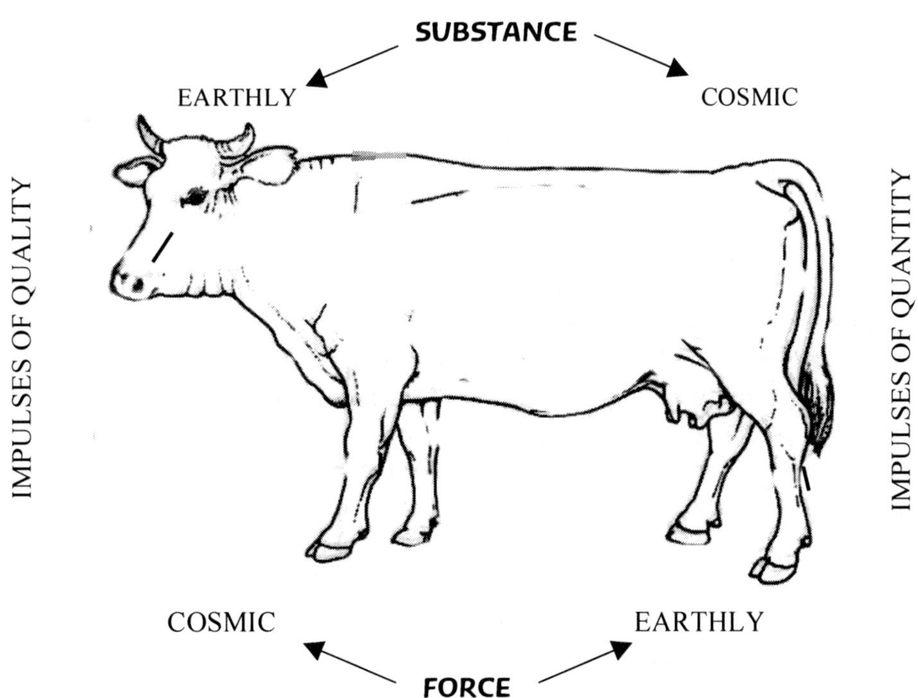

To complete this image - although Dr Steiner will say this later in the lecture - the physical body of man and animal that is located below the neck is nourished by what we call *cosmic substance*, a substance that is finely distributed throughout the surrounding atmosphere and that we take in through our senses. Another current brings this comic substance down into the body. This current arises because the ingested earthly substance that remains after building up the head 'bounces' back from the cranium. This generates a current capable of descending whilst bringing with it what is captured from the outside through our senses. Contrary to what happens in the head where the earthly substance can condense properly only in the presence of the cosmic forces of the universe, the cosmic forces conveyed downwards can properly build up the physical body only through the involvement of earthly forces.

To summarise: the head is composed of earthly forces thanks to the action of the cosmic forces that shape them. The remaining part of the body is made of cosmic substances thanks to the earthly forces that shape them.

We can say a little more about the form of the body. The cosmic forces shape the earthly substances into cosmic forms – hence the 'ball' of the head or at least the cranial cap. Reciprocally, the earthly forces shape the cosmic substance into earthly forms - radial and diversifying. This is demonstrated by the very fact that the head has a cosmic shape - round - and consists of a single body, while the rest of the body springs from a single trunk and spreads as it proceeds towards the periphery, according to the earthly principle of division and fragmentation.

At this point we may ask where are the so-called cosmic substances. Well, if we remember what we said when commenting on the seventh lecture, part of the cosmic substance that we absorb is the result of the dematerialisation of the Earth minerals that have been brought into flow by the trees and then released into the air. These also serve as a reminder for the forces that come from the cosmos. Do not be astonished that part of the free astrality is derived from the mineral world because matter is only light condensed into darkness, which can be processed through the trees so it is free to be light once more.

To help us understand what we have said and what will come later in these meetings we can draw a picture of the forces and label those that contribute to substantiation. We also note (picture below) that the earthly forces act in the lower body, while the cosmic forces act on the head, which is why it does not hang down as do the other 'limbs', but it turns upward.

We can easily understand how important it is to have food of the highest quality because this makes it possible to create substance for a brain that is, in turn, able to direct the cosmic substances to build a strong and healthy physical body. The possibility of forming a good brain is also a prerequisite for clear thinking and, therefore, a strong consciousness. Moreover a sufficiently robust and agile consciousness allows us to act in an appropriate way in all human activities and is the basis for a proper way to practice agriculture. We recall that in the past we were able to say that such a healthy consciousness is necessary to prepare good homeopathic products and even put right any shortcomings or deficiencies caused by incorrect use. Also in this case we can activate a positive spiral; a good nutrition enables a robust consciousness, which in turn makes it possible to create high quality products that form the basis for an even better nutrition … and so on.

We have to consider the act of thinking. It is good to remember that thoughts are not made by the human brain as might seem apparent, but that the brain makes thoughts perceptible to us from their irradiation out of the spiritual world. The importance of sustaining a brain by healthy eating is emphasised by this knowledge. The quality of the brain determines the quality of our thoughts. A healthy brain makes it possible to connect with beings (the Angels) that bring evolution forward. The reverse should also be clear.

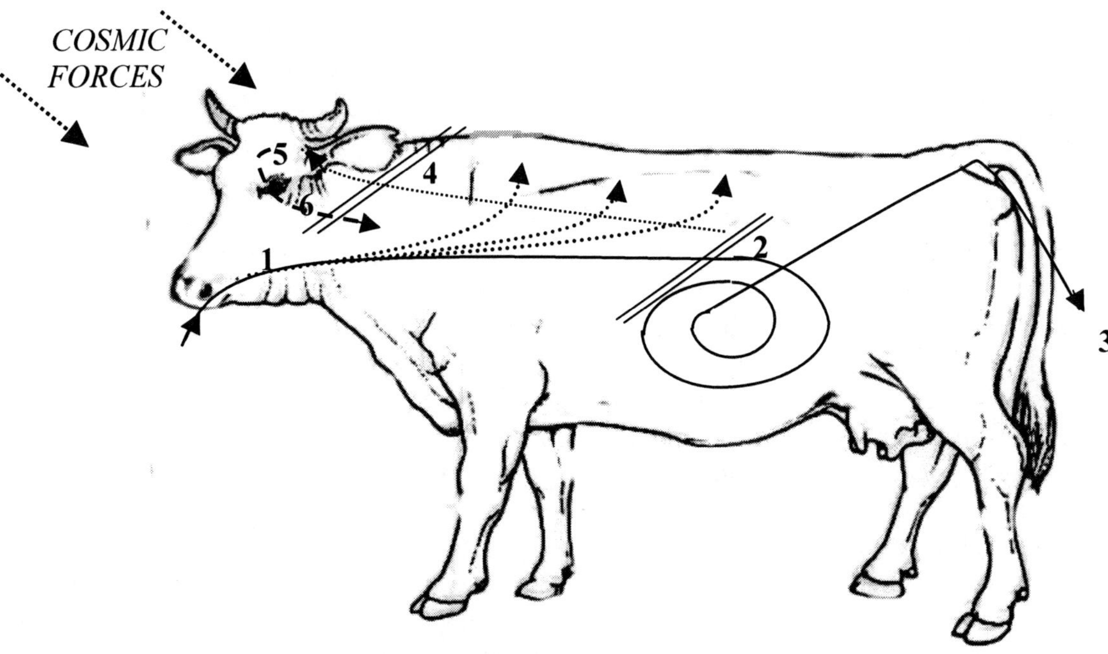

1 = INCOMING EARTHLY NUTRITION
2 = DIGESTED THROUGH THE FOUR ELEMENTS
3 = EXCRETION OF THE INABSORBABLE PART
4 = ETHERIC UPSTREAMING TO THE HEAD WITH SUBSTANTIATION IN UNION WITH THE COSMIC FORCES
5 = REFLECTION FROM THE CRANIUM
6 = REFLECTED STREAM THAT BEARS THE COSMIC NUTRITION

To fully understand what we have just said we must learn to distinguish between the ability to think and the power of thought. The ability to think is an attribute of the soul, and is what allows you to understand a thought. The power of thought is the ability to concentrate on a thought and process it over a long period of time until one grasps its intimate truth. The latter belongs to the etheric brain and is a capacity guided by the will of the I. The ability without the force is likely to lead to vacuous fantasising that leads nowhere, especially if the awareness is not sufficiently trained. When it is accompanied by force it can bring the highest communication with the spiritual world.

A poor quality diet affects all of this, even when we eat food that cannot be etherised. (This goes beyond the intestinal membrane and is deposited in fatty tissue, forming the seat of many diseases.) We believe we can go back to what we said earlier about the origin of cosmic substances, to present a further dynamic for reflection. Cosmic substances of earthly origin are derived from dematerialisation of minerals by trees. We can deduce that the widespread deforestation that man has been inflicting for decades, has consequences that are rarely thought through with clarity. It is culturally accepted that deforestation has consequences upon the reduced oxygen in the atmosphere. (In reality this is wrong because 90% of the oxygen in the atmosphere is not produced by trees, but by algae in the sea.) The most serious consequence of the gradual elimination of forests is the reduced production of cosmic substance of terrestrial origin and, therefore, in the diminished availability of this subtle nutrition, that we have called cosmic nutrition.

The eleventh paragraph ("*That is why it is ...*") reiterates the importance of allowing the animal to live in a pasture and stresses that this is the only way it can choose the type of grass that is more congenial in the various times of the year, depending on the forces present in the cosmos with which it can connect and resonate. An animal raised in a dark stable, unable to enjoy a connection with the forces of the cosmos, will generate progeny that gradually lose their propensity to process the cosmic forces. Were such an animal set free it would no longer be able to choose the best food for itself and will increasingly fail to distinguish good plants from those that are harmful. Just as the loss of the ability to connect with the forces of the cosmos requires some generations to be completed, in the same way reconnection also requires some generations. The impossibility to process cosmic forces, in fact, undermines the formation of a healthy brain, and this leads to the inability to adequately engender the rest of the body through the cosmic nutrition.

In the twelfth paragraph ("*Let's go on now...*") Steiner poses the following question: "*What is the purpose of the brain?*" We have already partially answered this question because the brain, as well as its role in the subtle nutrition process, serves as a basis for the Ego. This is of course the case for humans, because the animal "*has no ego - it's brain is only on the way to ego-formation.*"

For completeness we can add that the human I works mainly at the front of the brain, so much so that the bones located on either side of the front of the skull are called the 'temples', precisely because they form a kind of temple for the Ego. Without going into this too deeply, because that would lead us away from the present subject, we can bring to mind that man has three temples: the first is the one we have just mentioned which is in the head, the second is in the heart and the third at the base of the spine, in the area called the perineum, a term that derives from the Greek and means 'near the temple'. We believe it is no coincidence that the end of the spine is also known as the 'sacrum". There are three inner temples because they serve the forces of the I, the soul and the body. In the third temple, dedicated to corporeality, a physical basis of Life is retained. This consists of eight cells that are none other than half of the first sixteen cells that are formed in the developing human body. These eight cells, unlike the other eight, do not continue to multiply to form the body, but remain unchanged throughout life in the perineum, as representatives of the germs of Life.

The paragraph concludes with another question, "*So how does the brain come about?*" The thirteenth paragraph ("*Consider the body's ..*") contains the answer. Once again the point is that the brain is essentially excluded from terrestrial organic processes, and is removed to serve as the basis of the I. Steiner then adds more clarification of the process: "*Now, on the basis of the process in which foodstuffs are consumed, digested, and distributed by the metabolic-limb system, a certain amount of earthly matter is able to reach the head and the brain, and in this way a certain amount of earthly substance is actually deposited in the brain. But the foodstuffs are not only deposited in the brain, they are also excreted by the intestines along the way. Whatever cannot be further assimilated is deposited in the intestines.*" We believe that these sentences are sufficiently clear and are compatible with our own comments. We consider that there is a correspondence with what we have put in the drawing in the preceding pages.

These phrases, however, raise another question, namely, what is it that is really extracted from the animals' fodder? This is not only the most 'noble' of the material component of plants, but also the etheric component and astral forces and traces of the plants' group soul. Obviously the forces of the I present in the animals are much lower than those found in humans, so a person is able to extract more of these subtle elements from their food than animals, and this has a very important consequence. This is why human excrement is not suitable for composting and fertilisation, in contrast to the manure of animals. Manure is not so important for the amount of organic matter that it brings to the soil, but for the subtle components that remain in the dung after being excreted from the body of the animal. We believe it is comprehensible that the forces of the plants' group I that remain in the animal manure can serve again once they are returned to the soil as fertiliser. These forces nurture links with the Group I of the plants that grow on that same ground, allowing a further improvement of quality - like the virtuous circle that has been reported at the beginning of this meeting. The astral component remaining in the fertiliser favours the influence of nitrogen, which we know to be the bearer of cosmic imagination. (This also gives us an opportunity to say that if we are unable to get sufficient manure, we can still establish a healthy and strong connection with the Group I, using our seed baths and, through spraying our 'Pro Nitrogen' preparation. This particular preparation increases nitrogen in our soils in the period in which it is naturally and abundantly granted from the cosmos, and so forth.)

From what we have said so far the relationship between the brain and the intestine should be clear. Steiner says: "*The substance of the brain is simply intestinal content taken as far as possible, whereas premature brain deposits pass out through the intestine. The contents of the intestine, as regards their processes, are very much related to the contents of the brain.*"

Thus, we arrive at the fourteenth paragraph. ("*It would be crass to say ...*") Here Rudolf Steiner focuses on a particular dynamic that we have already anticipated, which is that when the brain matter is formed of good earthly substance it serves as a "*basis for the development of the I*". He reiterates that the animal, being devoid of an individuality or I, has a reduced ability to extract the subtle components from their food, leaving more in the manure. He emphasises that this is the reason that animal manure, as opposed to 'humanure', can be used as a beneficial fertiliser. At the end of the paragraph, confirming what was said in previous lines, Steiner points out that manure is placed on the ground, which is the place where the root grows "*... the plant's ego-potential develops down here in interaction with the manure*". By this point it should be obvious that fertilisation is important primarily for the ability to encourage a connection with the I.

Now we can once again turn our attention to the sketch that we made at the beginning of our deliberations, and insert information about nutrition that Steiner did not give during the agriculture course but that he provided on other occasions as recorded in many of his other lectures.

What we eat, which is still full of etheric forces during mastication (the first stage of the digestive process), undergoes a separation through which the physical part of the food is detached from its etheric. The physical component follows the path that we have already examined. The etheric component, which preserves the memory of its previous state, would tend to develop the same processes that developed in the plant when it was embodied, and therefore must be destroyed by our etheric body. Now, we should know that the etheric body follows the etheric laws that are completely complementary but opposite to those that regulate the physical body. Thus engaging in the disintegration of the etheric body of what we have swallowed strengthens our own etheric body, contrary to what happens to our physical body that is fatigued by engaging in activity. Therefore, the more we eat food that is rich in etheric forces, the more our own etheric body must engage to overcome it, and the more will our 'etheric muscles' be reinforced. This is very important because our etheric body is responsible for all our internal movements, and also for the strength with which we move outwardly and which make our activities so harmonious. The internal movements are those that accompany all our vital functions. The importance, therefore, to have food grown in a healthy way (at least biodynamically) is emphasised.

Of course, to cope with a food that is very rich in vital forces, our etheric body must be sufficiently strong to oppose it. This is why sick people are recommended to cook their vegetables, because during cooking the etheric body of the plant undergoes a pronounced weakening. The same recommendation can be made to healthy people for feeding in the evenings, when an intense physical activity has absorbed so many of our forces, temporarily weakening the etheric body.

A part of our food also goes to increase our ability to move, which we could also call our 'will'. We believe it is not a coincidence that the younger generations, which have been fed so much industrially prepared food or catered for in fast food establishments, prove so inactive and amorphous. The will is not only that which leads us to engage in active muscular efforts, but also one that allows us to remain focused on a thought for a long time. Again this points to the importance of a good and healthy diet.

There is a further aspect of nutrition that we want to emphasise. We have said that the physical component of food is taken down into the stomach and intestines and that, during this journey, part undergoes a process of etherisation. This food is then conveyed upward to become the brain substance. But here its path does not end: as we have already mentioned, what is not condensed in the head is sent back down in a form somewhat like a message, and brings down with it that which has entered the head through the sense organs (the so-called cosmic substance), and this then becomes substance in our physical body. But throughout this process, during the ascent and descent, it must receive a highly personal imprint. So during the food's ascent the etherised food receives the imprint of our physical body, our etheric body, and our astral and I, so that the brain is formed in its diverse parts in a way that is compatible with our whole being. These 'fingerprints' are maintained as in a message, so that what comes into us as a cosmic force is condensed on the way back down in a form that is relevant to our totality, and we can continue to be ourselves in spite of being reformed from continually replenished substances.

Now we believe we have demonstrated that the brain is our main organ for the qualitative aspects of nutrition. Do not forget that the word nutrition ('alimentazione' in Italian) can be split into three parts: *al* - *mento* - *azione*. The last part ('*azione*' = action) needs no further clarification but the first parts may. The prefix '*al*' is an Arabic monosyllable that emphasises connections to the divine or spiritual. The central part indicates that nutrition is defined by two poles: the chin (It: '*mento*') concerns the coarse aspect and the mind (It: '*mente*') the subtle. In conclusion, eating is that act in which the physical and mental are strongly connected, allowing a relationship with the divine, understood in a broad sense as that which comes down from the cosmos. In all this, of course, the brain plays a crucial role.

Twenty-eighth Meeting

Before we continue reading and remarking upon the eighth and last lecture of Dr Steiner's agriculture course, it is perhaps appropriate to recall the context in which what we are about to read was delivered. We can bring to mind that Count Keyserlingk, acting in this instance as a spokesman for the Anthroposophical movement, had asked for and obtained the agreement of Dr Steiner to give the agriculture course to address three fundamental problems that were afflicting the farmers. These were the loss of vitality of seeds, the decreased ability of agricultural products to feed people adequately, and the loss of fertility in the livestock. To answer these, Steiner speaks in the first lectures about the plant, the forces acting on them and their nutritional value. Later, in the central lectures, he reveals the biodynamic preparations - the tools that harmonise vital processes in the plant by making an effective connection between the plants, the forces of the cosmos and the Earth. Finally, in the sixth and seventh lectures, he speaks of plant diseases and of agriculture as the basis for health. In this last lecture, the eighth, he finally concentrates specifically upon animals. He initially gave the new foundations, then gave indications of how it is possible to intervene in the mineral and vegetable kingdoms thanks to the tools outlined in the central lectures, and now he has come to the animal kingdom.

In the fifteenth paragraph ("*A farm of this kind ...*") Steiner emphasises the "*wonderful interactions*" between plant and animal. In other words he reiterates what we have already had the opportunity to say about the fact that a plant grown in the right way adequately supports, as a food, the development of the animal in all its aspects, including the ability to connect to the individual principle of its species. Such an animal will then produce manure "*which in turn allow the plants to grow properly out from their roots in alignment with the force of gravity.*" We believe that the concept is easily understandable. What Steiner does not explain here is why this is happening. We will try to help in this with the drawing below.

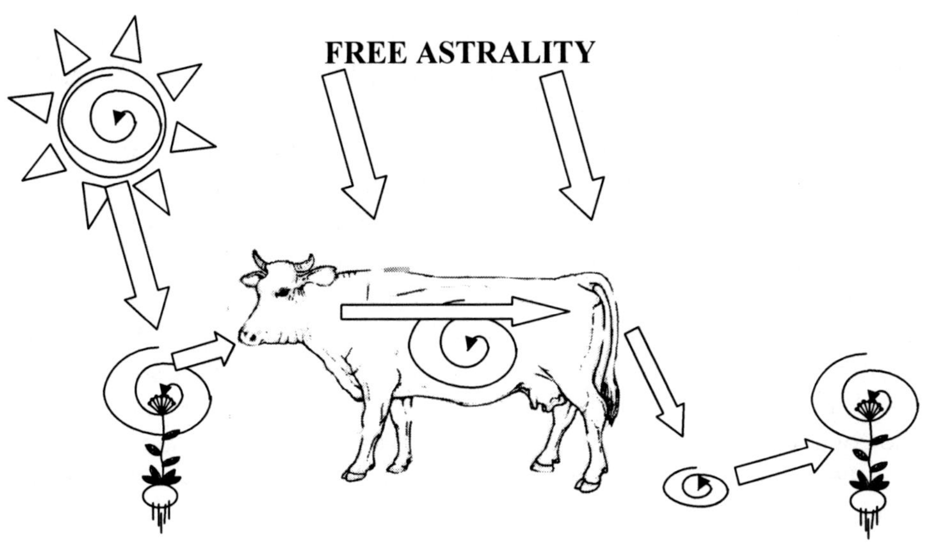

The second sentence of this fifteenth paragraph refers to the presence of forest and orchard that develops the astral element. We have represented these important components of agriculture with an annual plant that grows above the ground. Following the flow of this paragraph we also represent the animal that eats these plants, and the expulsion of manure by the same animal, which then supports the plant.

We think it is crucial to clarify "who" is the arrow that we have generically called *free astrality.* Well, this arrow represents the Sun, which this time is not indicated with its usual symbol, but in a way that highlights its intimate nature. The Sun, as we have already had occasion to say, is really a vortex and so we have drawn it in this way. This same vortex is echoed in phyllotaxis - the way that leaves grow up a stem - and also in intestines, and finally in manure which passes into the soil and finally arrives at the roots of plants. This sketch, in other words, should enable us all to understand that agriculture is an expression of the power of the Sun. Moreover, the Sun is the representative in our solar system of the world of stars, which is the visible expression of the spiritual world. The forces by which the plant is connected to the Earth through the roots are spiritual forces. Indeed, as the top of the plant is connected to the daytime Sun, the roots grow in the direction of the nighttime Sun. The Sun, at night, acting through the Earth, brings those forces that grow roots. We hope that no one has forgotten that invitation - when we do not know what to call certain forces - to call them the forces of the Sun! These same forces, if we go back in memory to the fourth lecture, are those brought by preparations 500 and 501 that both undergo a spiral dynamisation before being sprayed on the land. To bring the forces of the Sun into our fields one can also adopt rhythms that resonate with the Sun, and grow 'companion plants' which bring solar forces, such as the apple, citrus, olives, hawthorn, and the dog rose.

In modern [Italian] agriculture the animal has almost disappeared and so the whole cycle that we represented in the sketch cannot always be complete. There remains the possibility to resort to the practices mentioned above when trying to enhance agricultural effectiveness.

Incidentally, note that the farm animal once had another crucial role. Because it possesses an astral body, it was able to absorb excess astrality around the farm, such as that produced by the farmer and his family, and 'earth' this through their hooves. Today this is no longer possible because animals became un-economic, but all this had negative consequences on the health of agriculture. Returning to preparations 500 and 501, we can say that the 500 is the light that entered the darkness and became Life, while the 501 is still light, understood to be like consciousness.

We believe it is important to note that Steiner referred to the animals' ego forces using the adjective 'proper' [It: *giusto*] and refers to the growth that results in the plant uses the adverb 'properly' [It: *giustamente*]. Evidently Steiner wants to refer to spiritual logic thus elevating the sentence from its common meaning to far higher levels. Besides the last sentence reads: "*This is a wonderful interaction, and we must try to come to an ever better understanding of it.*" This implies that we can gradually and continually raise our levels of understanding to embrace the whole problem of nutrition and of how the Light, which dies towards the Earth, may rise again to supply forces for the conquest of freedom and love. These words emphasise once more the importance of the quality of food and therefore also the best practices within breeding and agriculture.

The sixteenth paragraph begins with the following sentence: "*Because of this interaction, you see, the farm is a kind of individuality.*" There is a clear reference to the agricultural organism and implies the force of an I, without which an organism cannot exist. In the third sentence of the paragraph Steiner says "*You must arrange things so that the cycle becomes self-sustaining.*" Exclude animals from the farm and one must use manure that comes from animals of a different property and that is appropriate to the needs of that different farm. Of course all this must be guided by the awareness that the farmer must choose "*the right number and kinds of animals on a farm so that you get enough of the right manure.*" And he alludes to the consciousness of the farmer in the last sentence: "*And you must also make sure that you plant what your animals instinctively want to eat.*" Indeed, the choice of forage involves the understanding of the links we have already mentioned between plants and animals. We note that the reference to the I continues even in the last sentence because instinct is the way the group I guides all the animals under its sway.

The seventeenth paragraph ("*Setting up experiments …*") repeats the idea of the self-contained farm individuality, at least as an ideal towards which one must aspire. In fact, since 1900, as recognised by Steiner, the world economic order made complete independence impossible. Today the situation is clearly even worse and it is therefore necessary to accept a series of compromises, but we must start from the awareness of what would be the "*proper*" way to set up the property and constantly strive to come as close to it as is feasible - though not fanatically.

Paragraph eighteen ("*So, by now we should …*") finds Steiner setting up what he will elaborate upon in the following paragraphs: " *... we should be able to discover in concrete terms how the animal relates to the plant organism*", because the problem of fodder can only be dealt with on that basis. We note that at this stage he uses the word "*concrete*" to indicate the famous *nexus* that we find on the spiritual level. *Concrete* apparently refers to the fact that in practice there is no other way to resolve the problem.

In the nineteenth paragraph ("*Look at the root...* ") Steiner encourages the observation of plants in their relationship with the ego or I forces which the manure has brought to the soil, and concludes the paragraph by saying: "*This process is helped if it can find the right amount of salts in the soil.*" With this phrase we want to highlight the importance of the activities of the elemental beings of the Earth, the gnomes. These gnomes have the ability to act effectively only through the minerals present in the soil and salts occuring in the form of crystals, which are among those minerals most associated with life. They are the only minerals that can grow. Remember that gnomes are emanations of Archai, also called the Spirits of Personality. From this point of view, the conventional method of cultivation shows some understanding that a successful fertilisation must contain some salt. Unfortunately this has been directed towards the use of synthetic salts. If only natural mineral salts had been used we would have gotten to the heart of the problem. Do not forget that Steiner advised the use of potassium sulphate to enhance land lacking in organic matter, and everyone here knows or can guess who also suggested that *digitalis purpurea* (foxglove) should also be grown in those areas. Indeed, in view of the fact that the physical plane resounds with the astral plane, Steiner has suggested that a poisonous substance (thus an astralised substance) should accompany the supply of minerals on the physical level, to put the latter into circulation and to assist in its rapid incorporation into the organic world.

In this regard we invite anyone who is in a situation where some exceptional event has completely upset the vital balance to note that nature takes steps to restore conditions so that life can return. Take an example from one of the participants in one of our summer courses that was held in a mountainous village. A landslide had swept away all the vegetation in its path, and it was curious to note that the ground at the top was completely dead, but little by little as you came down there were first some signs of life and at the base of the landslip there was a rich herbaceous vegetation. Careful observation allowed one to understand the processes that nature had put in place so that this could happen. At the top were two plants: coltsfoot and a mountain orchid. There were also many grasses (which we will understand better after considering the twenty-first paragraph of this lecture). These two plants are very astralised, the first of which is used in medicine as a remedy for lung diseases, while the second stimulates blood circulation. These two plants are able to stimulate the rhythmic system. Coltsfoot puts the mineral component of the land in motion (in breathing it helps the expulsion of carbon dioxide and thus carbon) and the orchid activates that little bit of life that was there. In those conditions it would have been useless to bring in organic matter, because this would have remained locked in a state of latent life. Immediately below was hornbeam, which we know is linked to Mercury, a planet with a fast cycle and therefore also capable of bringing the forces of movement. Underneath were the burdock, nettle and chamomile, which carried out the task of purifying the excess astrality so, from the activity of the pioneer plants, only the vital aspect remains so that the lower fertile land was rich in humus. We treasure the chance to be able to watch nature as she offers lessons of incredible scope. We believe that what we have described is a basis for other work in extreme conditions, such as, for example bringing life to the deserts.

Now, going back to the beginning of the nineteenth paragraph, we can focus on the fact that Steiner mentions the elemental beings; we can notice that he talks about the "*right amount of salts*", therefore he once again uses the term "*right*" to draw the attention to a higher level. Besides, the soil's composition and the presence of salts within it may not be modified by the farmer, who may instead be able to favour the action of the elemental beings, whose presence in the soil is "right".

At this point we could also ask what the difference is on a spiritual level between a mineral and a salt. Well the mineral spiritually represents the forces of the Father, while the salt, which is the union between a base and an acid and can therefore be considered *mercur*, is the Son. The action of the Son brings life through the light of the Holy Spirit. In this way we can say that we understand the action of potassium sulphate on the soil on a different level. We can also consider the fact that potassium is the representative of the constellation of the Virgin and sulphur of Pisces. We now know that the Pisces-Virgo axis is an essential reference for any healing, because through the Virgin one can control the influence of Lucifer and through Pisces one can balance the influence of Ahriman, and through the link between Fishes and the Virgin one can prevent the cooperation between the Ahrimanic forces (from Sagittarius) and Lucifer (from the Twins), in this way inhibiting the action of the third force of evil - the Asuras.

The nineteenth paragraph then suggests roots as the nutritional base for an animal to whose head we would like to guide earthly substance. We have already alluded to the fact that the head is the centre of the nerve-sense system and is the receptor of the cosmic substance that, on being carried back down, will allow the formation of the remaining part of the physical body. But Steiner says "*You can imagine someone putting it like this: 'I must give root fodder to an animal that needs substances conducted to its head, so that it can become as active as possible in its senses, so that it can develop a cosmic relationship to its cosmic surroundings.' Don't you think immediately of the calf and the carrot?*" The carrot can be seen as a ray of light confined in the soil and thus is able not only to feed the head as substance, but also to help create a subtle connection with the forces of the cosmos.

Of course this also happens in humans, where the carrot helps to develop the forces of discernment that must guide all the activities of farmers. An inappropriate level of consciousness is the main cause of a series of plant pathologies. A lack of adequate discernment is the cause of an infestation by wooly aphids. We believe that the widespread sowing of carrot in a field where there is a history of wooly aphids will not fail to have a significant therapeutic effect. One may consider the carrot to be useful whenever symptoms are detected indicating lack of Light in the soil. This failure could be inferred, for example, by the presence of the notoriously blind moles, so one could imagine that the carrot is a good remedy for such appearances. The arguments that we have made are in line with what Steiner has just said: to "*discover in concrete terms how the animal organism relates to the plant organism.*"

If we wanted to continue to look at other instances in which the forces of the carrot could be beneficial perhaps it would take too much time from the work we have set ourselves and we could go too much by the text. Let us therefore use another example: we know that cosmic nutrition comes through the senses - mainly our sight, hearing and smell - and we also know that cosmic images come to us through cosmic nutrition. We also know that these images are widespread in the soil, but also within us, through nitrogen. In particular nitrogen carries cosmic images in the blood to the heart and particularly in the atria. We may at this point think of the modern prevalence of cardiac diseases and malformations in humans, making cardiovascular diseases the main cause of death in many parts. Is this a reflection of the inability to receive cosmic images due to inappropriate life styles and a poor diet? So, we can suggest that people undertake an artistic activity and enjoy long walks in the woods, but it can be of enormous help to include carrots in the diet, ideally grown in the way we propose.

To conclude our sidetrack about carrots we will add that the carrot is bound to the light of Jupiter and the forces of Capricorn. Capricorn is the gateway for all pests and diseases and, given this, it seems perfectly logical that there is no better way to stop pests than to shine light in their port of entry. Jupiter is the "memory" in our system of the Old Sun, in which the spirits of Wisdom acted, and the Light of Jupiter is the light of cosmic wisdom we know as the Cosmic Sophia. We remember that the carrot is rich in meteoric iron, which is the iron from which the sword of Michael is forged. We do not see how this deployment of celestial forces against parasites can be improved. At this point, if we say that the carrot is 'sacred' we believe we have made an absolutely acceptable claim.

This brings us to the twentieth paragraph ("*Let us continue ...*") in which Steiner says that the nutritional process may not end with the formation of the head, but it is also necessary that the same process enables the formation of the remaining part of the body. The term '*will*' is used here, in the sense of cosmic will, which is also known as the free Life, unbound ethers, or Zoe. Steiner continues: "*The head must now be able to work as force, as will, and thus also be able to engender forces in the organism, so that these in turn can be worked into the organism.*" In practice Steiner says what was said here earlier - that the body of the animal and humans, is formed by the free Life that enters into us through organs linked to the nerve-sense system. But to facilitate this return process there is a need for "*a second kind of fodder, which will let a particular part of the body - in this case the head - to work upon the rest of the body in the right way.*" We find, once again, the word '*right*' because here is another reference to the relationship between plant and animal worlds and thus to a spiritual logic.

Paragraph twenty-one ("*So, suppose ...*") reveals what this second nourishment might be. But first, in the third sentence, Rudolf Steiner explains the basis upon which the comments are made about the right choice of fodder: if we provide something that helps cosmic substance from the head to radiate throughout the body, we must try something that "*naturally has a ray-like form, or else something in which this radiating tendency is gathered together and concentrated.*" An example is linseed or other products that show the same force. Also suitable for this purpose are long-stemmed herbs such as grasses then transformed into hay. Now we can better understand why the presence of grasses complemented the coltsfoot and orchids, in the example of the landslide we used earlier.

If we have really grasped Rudolf Steiner's principles then we can also be creative with further thoughts on that basis. For example, we could think that if the carrot mainly serves to form an organ which is capable of incorporating cosmic nutrition and that the linseed radiates these forces to give substance to the physical body, a potentised product based on these substances could significantly reduce the need to eat physical food and could therefore be a remedy for hunger in the world. A similar product could also be used to reduce the need for animals to take fodder and consequently it would reduce the cost of keeping livestock. Even for plants a product made as intimated could make a living possible from very poor soils. Obviously such a product will produce limited effects in the short term, but continued use for three successive generations could imprint this capacity as a permanent feature. Of course, the effect might be more evident in animals that already have a good ability to nourish themselves from the cosmic forces. For example, it will be easier to achieve a good response from a goat rather than a sheep. This stems from the observation that the ratio between ingested forage and milk produced is much lower in the goat than the sheep. This demonstrates the goat's greater capacity to exploit subtle nutrition. Moreover it was possible to reach the same conclusion by examining the foraging habits of the two animals. Sheep browse directly from the ground, while goats prefer to take food only from a certain height above the ground. Sheep mainly eat fresh grass even when it is very humid. Goats do not tolerate wet grass and can even eat thorns.

Paragraph twenty-two ("*Let's carry on and look ...*") begins with an invitation to make a further step forward on the same subject and try to understand how to act in the case that you want an animal strengthened in the mid-zone of rhythm. Steiner himself clarifies that this is useful in dairy cattle. Note that Steiner immediately clarifies that milk is not just a product of the animal's metabolism but of the rhythmic part that he defines - referring to the production of milk – as the region "... *where the head or nerve-sense system develops in the direction of the respiration, and where the metabolic system develops in the direction of the rhythmic system; that is, where the two interpenetrate.*" The mistake of regarding milk as a product of the rear part of the animal results from a superficial observation of animals such as cows and goats which have their udders towards that part of the body. Many other animals such as pigs, dogs, cats and more produce milk right into the thoracic area. We think that the migration of mammary glands toward the rear of the body has been caused by a constant search for increased production, which has brought about this metamorphosis of the animals.

Women's breasts are even more forwards/upwards - towards the area of the heart - and this is due to the fact that they will nourish a child who will be the bearer of an individual I. This requirement clearly does not exist in animals whose milk bears a stronger metabolic quality, while human milk, as we have said, is moved more towards the rhythmic area.

It is worth clarifying the passage in which Steiner presents milk as a "*substance on its way to becoming a sexual secretion ... transformed in its encounter with the head forces.*" If we follow the path which food takes through the metabolic area we can see that it first passes through the region of the Venus forces, which generally preside over the process of the secretion of gastric juices, and then the area where the forces of Mercury act more resulting in intestinal peristalsis. These are complex contractions in the annular musculature of the digestive tract that propel the food along the intestines. Finally the remaining forces of the food are captured within the area where the Moon acts, which is turned into sexual energy. But the Moon does not intervene directly in the digestive process, which has been exhausted by the action of Mercury, where the food is brought to the phase of Fire and can cross the intestinal membrane and become Warmth. Lunar forces may capture part of those forces and use them for sexual activity. But if we want to increase milk production, we must ensure that these remaining forces are captured by the lunar forces in however small a degree in order to maximise the forces that will be channeled towards the central area. We already know that the best fodder plants in this regard are those that are very rich in the forces of Mercury, such as clover and alfalfa. These demonstrate this wealth by their extraordinary ability to 'come again', or grow again after being cut.

What seems strange is that milk, which after all is a white liquid, is not so very much attached to the Moon, but in fact represents a kind of sublimation of the Moon's forces, like a polar action. It is no coincidence that women have suspended menstrual and ovulation cycles for a few months, beginning when milk production comes to feed the baby.

The twenty fourth paragraph ("*For processes of this kind ...*") now says that to encourage the process of milk-formation one must not use a root yet, but "*a type of fodder that does not work as strongly in the direction of the head as roots do.*" Steiner tells us that since milk, as we have seen, must remain related to the sexual force, we must make sure that the fodder "*should not contain too much astrality – not too much of the blossoming and fruiting tendency.*"

We believe that this sentence should be clarified. When we said that lunar forces capture and bring down the forces of food that should otherwise go back to the head, in fact we made reference to Lucifer because we must keep in mind that Lucifer was an angel and that the angels are also called "Sons of Life". Moreover the forces taken by Lucifer are used for activities related to the transmission of biological life. Well, we know that Lucifer is a kind of astral being acting in the etheric, one who acts through the world of feelings and passions. If the fodder fed to animals was too rich in astrality, it would promote this Luciferic action and thus encourage the diversion of forces to the sphere of sexuality.

We know that fruits act on the metabolic system, while leaves affect the rhythmic system, so we will have to choose "*what is green and leafy.*" Steiner gives us another valuable indication when he said: "*we should not take too much of what is or tends to the flower or fruit.*" In other words, this means that in addition to using green grass, we must also take care that it is young, that it is far from flowering, as Steiner says "*herbs that grow*", meaning those herbs in a vigourous phase of growth and we know that the arrival of flowering marks the end of the growth of the plant and the start of its astralisation.

Twenty-Ninth Meeting

Again, we are reading and commenting on the eighth Lecture of the agriculture course, held by Steiner on June 16th, 1924 at Koberwitz. We will carry on from the point where we left off in the last meeting - the twenty-third paragraph (“*For processes of this kind* ...”).

Steiner poses an interesting problem: how can we establish the right way to feed a cow - normally fed with leaves and various species of grass - if we want to increase the production of milk? Note that we will always remain within the potential of the animal, which is to say that we will not support the practice of forcing its innate nature. Again Steiner suggests the solution and says "*I use plants that take the fruiting process – the process that occurs in the flowers and in fruits – and bring it into the leafing process.*" In other words, we are talking about plants that do not fully express their fruiting as is commonly done, but retain and express it at an earlier stage, in the formation of the leaf. And these plants "*are legumes, especially clover*". Indeed, a more careful observation of legumes allows us to observe that the part of them that we generally take for food is only the seeds, and that there is no fruit-pulp as in normal fruit. Moreover, the seeds grow protected by pods that look very similar to the leaves. We could say that these plants prolong their leaf stage up to the formation of seed. Also, we can notice that once the plant has produced the seeds, the plant has not exhausted its vigourous growth phase, but continues to produce stem and other leaves. Moreover, this flowering and fruiting occurs at the same time as the leaf-growth, and it is not a disruption of the normal growth process of the plants, which usually follows the order - root, leaf, flower and fruit-seed. In this ‘abnormal’ sequence of growth processes we also find the explanation of the formation of alkaloids in legumes, as the early flowering and fruiting comes at a time when the plant is still fully in the leaf phase. The leaf clearly shows the effect of a strong astralisation, of a Luciferic action.

Clover – as much as anything because of its ability to revive its full leaf stage even after repeated grazing - is the preferred legume for the purpose in question. Given finally the relationship of milk with animal astrality, we can better understand how the pulses in general, and clover in particular, are able to support milk production with their astralisation.

Further on we find an important observation about the fact that a legume-based diet does not have so evident an effect in the cow to which it is administered, but more in the heifer that will emerge from it. Indeed, the improvement of the function concerned is manifested in the next generation. We can add that if we want the function that we are aiming to improve to become a hereditary characteristic, it will be necessary to prolong the ‘treatment’ for three consecutive generations. What we have just said also applies to the plant world and this allows us to understand how important it is for farmers to start producing their crops from their own seeds.

The twenty-fourth paragraph (“*Now in this area* ...”) highlights the need for any intervention to be done with awareness. We believe we can grasp, once again, the reference to the links between the plant and animal worlds. Without this knowledge we run the risk of adopting a series of contradictory activities that will neutralise the effect of one with the other. In addition to understanding what we are doing we must also nurture an appropriate attitude, which is perhaps even more important than understanding. Do not forget that Steiner said that good doctors should prepare a remedy with their own hands and with love for their patient. The doctor should then administer the remedy himself, so that the patient can have confidence and be totally open to the healing power of the remedy itself.

Paragraph twenty five ("*It is similar when ...*") continues on the same line talking about forage but also stresses that without an adequate understanding one may well try to put together different ingredients, in the hope that the choices made will prove their worth empirically. But attempts in this direction often end up in such a confusion that they are useless. The loss of knowledge of the links between the various forces makes it impossible to tackle the problem with rationality. Steiner concludes the paragraph: "*This situation can only be dealt with rationally by thinking along the lines I have indicated. In that way the question of animal nutrition is considerably simplified, so that it is possible to have an overview of it.*"

The twenty-sixth paragraph ("*You can understand the way ...*") should present no significant difficulties.

The twenty-seventh paragraph ("*Let's take another example ...*") considers - well - another example. The first three sentences, indeed, appear to be quite self-contained, but basically say that all parts of the plant are nothing more than the metamorphosis of the leaf, so the plant has a common disposition in every part it, although these specialise only at a certain stage of growth. To understand this better we could refer to stem cells which contain a basic message with all the possibilities of the various specialised cells of various organs of the body, so much so that they can be used in different parts of the body to regenerate damaged tissues or organs. Another example is the famous three embryonic layers (endoderm, mesoderm and ectoderm) from which our whole body has been built. Obviously these are general messages that combine to manifest a wonderful diversification and specialisation of all the different parts of our body.

In the first sentence of the paragraph Steiner invited us to look "*to the flower, and to the fruiting processes within the flower.*" In the fourth sentence he points out that a new plant is normally grown from seed: "*you must take the germ of the fruit that appears in the flower and put this into the ground.*" For the potato instead we use the eye of the tuber, because the potato rarely gets to produce fruit and so does not provide seed, which is the normal part of the plant for the reproductive function. Thus it is necessary to use other parts of the plant that do have this function, which, in this case, is the eye of the tuber. The farmer excises the eye of the potato trying to remove the least possible amount of pulp with it so that it becomes 'almost' a seed. By so doing varietal improvement is much faster. It is no coincidence that even the standard agronomic practice recommends using small potatoes as seed. Note that all plant reproduction performed in the laboratory, including meristem multiplication, exploit this 'common disposition'.

The paragraph concludes with the observation that it is possible to "*enhance an incomplete fruiting process, however, by means of processes that are outwardly similar to outer combustion.*" Steiner wants us to note that the fruiting process is a kind of death, and this is particularly evident in annual plants. The fruit is ripe when it detaches from the plant and dies. But this process of death is not fully carried out because death does not affect the seed, which remains vital. The aging process is similar to a process of combustion that in some way may be reproduced by people.

Those who touch a fruit that is still on the plant can easily perceive the effect of cooking fruit in the Sun. You will notice how a ripe fruit (for example, a tomato) is warm to the touch on a sunny summer day.

The twenty-seventh paragraph holds an example to elaborate this scheme of animal nutrition. Let's start from those parts of the plant which can be strengthened in their quality " ... *if you first spread them out in the Sun to steam a bit.*" Indeed the hay is left in the Sun before it is dried in the barns. The former is drying but should not be taken too far, otherwise the next process that takes place in the barn would not be possible. The action of the Sun should only cause evaporation of excessive moisture while at the same time enriching the fodder with solar life forces. "*There the latent tendency is taken a little further toward fruiting.*" We could also say that parts of the plant that do not specialise in the production of fruit, but which have this possibility as a potential, can carry this forward by drying in the Sun and obtain a nutritional value similar to that of the fruit. The understanding of these effects prompted people to begin to cook food. Cooking is a process that outwardly, as stated at the end of the previous paragraph, is "*similar to outer combustion.*" Moreover, the Sun's warmth matures and develops the fruits of plants, not the green part which needs further cooking to manifest this latent possibility.

Indeed, "*Even when the flower and seed parts of the plant are given as fodder without being cooked, they work on an animals digestive system, primarily by means of the forces they contain rather than through their substantiality.*" We know that the flower is made up of cosmic substance and earthly forces; the green parts must later be brought actively into the processes of maturation. This is why the hay is so important in the feeding of cattle. Clearly cooking always improves the ability to nurture the metabolic system, as do fruits. It is worth recalling that the metabolic system corresponds to the flower in the plant and in animals it is responsible for fattening them up. We can also understand why at the time for fattening pigs, they were given the so-called 'swill' to eat, which was just a mixture of domestic left-overs such as fruit that was no longer usable for people and flour, appropriately boiled.

The twenty-ninth paragraph ("*Think of the animals ...*") asks us to consider the difference between animals that live in the mountains and those that live on the plains. It is clear that those that live in the mountains must work harder, if only because of the slopes that must be overcome. The effort required develops will forces, forces that would otherwise have to be generated through a proper diet. The fact that the lack of such forces cause the poor performance on the part of these animals, both in growth or in the production of milk or fat, is justified by the fact that the forces will operate in the region of metabolism, which is the same region from which the forces originate that emphasise the production of milk and fat, and the strength to pull heavy loads.

At this point, curiously, Steiner argues that for animals grazing in the mountains; "*you must therefore make sure that enough of their fodder comes from the aromatic alpine herbs, in which nature itself has enhanced the flowering and fruiting parts through the Sun's outer cooking process.*" So he says that these plants do not need the further process of drying in the Sun, because nature has already brought about more than the normal solar cooking, bringing up the processing of sap into the 'fire' of essential oils. Remember that the production of oil is favorably affected by woodland, which is the bearer of the I forces.

If we remember what we said about the influence of trees we should be clear that the action of forests induce great vigour, sometimes even enough to prevent flowering. But we said that the woods brings forces of aromatic quality to aromatics, even stimulating the production of oils. At first glance it seems that we have contradicted ourselves. Actually this is not the case, because the woods, being the bearer of the I forces will be able to diversify its operations according to the situation. We need not be surprised that a wood stimulates the etheric body of normal plants, which it can do because it is the bearer of the I forces, reinforcing the processes of vigour. However, in aromatic herbs it acts at the level of the I. There could, however, be a question at this point because the forest does not support flowering or fruiting. The reason must be sought in the fact that the forest is primarily and naturally particularly rich in forces of the I, secondly it is rich in the etheric forces and, finally, also has a good physicality. The astrality is particularly lacking, so much so that it can never produce a showy flowering. Here it is the case that, given this shortage, when it is confronted with plants rich in astrality like fruit-bearers, it has no choice but to absorb what they lack, yielding etheric forces, known to hamper the process of flowering.

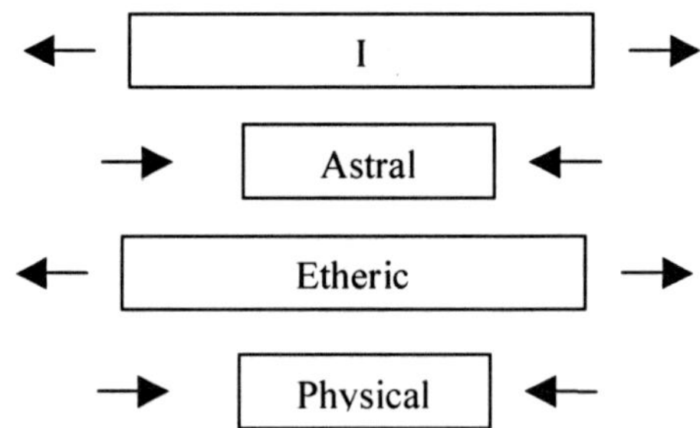

The schema shows this dynamic: the excess of "I" (forest) compresses the astrality (fruit bearing), enhancing the ethericity (annuals).

After speaking of aromatic herbs and their ability to carry the process of cooking by the Sun well beyond the level possible in other plants, Steiner spoke again of the latter saying that they can also be used to bring power to limbs, as long as they undergo treatment based on artificial cooking: boiling.

We might ask ourselves why Dr Steiner speaks of boiling or simmering and not about other types of cooking. We could say that the reason lies in the fact that in the case of boiling, cooking brings in the water element and therefore the presence of the the alchemical ether together with the heat. Especially when boiling there is the creation of movement, which is a characteristic of life. What we are talking about is a unique process, not comparable even to the 'cooking' of hay in the sunshine. The process in boiling is that which the alchemists called 'wet', in which water and air are united, while the path followed by the hay is 'dry', in which the Spirit acts directly in the field. We could even say that when boiling we are in the realm of Bios or biology, while hay intervenes in the action of Zoe, the Life spread throughout the cosmos. Note that making silage with green plants remains in the world of Bios. Only hay bears the forces of Zoe. This food is used to feed the animals in the winter season when the animals remain closed inside and are not allowed to go to pasture and draw directly from the cosmic nutrition. Hay also decreases the amount of milk produced but strongly increases the percentage of fat and, therefore, quality.

Returning to boiling, we all know that this happens at a temperature of 100° C, a temperature that is indicated by a number that has a particular meaning. To understand this fact we need to refer to the Gospel passage in which the Evangelist relates the 'parable of the sower' (Matthew 13.1). In verse 13.8 we find that the seed which '*fell into good ground, and brought forth fruit, some an hundredfold, some sixtyfold, some thirtyfold. Who hath ears to hear, let him hear.*' 30 represents the mysteries of space (the angle of 30 degrees is linked to Light), 60 is central to mysteries of time (the sexagesimal system by which time is measured) and 100 the mysteries of matter. Remember that Steiner said it is a great mystery that the spirit has become matter, but even greater is the mystery of matter returning to become spirit.

100 then is the number that represents the completion of matter. In Steiner's lecture series presented as "Man as Symphony of the Creative Word"[57] we can read that the mysteries of matter are enclosed in the cow and whoever knows the secret of the cow knows the secret of number, quantity and measure. Moreover, the cow is the only animal that has a relationship between the weight of the blood and body weight total of 1:12, expressing a particular link with the Zodiac, which is the origin of matter. We have repeatedly stated that any constellation of the Zodiac has one or more substances as an agent in the physical world. From everything that we have said it should be obvious that between cattle, the process of boiling, and the mysteries of matter, there is a special bond.

Enlarging upon these considerations, we may also say that the centesimal dilution within homeopathy enables one to gather the free peripheral forces and bring them to incarnation. We know that homeopathic pharmacists also make other dilutions including the decimal and LM potencies. The decimal dilution is used to stimulate the forces latent within the patient, and the LM potencies open up to the Light of the macrocosm (Zoe). From what we have just said you can deduce that the decimal dilutions should be particularly suitable for humans, as we have an I within us and therefore the spiritual element, while the centesimal potencies are essential for the plant world which must be open to the spiritual forces from the cosmos. The threshold of 100 degrees, however, also brings purification to matter on the physical level, which then enables access to higher forces.

The thirtieth paragraph ("*And it would also ...*") starts from the consideration of what we have said for the animal and whether we can support the metabolic part of the body with an adequate diet, and asks if this might also be useful for humans. Indeed, says Steiner, there is the danger that one who tends to follow a spiritual development along the path of mysticism is prompted to accelerate this process by adopting a raw-food diet, with the consequence that if one part makes progress the physical part on the other hand becomes increasingly weak and indolent.

The raw diet still has the great advantage of activating the nitrogen and then making the body capable of receiving cosmic images. To assist us in understanding the effects of a meat diet we can try to explain using the following diagram below.

When we eat meat, which is ascribable to the Air element, nitrogen will certainly be ingested too and this will take us towards Water and then continue up towards Earth (the incarnative cycle of Bios). In fact, this forms deposits of uric acid which are the cause of the unpleasant pains that affect men of modern civilisation such as rheumatism and gout.

Vegetarian food, by contrast, starts with food that is of Earth and through chewing, or boiling, shifts in the direction of Water (excarnating of Bios and incarnating Zoe), taking a path that we will recognise as a cycle of life (counterclockwise) opposite to

[57] *Lecture 3* – "No other animal has the same proportion between the blood-weight and the entire body-weight as the cow; other animals have either less or more blood than the cow in proportion to the weight of the body. And weight has to do with gravity and the blood with egoity; not with the ego, for this is only possessed by man, but with egoity, with separate existence. The blood also makes the animal, animal — the higher animal at least. And I must say that the cow has solved the world-problem as to the right proportion between the weight of the blood and the weight of the whole body — when there is the wish to be as thoroughly animal as possible."

the one that was just mentioned which is the cycle of death (anticlockwise).

Vegetarian food then lets you dissolve the deposited uric acid and brings it back into motion. But beyond the dissolution of uric acid, what is important is that you activate a path through which the material can return to become spirit. If vegetarian food is interpreted as meaning raw, the journey along the cycle of life will remain confined to the biological aspect, that is a diet among the Earth and Water elements. The introduction of a certain amount of cooked plant foods makes it easier to continue the journey towards the cosmic pole - to the Air and Fire elements.

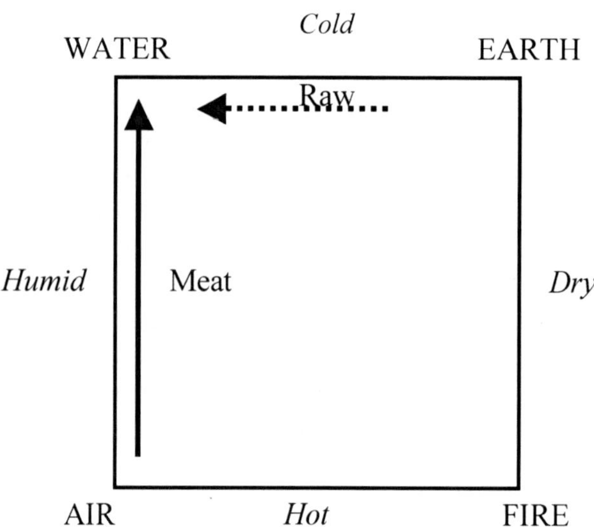

There is another aspect of raw food to consider and that is the constitution of those who undertake this diet. If the person who eats only raw food is sufficiently strong they will be further reinforced by this diet. Conversely if the person is essentially weak they will be further weakened. The ingestion of raw food means that the etheric body of the food is still essentially intact. Our etheric body must destroy the etheric body of the plant that we eat, as we have often said. Now if our etheric body is strong enough it will be able to make a substantial commitment in performing this function so it will be strengthened and reinvigourated. Conversely if our etheric body is too weak to be able to counter the strong ethereal body of the plant effectively, this will cause a further weakening of the eater.

If we consider children's nutrition we must consider various foods. In general the child does not have sufficient forces to deal with the type of diet that is suitable for adults. As long as the child is inside its mother it receives the nutrients they need, and in the first months of post-partum life it is nourished with breast milk which is particularly rich in formative forces but is easy digestible at the same time. It is not therefore sensible to wean the infant with foods that are difficult to digest such as raw vegetables. Weaning onto cooked food should be carried out gradually. It is no coincidence that in societies that are defined as 'primitive' food for weaning is often premasticated by the mother. This is not only to make it more tender, because the infant has not yet produced milk teeth, but mainly because chewing is already starting the digestive process.

Steiner also presents the opposite case - that is of a very vigourous grown-up who wants to adopt a raw food diet. Even with this strength the same processes will apply, but thanks to the person's vitality there will be no damage, and this may bring into movement those deposited substances caused by the former regime – those that may otherwise have generated gout and rheumatism. Indeed the grown up can be further strengthened.

It may seem a bit odd that Steiner defines choosing to become mystical as eccentric. This comes from the recognition that the mystical path is no longer suited to people's needs. Mysticism made some sense when people lived in a polytheistic culture in which divine powers were attributed to statues and images, and they recognised the impression of the incarnation of divinity in those same objects. In this context it was appropriate for seekers to detach themselves from the world and to try to find the same God within themselves. Today it seems more responsible to the needs of the world to engage in worldly activity, because the condition to which we have reduced our planet calls for immediate action to try to reverse its course.

The thirty-first paragraph ("*These things all have ...*") refers to the different issues and states that "*there are no hard and fast rules,,*" but that *"we have to learn to apply them differently in individual instances.*" Then Steiner reiterated the advantages of a vegetarian diet. This paragraph seems absolutely clear so we will refrain from further comment.

The opening thoughts in paragraph thirty-two ("*And now, on the subject ...*") are likely to be a cause for surprise for many people, because it contradicts the common image of pigs. The paragraph starts with the observation that what makes animals fat is not the food that is ingested physically but the cosmic substance that is assimilated. It is interesting that Steiner uses the expression "*filling a sack*" because it allows us to understand that the gross food serves to enlarge this metaphorical sack so that it can carry the greatest possible amount of cosmic substance.

At this point, the pigs are called "*heavenly creatures*" because, "*their fat bodies consist entirely of cosmic substance*" and "*what they eat is only needed in order to distribute this cosmic substance to the different parts of their bodies*".

This fattening is enhanced "*when you give them fodder with an inherent fruiting tendency - preferably treated further by boiling or steaming - and also when you give them fodder where the inherent fruiting tendency has been enhanced by cultivation.*"

The discussion about fattening animals for bulk is not completed by this and we will see in the next few paragraphs, in our next meeting, what is necessary to consider in the diet of the animal for milk production and also food for the head when we need Earthly substance.

Thirtieth meeting (January 8, 2005)

We have almost reached the end of the Koberwitz course of 1924 and of all our labours that have lasted for thirty meetings. There remain only a few paragraphs upon which to comment and this still requires a few hours of toil. Then we believe that we can present to Rudolf Steiner the result of our efforts as a sign of gratitude for the magnificence of what he left to humanity in the last Pentecost of his life.

In our previous meeting we ran out of time to finish considering the last part of the thirty-second paragraph of the eighth lecture ("*And now on the subject ...*"). We were talking about fattening up animals and more specifically about the fact that animals fatten upon the fruiting parts of the plants. We have also said that these parts are still more effective for our requirements if they are cooked by the Sun or with an additional treatment through boiling or steam cooking.

Steiner now leads us back to a broader vision of the plant recalling that the fruiting process, which finds its completion in the latter part of the life of the plant, in fact occurs - even if not fully expressed - in all the earlier stages of development. What we are saying has already been clearly stated by Rudolf Steiner in the twenty-seventh paragraph where we can read: "*It is characteristic of plants - and that is why Goethe was so fond of them - that the potential to develop any of their parts is present throughout their body.*" Only by taking this into account can we understand why Steiner introduces the example of the beet at this point in the lecture: the beet is a root so it should be suitable to nourish the head of the animal. In the light of what has just been said, fruiting forces are also present in the root, and man has insisted on the presence of an enlarged root and especially in an intensification of its qualitative processes to the point of producing sugar.

Something similar occurred with the potato. The potato is not a root but a rhizome, which is a swelling of the lower part of the plant. This enlargement is the result of flowering forces retained in the lower part of the plant. The fact that these flowering forces are of an astral nature which act in the developing stages of the plant - earlier than they should in the 'normal' sequence of events - is responsible for the formation of alkaloids. These are strongly astralised substances that we recognise as toxic. Now, since we recommend a certain restraint in the consumption of potatoes, we should ask ourselves a similar question about the beet and products derived from it: can these be recommended as food for human consumption? The case of beet is different from that of the potato because it is the astral flowering forces that are detained in the potato, while the beet retains the spiritual forces associated with fruit (rich in sugars and hydrogen, a substance which bears the spiritual forces). Therefore we can say that in the case of the beet the spiritual-Fire acts in the root-Earth and this corresponds to the 'dry way' of alchemists. In other words if a person is 'ready' to take this route of initiation they can be nourished by the beet. Otherwise is it better if they do not eat them.

Since we have moved our attention from animals to humans, we may ask ourselves whether the ability to evenly distribute the cosmic matter (within the body) is greater in males or females. Considering that we are dealing with the distribution of cosmic matter, we should say that males in this case have a certain advantage since the I forces within them are generally more active than in females and therefore they may distribute the cosmic matter more evenly in their body, since it is thoroughly permeated by the forces of the I organisation. Women instead tend to have a greater presence of soul forces within them. In nature, the soul forces are not evenly distributed: this is why the various species of animals are distributed irregularly on Earth. Man is a microcosm and so it should not be a surprise that the soul forces are also irregularly distributed within us. The thighs are the part of our body that represent our 'animal nature' (from which comes the myth of Dionysus' pregnancy in Zeus' thigh). This is the reason why many women carry weight in this part of their body.

The thirty-third paragraph ("*So, what should we*...") repeats the principle from the beginning of the previous paragraph. It is still stressed that animal fat is cosmic substance and it is therefore necessary that the substance can flow copiously inside the animal and can then be spread throughout the body. Based on the assumption that the action of distribution is favoured by "*something with a strong fruiting tendency, which has also had some further treatment*", it is now specified that "*this condition is largely met by certain kinds of oil-seed cakes and the like.*"

The cosmic substance is captured by the head of the animal, as a centre of the nerve-sense system. The head, however, is formed by terrestrial substance and therefore we must be careful that, albeit in small quantities, "*since in this case the needs of the head are minimal*", we add food containing such substances. "*Thus we should add something root-like to the fodder of these animals, even if only in very small amounts.*"

In paragraph thirty-four ("*In general we can say*...") Steiner sums up a bit what he has said so far, when he says, "*that what is root-like has a function in relation to the head, what is flower-like has a function in relation to the metabolic-limb system, and what is leaf-like has a function in relation to the substantiality of the human rhythmic system.*" After that he mentions the indispensability of salt. In other words Steiner mentions food for the various constituent bodies and finally ascribes salt as the nourishment of the Ego that, as an organisation, presides over the healthy performance of all other functions and the proper conduct of all processes.

Salt has a cubic crystal form and as it is a crystal it belongs to the dead merely physical world. If, however, we assume that it derives from the 'crushing' of a sphere worked upon from six different directions, we can say that part of the ball is invisible but remains connected to the six faces of the cube as if they are still floating on it - the etheric-vital aspect of the salt.

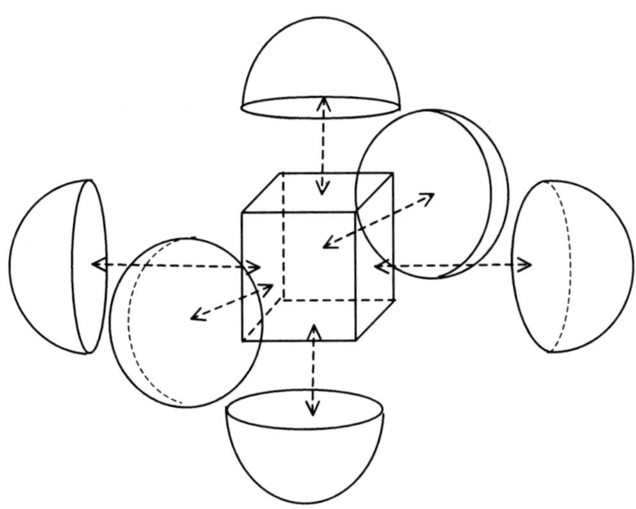

In humans and in animals the saline aspect finds expression in bone which, like a crystal of salt, represents the dead part of our body. However, their contractive force governed by calcium salts - calcium carbonate and phosphate - is kept integral by our subtle bodies. In the agricultural world the same function is performed by lime in the soil, which, while appearing dead from the organic point of view, plays an important role in the retention of imponderables. It is not by chance that olive trees prefer to grow in calcareous soils to promote the processes of production and maintenance of oil in *drupacea* (stone fruit) which appeal to the forces of the constellation Libra that has calcium as its earthly ambassador. We normally attribute calcium with the Ahrimanic force and from a certain point of view this is correct because the act of extreme contraction that this element expresses is considered contrary to life. On the other hand it is also true that one can find physical forms of calcium that allow us to see it in a different light. Do not forget that Mary Magdalene saved oil of nardo (from a plant that grows in the Himalayas) in an amphora of alabaster with which she sprinkled the body of Jesus. Alabaster was used in the past for making windows in churches, because it can be made very thin allowing the passage of light, symbolising the way that the forces of light will have their victory over the Ahrimanic forces of darkness. It was pure calcium carbonate. On the other hand it is also true that we normally say that matter is the kingdom of Ahriman, but it is equally true that matter may be defined as *spiritual light shattered into visibility*. The fact that matter in the process known as 'the fall' has been pervaded by the forces of evil should not make us forget that any substance which is derived from the spirit world brings with it, in germ, the possibility of resurrection.

The next paragraph is the thirty-fifth ("*Now there is still one* ..."). Here Steiner introduces a series of considerations on a vegetable that had only recently become part of the human diet. This is the tomato. These considerations, upon which we will enlarge, are still valid now. Today the tomato is the subject of a broad consensus in the world of consumers and is the subject of special attention from doctors and nutritionists who attribute remarkable health-giving qualities to the tomato; the tomato is considered as a general anti-tumour remedy. Clearly, those who support these theories have never exerted themselves to understand the so-called '*nexus*' between the plant-world and the animal-world, or between plants and man.

The sentence that begins the thirty-sixth paragraph leaves little room for doubt concerning the effect of tomato. We read that tomatoes, "*have a significant effect on everything that tends to separate itself from the organism and develop an independent organisation within the body*".

It seems to us that the sentence is too explicit in stating that the tomato has the general capacity to support processes that are developed in independence from the body. This particular dynamic should not be seen only in its negative aspect, so we can appreciate the link between tomatoes and the liver, which is the organ that acts most independently. "*... under certain circumstances, adding tomatoes to the diet can have a beneficial effect on an unhealthy liver*".

We must, however, ensure that this virtuous aspect is active and not canceled out by the potentially harmful action. This is achieved by judicious consumption. We will explain more clearly: the liver is the organ that specialises in connecting with the etheric body and this strong etheric character is highlighted by the extent that the liver is able to regenerate itself. It is interesting to note that the tomato also has this enormous etheric capacity - wherever it touches the ground it makes new roots. Anyone who has seen a tomato plant is sure to have noticed that the tomato has adventitious roots, namely roots emerging from the plants' stem above the soil, as if the Earth's life is insufficient to contain the roots. Well this link between the liver and the etheric presupposes a certain link with the world of rhythm. In order that the tomato can have a therapeutic stimulus upon the liver the tomato should be eaten in small quantities, outside the normal rhythm of diet, or between meals.

The fact that the tomato is a very special plant and that its fruit should not be consumed like any other can also be inferred from another kind of observation. The tomato shows, as we have already said, an extraordinary etheric vigour in its lower parts. On other occasions we called this the 'Earthly pole' of the plant. In contrast the opposite pole produces a fruit that we can tell by its colour and shape is highly 'cosmic'. This strong - even excessive - contrast cannot go unnoticed, because we know that health is always a median between poles.

In order to mitigate the potential damage, the tomato should be used in the regular diet only when it is absolutely ripe and then it should be consumed without being cooked much, an operation which would otherwise enhance the already excessive heat forces even further.

Early flowering in tomatoes betrays excessive astrality. The plants in question do not bloom at the top, but along the stem in between two leaf stems. We know that an orderly development of the plant would produce the flowering stage when the leaf stage is finished, as a metamorphosis of the flowering. This anticipatory flowering is the result of the same dynamic that creates alkaloids.

A later sentence removes any doubt about the anti-tumoural potential of the tomato. We can read: "*let me say this in parenthesis - people diagnosed as having cancer should immediately be forbidden to eat tomatoes, because cancer from its very inception makes a certain part of the human or animal body independent of the rest of the organism.*"

The thirty-seventh paragraph ("*But now we need to ask ...*") explains the reason for all the things that we have said about the tomato. Tomatoes are the only plants that "*feel most at home when they are given manure or compost that is as close as possible to the form in which it comes from the animal or other source.*" This means that the manure has not yet been incorporated in a composting process whose primary purpose is to weaken the self-imprint of the original animals and to favour the activities of a new I which is that of the compost heap. This ego harmonises the processes and above all organises the vitality of the heap so that it is compatible with the vitality of the soil, which is not a turbulent life. Still, the tomato is the only plant that likes compost made with other tomato plants, as if it were a cannibal.

In its place of origin this plant has a completely different effect. The tomato originates from the Americas where the dominant forces are contractive Earthly forces. In those conditions the vitality of young plants is sorely tested and the tomato's strong etheric carries the plants beyond the initial phase of life. The etheric can be stimulated by an astral component such as an alkaloid that acts as a drug and allows the plant to overcome the limits imposed by its natural environment. The indigenous people in those places are also very earthly by nature and because of this they can safely eat plants rich in alkaloids.

The fact that the tomato plant is so very full of etheric and astral forces critically affects its ability to fully express itself in its two other bodies: the spiritual and physical. Indeed, the tomato does not have the capacity to keep itself upright and even struggles to maintain a solid physical substantiality. Of course, the I weakness is reflected in the fact that even those who eat tomatoes, thereby encourage autonomous forms that are beyond the organisational control of the I. "*This is the reason that they can influence what works independently within the human or animal organism*".

The thirty-eighth paragraph begins by saying that, "*in this respect, potatoes are somewhat similar*". The potato also originated in the Americas and is rich in alkaloids. The second sentence of the paragraph seems particularly significant because it states: "*They too act extremely independently, that is to say, they tend to pass very easily through the whole digestive process and then enter the brain and make it independent; they make it independent even of the influences of the other organs in the human body.*" In other words, the head no longer has good connections with the rest of the body and can 'escape' upward (as carbohydrate which is the food-stuff connected with the upper central part of the brain) and be subject to Luciferic forces who bring the fantastic thoughts, or be prey, as a reaction (or karma) to Ahrimanic earthly forces bringing the materialist thoughts (decay of the fore-brain). So: "*we should eat only enough potatoes so that our brain and our head in general are stimulated.*"

The thirty-ninth paragraph begins with a very significant sentence: "*In these lectures, I have only been able to supply certain guidelines, of course, but I am sure that they will provide a foundation for many different experiments extending over a long period of time, and then they will lead to brilliant results if worked into your agronomical practices on an experimental basis. That should be a guideline for dealing with the material presented in this course.*" The fact that we read about "*certain guidelines ... a foundation for many different experiments,*" shows that Steiner hoped that what he communicated could be developed for the future through experiments.

The paragraph continues with the recommendation that only those directly involved in agriculture should speak of the contents of the course. This had already been expressed at the beginning of course, in the seventh paragraph of the first lecture and can be called a 'principle of responsibility'. The next paragraph ("*Because of the admirable tolerance ...*") is a call not to immediately disclose the contents of the course. This invitation now, in the age of communication, when any news or information can be available in a few seconds in any part of the globe, may not be immediately understood. Dr Steiner gave an explanation in Dornach in his report of June 20th 1924, on his return from Koberwitz. The eighth paragraph of the report says: "*One condition for success, however, was strongly and repeatedly emphasised: for the time being, the content of the course must remain the spiritual property of the circle of practicing farmers. Although some people only casually interested in agriculture were also present at the course, they were not permitted to join the circle and were expressly instructed not to fall into the usual anthroposophical habit of immediately talking about everything with everyone. These things will only be able to live up to their true potential if the content of the course remains in the hands of the specialists and is tested by the farmers. Some things will require fours years to try out. In the meantime, the practical pointers that were given are not supposed to stray outside the agricultural community. These things are meant to enter right into practical life, so it does no good just to talk about them. Anyone who heard these things and goes around talking about them will be doing them an injustice.*"

Apart from the need to submit the teachings received during the Course to experimentation, which would have meant all those working would have made those things that they have just heard their own and establishing a continuity of consciousness, there was also another reason why Steiner urged all the delegates to avoid broadcasting the contents of the course. It is a principle that has always inspired members of the Rosicrucian world and consisted of acknowledging the need for the content of any new insight or new truth to remain undisclosed to the public until it has been matured within the group for at least 100 years. Recall that the Rosicrucian movement was characterised by the belief that spiritual knowledge should not be maintained at the level of 'think tanks', but had to be used for an inner development that sustained the practical life, activity and evolution of all natural kingdoms: this was the way and door into practical life.

We think it should be obvious that any spiritual truth must be properly understood and experienced before being exposed to third parties to avoid speaking without competence. The period of 100 years, however, was not any arbitrary period because it was the interval between successive incarnations of the Bodhisattva. Remember also that the Bodhisattvas are lofty spiritual beings whose mission is to bring the teachings received directly from Christ to men. The Bodhisattvas act through repeated incarnations and when the mission of one of them is complete, or the teaching of which he was the bearer becomes the general property of humanity, the Bodhisattva ascends to the rank of a Buddha. Such a one no longer needs to be incarnated and leaves the task of helping to the next Bodhisattva. The Bodhisattva currently acting in humanity is the one who will be called the Maitreya Buddha and who succeeded the Gautama Buddha who last incarnated six centuries before Jesus Christ.

Today all human activity is caught in an accelerating whirlpool, so the incubation of teaching is considered complete when kept confidential and subject to verification and testing for a period of a minimum of three years. However, the text that we are commenting upon was publicly available for the first time in the seventies and before that had only a small circulation of typed copies in the anthroposophical world, limited and individually numbered, which were registered and assigned to those who wanted to study them and who would take responsibility for maintaining their confidentiality.

We believe, however, that the primary reason for such confidentiality regarding the agriculture course should be sought in the fact that it contains great truths about the laws of life, with which it is possible to act with intense force for the good of nature and man, but one can also do great damage: what determines the outcome is, once again, morality.

The forty-first and last paragraph contains thanks to Count and Countess Keyserlingk for their hospitality, and thanks and greetings to all the delegates. We can do no better than follow the example of Rudolf Steiner and thank all those who, for four years, have had the patience to follow us, and we hope that our comments will find application in the practical agricultural life.

Publications available from l'Albero della Vita

All titles are by Enzo Nastati except as indicated. The English title of the publications that are currently translated follow the Italian details and are underlined. English copies are available from: Considera, 1 Prospect House, Hitchings, Blakeney, Gloucestershire, GL15 4BJ, UK or via mark@considera.org. Clearly there are many other publications that could be translated given funds or volunteer translators. Please contact l'Albero della Vita or Considera should you be able to assist with either.

Many publications are also available in Spanish. Non-English titles are available from l'Albero della Vita. The list overleaf is comprehensive at the start of 2009.

l'ALBERO della VITA
ASSOCIAZIONE INTERNAZIONALE per la RICERCA SPERIMENTAZIONE e DIVULGAZIONE delle DISCIPLINE ECO – COMPATIBILI
Via Villaorba, 19 – 33033 Beano di Codroipo (Ud) - Italia
tel. 0432 905724 fax 178 274 9667 e-mail segreteria@albios.it

AGRICULTURAL COLLECTION

1) *Introduzione al metodo omeodinamico*; [**Introduction to the Homeodynamic Method**]
2) *Nove incontri con l'agricoltura biodinamica:* corso base; [**Nine meetings about Biodynamics** – in translation.]
3) *Manuale di coltivazione omeo-dinamica:* l'applicazione del metodo nelle varie colture; [**Homeodynamic Agriculture Handbook**]
4) *Calendario Agricolo Astronomico:* con le indicazioni degli aspetti planetari e zodiacali per le semine, lavorazioni e trattamenti (pag. 35 più 13 tavole aggiornate ogni anno); [**Astronomical Agricultural Calendar** – available each year]
5) *La Concimazione: ovvero i modi per portare vita alla terra*; **[Fertilisation]**
6) *Quattro aspetti dell'organismo agricolo:* le indicazioni per riconoscere e progettare un organismo agricolo secondo le sue esigenze;
7) *L'organismo agricolo:la multifunzionalità e l'allevamento ovino*, di **Enrico Meineri**
8) *Azioni delle 13 Sante Notti nel piano Fisico, Eterico, Animico e Spirituale*; [**The Influence of the 13 Holy Nights on the Physical, Etheric, Astral and Spiritual Planes**]
9) *Le Consociazioni vegetali su base zodiacale*; **[Zodiacal Plant associations.]**
10) *Coltivare i cereali con il metodo omeodinamico*; **[Cultivation of Cereals]**
11) *Principi di frutticoltura secondo il metodo agricolo omeodinamico*; **[Orchards]**
12) *Coltivare le officinali con il metodo omeodinamico*; **[Medicinal plants]**
13) *Coltivare la vite con il metodo omeodinamico*;
14) *Coltivare in orto e serra secondo il metodo omeodinamico.*
15) *Coltivare in aridocoltura con il metodo omeodinamico*; **[Dryland Farming]**
16) *Comprendere e curare le parassitosi:* parte prima. Un percorso per comprendere i parassiti ed i mezzi di cura; **[Understanding Pests and their Control pt 1]**
17) *Comprendere e curare le parassitosi:* parte seconda. Come affrontare il problema dei parassiti con la meditazione; **[Understanding Pests and their Control pt 2]**
18) *Il Giardino come Paradiso*: indicazioni spirituali per la progettazione; [**Paradisical Gardens** – in translation]
19) *La rigenerazione della semente e le nuove piante secondo la scienza dello spirito*; [**Regeneration of Seeds and New Plants**]
20) *Comprensione ed allevamento degli animali secondo il metodo omeodinamico*;
21) *Apicoltura ed evoluzione dell'ape:* una nuova visione dell'essere dell'ape, suo compito e sua cura; **[Apiculture and the Evolution of Bees]**
22) *Aiutiamo gli animali domestici:* comprensione degli istinti, bisogni e sviluppo di **R. Nastati**
23) *Il ciclo eterico della terra: indicazioni per un agire agricolo cosciente in armonia con le forze macrocosmiche;* [**The Etheric Cycle of the Earth**]
24) *Commento al Corso di agricoltura tenuto da Rudolf Steiner a Koberwitz nel 1924*;
25) *L'agricoltura come via di spiritualizzazione della Terra: la cristianizzazione della natura;*
26) *Il diffusore per prodotti omeodinamici: come favorire il fluire della vita dal cosmo alla terra* (NON DISPONIBILE);

SPIRITUAL SCIENTIFIC COLLECTION

27) *Glossario scientifico-spirituale:* una pratica guida per chi si addentra nella terminologia antroposofica, di **Fabio Montelatici**; [**A Spiritual Scientific Glossary**]
28) *Lo sviluppo del pensiero terapeutico*: metodo per l'educazione del pensiero al fine di riconoscere l'essenziale delle percezioni ed individuare il processo terapeutico;
29) *Libertà e Amore:* un approccio scientifico-spirituale alle mete dell'umanità; [**Freedom and Love**]
30) *Punto e sfera*: *aspetti individuali, sociali e spirituali nella relazione tra micro e macrocosmo*;
31) *Meteorologia*: un'indagine tra passato, presente e futuro, del rapporto tra uomo e macrocosmo e l'interagire con la sfera meteorologica; [**The Weather**]
32) *Dall'Astrologia all'Astronomia in noi:* una via verso l'interiorizzazione del mondo stellare;
33) *Esseri elementari:* loro origine, azione e trasformazione;
34) *Le Basi per una Nuova Omeopatia*; [**Foundations For A Development Of Potentization**]
35) *Santo Vangelo di Gesù Cristo*: tradotto dal greco e con indicazioni spirituali per una loro libera comprensione **Autori vari**;
36) *La missione di Michele tra la divina Sofia ed il Cristo* (in preparazione);
37) *Dalla terapia di Raffaele alla terapia di Michele*; [**From the Therapy of Raphael to the Therapy of Michael** – in translation]
38) *Il mistero della nascita ed il mistero della morte,*;
39) *Della vita dopo la morte di* ***R. Steiner***;
40) *Le preghiere della cristianità: il Pater, l'Ave ed il Gloria*; [**The Prayers of Chrsitianity – Lords Prayer, Hail Mary and Glory Be.**]
41) *Impariamo a pregare*: un cammino nel dialogo con Dio;
42) *Maria-Sofia*: un'indagine scientifico spirituale sull'azione di Maria nella Terra e nel Cosmo; [**Mary Sophia.**]
43) Il ritmo della Terra fisica: come agire per stimolare la vegetalizzazione della Terra;
44) *Il Natale nei Misteri del Tempo*;
45) *Da Pasqua a San Giovanni: un percorso nei Misteri della Luce*;
46) *Uso giusto e sbagliato della conoscenza esoterica (Confraternite segrete)* di **R. Steiner**;
47) *Il Vero, il Buono ed il Giusto:* i tre ideali morali dell'umanità;
48) *La venuta dell'Anticristo nella scienza dello spirito, nella sacra scrittura e nelle profezie*; [**The Coming of the Antichrist**]
49) *Avvento e Natale: raccolta di leggende, storie, saggi, scritti e tradizioni per meglio vivere le festività.* **Autori vari.**
50) *Indicazioni per mamme e bambini di* **R. Steiner** con commento di Enzo Nastati;
51) *L'anno solare come simbolo dell'anno cosmico di* **R. Steiner**;
52) *Io Sono* del **Conte di St. Germain** con commento di Enzo Nastati; [**I am**]
53) *I quattro eteri* di **Ernst Marti**;
54) *L'eterico* di **Ernst Marti**;
55) *Silice, calcare, argilla*: processi nel minerale, pianta animale ed uomo, diciassette indagini scientifico-spirituali di **Benesch e Wilde**;
56) *Le forze di vita e morte che intessono nella Natura: un preludio alla scienza della Vita, di* **E. Nastati e Fabio Montelatici**;
57) *Le forze formatrici negli organismi viventi (le cristallizzazioni con cloruro di rame, un metodo per comprendere le trasformazioni biologiche delle sostanze vegetali), di* **Magda Engquist**;
58) *Il cammino di iniziazione moderno attraverso il Libro delle Forme, della Vita e delle Coscienze*;
59) *Entrare nel Libro delle Forme attraverso i pensieri di Michele e la forza di Maria*;

60) *Interpretiamo insieme le fiabe ed i miti: Cappuccetto Rosso, Biancaneve, Perseo e Parsifal*;
61) *Antiche e nuove vie di Iniziazione:* dall'antica India al Cristo Eterico;
62) *I Sacramenti: misteri antichi ed Iniziazione cristiana*, di **E. Nastati** e **Fabio Montelatici**;
63) *La strada verso il fato:* un esame scientifico-spirituale del "Signore degli Anelli", di **Fabio Montelatici**;
64) *Il riscatto del Sole*: un'integrazione adeguata ai tempi all'interpretazione del "Signore degli Anelli", di **Fabio Montelatici**;
65) *Astronomia spirituale*: un percorso scientifico-spirituale dalla Via Lattea allo Zodiaco e dai pianeti alla Terra, di **Enzo Nastati**, **Fabio Montelatici** e **Achille Minisini**;
66) *Riassunto commentato di "Il corso dell'anno come via di iniziazione all'esperienza dell'entità del Cristo – un esame esoterico delle feste dell'anno" di S. O. Prokofieff*, a cura del **Gruppo di Trieste** dell'Albero della Vita (IN REVISIONE - NON DISPONIBILE);
67) *Igienistica dei 12 sensi dall'infanzia alla terza età*, di **Livio Casetti** (IN REVISIONE - NON DISPONIBILE);

ARCHITECTURAL COLLECTION

68) *Azione del colore nell'abitare*: con le indicazioni del come utilizzare il colore per risanare l'abitazione;
69) *Azione delle forme nell'abitare*: con le indicazioni del come utilizzare le forme per risanare l'abitazione;
70) *Indicazioni scientifico-spirituali per progettare e costruire secondo coscienza:* un'introduzione all'architettura in chiave spirituale;
71) *Corso sulla qualità dell'abitare*: 12 dispense con le indicazioni del come effettuare una analisi ambientale tra eteri, elementi e forze fisiche e sottonaturali e del come risanare l'abitazione utilizzando forme, colori e materiali appropriati.

NUTRITION COLLECTION

72) *Alimentazione cosmica e terrestre:* nei quattro regni della natura secondo la teoria degli eteri e degli elementi;
73) *La qualità nei trasformati alimentari*: pane, vino, succo d'uva, tofu, seitan, marmellate;
74) *Corso sulla qualità dell'alimentazione*: dieci dispense per conoscere i regni naturali, gli alimenti. la nutrizione cosmica e terrestre nell'uomo, l'alimentazione per i sette corpi, l'azione di temperatura, pentole, materiali, le trasformazioni alimentari;
75) *L'alimentazione del bambino in corpo, anima e spirito*, di **Enzo Nastati** e **Fabio Montelatici**.

RESEARCH COLLECTION

76) *Dossier Acqua*: l'acqua, la sua memoria, la sua vitalizzazione nel passato e nel presente con particolare riferimento al vitalizzatore "Aqua Viva" ed alle prove effettuate su di esso dall'Istituto EUREKA in relazione a diversi tipi di tubazione.
77) *Dossier Disinquinamento*: i risultati ottenuti dall'Istituto EUREKA nell'ambito del disinquinamento ambientale (piombo, metalli pesanti, atrazina, cloro, desalificazione, gasolio, molecole di sintesi, derattizzazione, insetti dannosi, acqua salmastra, ecc.) e del miglioramento dell'acqua; [**Pollution Remediation Dossier**]
78) *Dossier Agricoltura*: alcuni risultati ottenuti dai nostri soci sull'impiego del metodo omeodinamico in agricoltura, tra cui recupero di danni da gelata e da grandine, stimolazione dell'azotofissazione nelle leguminose, rigenerazione delle piante, ecc. [**Agricultural Dossier**]

L'ALBERO della VITA
INTERNATIONAL ASSOCIATION FOR RESEARCH
EXPERIMENTATION and INFORMATION on ECO~COMPATIBLE DISCIPLINES
Via Villaorba, 19 – 33033 Beano di Codroipo (Ud) – ITALY
Tel. 0432 905724 – Fax 178 274 9667 - E mail segreteria@albios.it

APPLICATION FOR AFFILIATE MEMBERSHIP OF THE ASSOCIATION 2009

To the President.

I would like to be an affiliate member of the association l'Albero della Vita.

DATE ....
FIRST NAME .... FAMILY NAME .... ……………….…..... ...
ADDRESS ....
... ...
... ... ……………… ……
NATIONALITY ... ... TEL ....
FAX ... ...
MOBILE …………… ... . E-MAIL ... ...
OCCUPATION:

□ Farmer	□ Office worker	□ Doctor	□ Therapist
□ Architect	□ Teacher	□ Retired	□ ……..

INTEREST ... ...
2009 RATES:
□ **Renew my subscription**: the annual rate is €33 and gives access to: telephone counseling every morning; participation in courses, symposia and conferences; homeodynamic preparations for the home and agriculture, handouts and books at favorable prices; annual Albios quarterly-magazine subscription, electronic monthly updates.
□ **I would like to join for the first time**: The rate is €22 for this first year and you will receive 2 free back-issues of Albios and all the services mentioned above. ***For those living outside Italy: first year rate is also € 33***.
□ **I would like to register as a Friend** of the project "A House for Michael". Minimum donation is €100.
Payment:
□ Bank account # IT 91G 07601 02200 000076697341 registered to l'Albero della Vita, attaching the receipt to the admission form, stating full and clearly the purpose for the payment. BIC CODE is: BPPIITRRXXX. POSTE ITALIANE (filiale di via Caboto, Trieste).

L'ALBERO DELLA VITA CAN OPERATE ONLY FOR ITS MEMBERS AND, THEREFORE, ONLY BY BECOMING A MEMBER YOU CAN SUSTAIN THE INITIATIVES

Privacy
Under L 675/96 it is your right to ask for the cessation of the storage or the updating of the data in our possession. In the meaning of act. 11 Ls 675/96 and in relationship to the information given by me in the requested data, I authorise the Association to treat my data for the management of members.

Signature ... ...

Translator's note.

I barely speak Italian! The Italian manuscript was passed through various net-based translation engines to receive back a mass of 'gobble-de-google'. After months of careful work and self-education I have hewn this book from that approximate mutilated block. The result is bound to be far from perfect, but I have to say that I am pleasantly surprised it is as readable as it is. I imagine it is about 95% spot on.

A second thing to say is that biodynamics is - tragically - still a niche interest, so there's little point in publishing thousands of copies. For this reason this book has been 'printed on demand' – but this gives a continuous opportunity to make improvements as they are suggested.

I am exposing this process to encourage those who speak Italian, English and biodynamics to speak up. Please let me know where this translation is wrong, poor, awkward, misleading, unclear etc. It can be improved if you think it worthwhile to pass on your suggestions.

I have pressed on - despite the urge to proof-read a few more times - to get this manuscript printed in time for the annual agricultural conference at Dornach this Candlemas - the subject of which is the Koberwitz course. I trust that the remarkable work behind Enzo Nastati's commentary will act as a powerful 'preparation' to bring the even greater gift of Dr Steiner's Whitsun labours of 1924 to practical fruition.

Mark Moodie
mark@considera.org

January 2009

I am very grateful that the above has brought forth some very useful input for this second print run. Thanks to Conrad Zirkwitz, Hugh Courtney, Wain Farrants and, once again, Mike Atherton - November 2009.

Commentary on Dr Rudolf Steiner's Agriculture Course
Enzo Nastati

ISBN: 978-0-9517890-6-3

Lightning Source UK Ltd.
Milton Keynes UK
UKOW01f0642120814

236791UK00003B/20/P